Problems in French History

Also by Martyn Cornick

INTELLECTUALS IN HISTORY: The *Nouvelle Revue française* under Jean Paulhan
THE FRENCH SECRET SERVICES (*with Peter Morris*)

Also by Ceri Crossley

CORRESPONDANCE GENERALE DE JULES MICHELET (*co-editor*)
FRENCH HISTORIANS AND ROMANTICISM

Problems in French History

Edited by

Martyn Cornick
Reader in Contemporary French Studies
University of Birmingham

and

Ceri Crossley
Professor of Nineteenth-Century French Studies
University of Birmingham

First published 2000 by
PALGRAVE
Houndmills, Basingstoke, Hampshire RG21 6XS and
175 Fifth Avenue, New York, N. Y. 10010
Companies and representatives throughout the world

PALGRAVE is the new global academic imprint of
St. Martin's Press LLC Scholarly and Reference Division and
Palgrave Publishers Ltd (formerly Macmillan Press Ltd).

ISBN 0–333–75381–X

This book is printed on paper suitable for recycling and
made from fully managed and sustained forest sources.

A catalogue record for this book is available
from the British Library.

Library of Congress Cataloging-in-Publication Data
Problems in French history / edited by Martyn Cornick and Ceri Crossley.
    p. cm.
  Includes bibliographical references and index.
  ISBN 0–333–75381–X (cloth)
    1. France—Civilization. I. Cornick, Martyn. II. Crossley, Ceri.
  DC33 .P75 2000
  944—dc21
                                                    00–040485

10   9   8   7   6   5   4   3   2   1
09 08 07 06 05 04 03 02 01 00

Printed and bound in Great Britain by
Antony Rowe Ltd, Chippenham, Wiltshire

Contents

List of Maps and Figures   vii

Notes on the Contributors   ix

Introduction   Martyn Cornick and Ceri Crossley   xiii

1   Catherine de' Medici: Saint or Sinner?   1
    Robert J. Knecht

2   Protestantism and Jacobinism in the Department of the
    Aveyron, 1789–1815   17
    Peter M. Jones

3   Chalk and Cheese or Bread and Butter:
    Political Culture and Ecology in the Seine-Inférieure during
    the Napoleonic Period   31
    John Dunne

4   Getting out the Vote: Electoral Participation in France,
    1789–1851   50
    Malcolm Crook

5   A Forgotten Socialist and Feminist: Ange Guépin   64
    Pamela Pilbeam

6   Attitudes towards Animals and Vegetarianism in
    Nineteenth-century France   81
    Ceri Crossley

7   Religion and Nationalism in Late Nineteenth-century France   104
    Brian Jenkins

8   Distorting Mirrors: Problems of French–British Perception
    in the *Fin-de-Siècle*   125
    Martyn Cornick

9   After Dreyfus and before the Entente: Patrick Geddes's
    Cultural Diplomacy at the Paris Exhibition of 1900   149
    Siân Reynolds

10   A Left-wing Intellectual of the 1890s: Georges Clemenceau   168
     David R. Watson

11   Seduction and Sedition: Otto Abetz and the French,
     1918–40   180
     Nicholas Atkin

12   *Les Années noires?* Clandestine Dancing in Occupied France   197
     Robert Gildea

13   Opening Pandora's Box: the Case of *Femmes tondues*   213
     Claire Duchen

14   Contemplating French Roots   233
     H. R. Kedward

15   From Greater France to Outer Suburbs: Postcolonial
     Minorities and the Republican Tradition   249
     Alec G. Hargreaves

16   Angels on the Point of a Needle: Counting Catholics in
     France and Spain   264
     Maurice Larkin

17   The Rule of Memory and the Historian's Craft in
     Contemporary France   285
     François Bédarida

Index   296

List of Maps and Figures

Maps

1   Turnout in the 1809 Elections.   35

2   The Departments of France in 1790 and 1848.   51

Figures

9.1   Groupe Français—De l'école internationale de
      l'exposition: Programme du Lundi, 10 Septembre au
      Samedi, 15 Septembre. (*Reproduced by permission of
      University of Strathclyde Archives*)   154

9.2   Paris International Assembly (Ecole internationale de
      l'exposition). Programme for the week 25–30 June 1900.
      (*Reproduced by permission of University of Strathclyde
      Archives*)   155

13.1 Capa 23 August 1944. First pictures of the capture
      of Chartres. Two French women collaborators,
      their hair shaved off, are led back to their home,
      while the population whistle loudly
      and jeer at them. (©*Magnum Photos*)   220

13.2 Capa 23 August 1944. Two women collaborators, one
      with a baby by a German father, are led back to their
      homes, after having their hair shaved off, amid jeers
      and whistles from the population. (©*Magnum Photos*)   221

13.3 Axis collaborationists denounced in Cherbourg. A French
      Resistance Party member explains the shorn hair of young
      women of Cherbourg accused of violating rules set down
      by the French Resistance Party for conduct with the
      Germans. The women were rounded up on Bastille Day
      morning, 14 July 1944, shorn, and paraded through
      the streets. (*Courtesy of the Imperial War Museum*)   225

13.4 Members of the French Resistance Party mete out
      punishment to all known 'Axis collaborationists'.

Housemaids, servants, etc. of the Germans were
gathered together, shorn of all their hair, and paraded
through the streets of Cherbourg, their bald heads the
emblem of their violation of the rules of the Party.
(*Courtesy of the Imperial War Museum*)   226

13.5   Collaborationist in Paris (no date). French civilians wave
the victory sign after shaving the head of a French woman
accused of collaborating with the Nazis. A swastika has
been painted on her forehead. (*Courtesy of the Imperial
War Museum*)   227

13.6   Montereau (no date). (*Courtesy of the Imperial War Museum*)   228

Notes on the Contributors

Nicholas Atkin is Lecturer in History at the University of Reading. He is the author of *Church and Schools in Vichy France, 1940–1944* (New York, 1991) and *Pétain* (London, 1998). He is also joint editor (with Frank Tallett) of several works, the latest of which is *The Right in France, 1789–1997* (London, 1997).

François Bédarida is former director of the CNRS Institut d'histoire du temps présent in Paris. He has published widely on British and French history, and among the latest works he has edited are *L'Histoire et le métier d'historien en France, 1945–1995* (Paris, 1995), *1938–1948, les Années de tourmente: de Munich à Prague: dictionnaire critique* (Paris, 1995) and (with J.-P. Azéma) *La France des années noires* (Paris, 1993).

Martyn Cornick is Reader in Contemporary French Studies at the University of Birmingham and is chair of the editorial board of *Modern and Contemporary France*. He has published widely in the fields of Franco-British perceptions and in French intellectual history, in particular a book on Jean Paulhan and the *Nouvelle Revue française* during the interwar period (Amsterdam, 1995).

Malcolm Crook is Professor of French History at Keele University. He has written extensively on the French Revolution and his most recent book is *Napoleon Comes to Power: Democracy and Dictatorship in Revolutionary France, 1795–1804* (Cardiff, 1998). He is currently exploring the development of electoral behaviour in France from 1789 to 1914.

Ceri Crossley is Professor of Nineteenth-Century French Studies at the University of Birmingham. He is the author of *French Historians and Romanticism* (London, 1993), and has published widely on aspects of nineteenth-century French thought and literature and also on Anglo-French cultural relations. He serves on the editorial committee of the newly formed society, *A la recherche de Michelet,* and is currently working on issues relating to natural history, medecine and science fiction in nineteenth-century France.

Claire Duchen was Senior Lecturer in French at the University of Sussex. She wrote extensively on feminism and women in postwar France and was the author of *Feminism in France from May '68 to Mitterrand* (London, 1986) and *Women's Rights and Women's Lives in France 1944–1968* (London, 1994). She recently co-edited *When the War was Over: Women, War and Peace in Europe, 1940–1952* (London, 1999).

John Dunne is Senior Lecturer in History at the University of Greenwich, and researches on Napoleonic politics and society, and French elites before and after the Revolution. His publications include a volume (with Jérôme Decoux) in the series *Grands Notables du Premier Empire* (Paris, 1993).

Robert Gildea is Reader in Modern History at the University of Oxford and a Fellow of Merton College. Among his publications are *The Past in French History* (New Haven, 1994) and *France since 1945* (Oxford, 1996). He is currently writing a study of the French under the German Occupation, with special reference to the Loire valley.

Alec G. Hargreaves is Professor of French and Francophone Studies in the Department of European Studies at Loughborough University. Among his recent publications are *Immigration, 'Race' and Ethnicity in Contemporary France* (London/New York, 1995) and (co-editor with Mark McKinney), *Post-Colonial Cultures in France* (London/New York, 1997).

Brian Jenkins is Professor of French Area Studies at the University of Portsmouth. He is the author of *Nationalism in France: Class and Nation since 1789* (London, 1990) and editor of the *Journal of European Area Studies*. He is currently preparing a book on *Nationalism in Europe: History and Theory* to be published by Macmillan.

Peter M. Jones is Professor of French History at the University of Birmingham. He has written extensively on the French Revolution and on rural France in the seventeenth and eighteenth centuries. His most recent book is *Reform and Revolution in France, 1774–1791* (Cambridge, 1995).

H. R. Kedward is Professor of History at the University of Sussex. Included among his major publications on France during the Second World War are *The Liberation of France: Image and Event* (editor with Nancy Wood) (Oxford, 1995), *In Search of the Maquis: Rural Resistance in*

Southern France, 1942–1944 (Oxford, 1993) and *Occupied France: Collaboration and Resistance 1940–1944* (Oxford, 1985).

Robert J. Knecht is Emeritus Professor of French History in the University of Birmingham. He has been Chairman of the Society for the Study of French History and is the author of several works on sixteenth- and seventeenth-century France, including *Renaissance Warrior and Patron: the Reign of Francis I* (Cambridge, 1994) and *Catherine de' Medici* (London, 1998).

Before retiring, **Maurice Larkin** was Richard Pares Professor of History in the University of Edinburgh (1976–99). His books include *France since the Popular Front* (Oxford, 1997), *Religion, Politics and Preferment in France since 1890* (Cambridge, 1995), *Man and Society in Nineteenth-Century Realism* (London, 1977) and *Church and State after the Dreyfus Affair* (London, 1974).

Pamela Pilbeam is Professor of French History at Royal Holloway, University of London. Her most recent publications are *The Constitutional Monarchy in France 1815–48* (London, 1999) and *Socialists before Marx: Women, Workers and the Social Question in France* (London, 2000). She has just completed a book on the French early socialists. Another recent publications is *Republicanism in Nineteenth-Century France 1814–71* (Basingstoke, 1995).

Siân Reynolds is Professor of French at the University of Stirling and was the second President of the Association of Modern and Contemporary France (1993–9). She works on both French and Scottish history, and her books include *Britannica's Typesetters: Women Compositors in Edwardian Edinburgh* (Edinburgh, 1989) and *France between the Wars: Gender and Politics* (London, 1996).

David R. Watson taught European History at the University of Dundee until retirement in 1996. He is the author of a biography of Clemenceau, and of many articles on French and European history, the latest of which is 'La Russie nouvelle', in M. Petricioli, *Une Occasion manquée? 1922: la reconstruction de l'Europe* (Berne, 1995).

Introduction

Martyn Cornick and Ceri Crossley

It is an honour to introduce this collection of essays offered to Emeritus Professor Douglas Johnson. Few have done as much to encourage and foster the disciplines of French history and French studies over the past half-century, so there could hardly be a more pleasurable task than to gather together these chapters from some of Professor Johnson's friends, colleagues and former students. Indeed, the students of France who owe an intellectual or professional debt to him are so numerous that had we been able to, we could have doubled the length of this book with contributions of the first rank.

We are aware of the criticisms that tend to be made of *Festschriften*, that they are sometimes said to contain disparate material and that they do not always succeed as a unified or coherent whole. Here, however, diversity should be seen as having positive value, for the very diversity of these contributions pays tribute to the unparalleled range of Douglas Johnson's interests, influences and expertise. After all, and to quote the words of Fernand Braudel, 'la France se nomme diversité'.[1] The title of the collection reflects the view that French history is never straightforward, that it is often problematic and full of paradoxes, contradictions and echoes. It has now become virtually a commonplace to say that in France the past is always present; yet that is precisely what makes the French story such a perennially inspiring subject. Douglas Johnson has always been able to impart this fascination to his students and fellow academic colleagues alike. The diversity of these essays also owes much to the fact that Douglas Johnson has helped French studies break free from what were sometimes narrow academic boundaries, away from the strictly literary or the solely historical to a state of affairs where academic things French may now be apprehended in their entirety. Today France is studied across a number of areas, including cultural and interdisciplinary studies, history, literature and political and social science, each of which boasts a lively association or journal, or both.

The essays offered here cover mainly the modern and contemporary periods of French studies. Some of the chapters cluster in given periods, such as the post-Revolutionary era, the *fin-de-siècle* and the Second World War. There are also a number of themes common to several of the contributions, including intolerance, the relationship between politics and religion, town and country tensions within France, politics and nature. Questions of identity – a subject which has long fascinated Douglas Johnson – are also explored in the form of studies of France and its Others (on England, Scotland, Italy, Spain and Germany), and reflexions on internal differences within France.

We have presented the contributions in loose chronological order, beginning with Robert Knecht, whose essay reassesses the evidence regarding the role of Catherine de' Medici in the Massacre of St Bartholomew's Day and examines the manner in which her reputation has been treated by propagandists, writers and historians. In a stimulating case study concentrating on towns in the southern Aveyron during and after the French Revolution, Peter Jones describes how the inherited grievances of local Protestants fuelled the fires of Jacobin intolerance. The idealism of 1789 failed to put an end to the antagonisms between local Protestant and Catholic factions. Sectarianism and gang violence were a prelude to a period of dictatorship in which Protestants sought revenge on their religious enemies.

John Dunne's interest is centred on a different region of France and at a slightly later period, that of the First Empire. His essay deals with Normandy and addresses the problems posed by the contrasting areas of the Pays de Caux (conservative) and the Pays de Bray (republican). To what extent did different types of farming lead to contrasting political allegiances? The nineteenth century is remarkable for the difficulties thrown up by attempts to turn the project of a democratic society into political practice. In his essay Malcolm Crook examines the extent of popular participation in the electoral process. He calls into question the notion that participation increased from a low level in the 1790s to reach a high point in 1848. Crook suggests that rather than accept that a general or uniform development took place, we should look instead at voting culture in France. The spirit of 1848 is central to the essay by Pamela Pilbeam, who describes the fascinating career of Ange Guépin of Nantes, unsuccessful politician but tireless socialist and republican activist. Guépin was a committed intellectual, a prominent figure who struggled to improve the conditions of the poor and raise the position of women.

Ceri Crossley traces the changing attitudes towards animals in France from the late eighteenth to the early twentieth century. His chapter draws attention to the extent to which the supporters of socialism and republicanism were often supporters of animal welfare and sometimes advocates of vegetarianism. In an essay which may be seen as a post-script to his previous work on nationalism, Brian Jenkins discusses the emergence of opposing models of nationalism in the post-Boulangist France of the last quarter of the nineteenth century. How did nationalism, in origin a left-wing movement arising in 1789, come to be reshaped, remoulded as a xenophobic, right-wing populism hostile to liberal democracy? For Jenkins the key element was religion. This was the moment when the Catholic Right abandoned traditional forms of royalism and espoused a mass politics grounded in militarism and anti-Semitism. In his essay Martyn Cornick focuses precisely on the xenophobia and the intolerance which marked not only French but also British political culture at the close of the nineteenth century. He explains how on both sides of the Channel the raising of issues of national and cultural identity led to the promotion of a discourse of superiority. Each nation sought in the other a defining opposition to itself while recognizing in difference a distorted image of identity.

In her essay devoted to Patrick Geddes, Siân Reynolds also focuses on a turn-of-the-century moment which was crucial for Franco-British relations. She describes how at a time of international tension Geddes forged intellectual links between Scotland and France. Geddes participated in the Paris Exhibition of 1900 and developed a network of collaborators which included numerous liberals and Dreyfusards. A further insight into the relations between politics and culture in the final decade of the nineteenth century is provided by David R. Watson, who focuses not on Clemenceau the statesman but on Clemenceau the journalist, novelist and intellectual. Watson's researches demonstrate that Clemenceau's influence was widespread in the years preceding the Dreyfus Affair and that his reputation in the broader intellectual sphere merits reassessment.

Closer to the contemporary period, Nicholas Atkin offers a reappraisal of the career of Otto Abetz, who was Hitler's ambassador in Paris and a close associate of Pierre Laval. Atkin illuminates the ways in which groupings designed to foster Franco-German reconciliation in the inter-war years were manipulated by Berlin, and he questions the account Abetz gave of his activities after the war. Robert Gildea, in an essay which challenges received views, examines the sub-culture of clandestine dances which flourished in the western departments during the

Occupation years. Drawing on archives, including *gendarmerie* reports, and on personal testimony, Gildea describes how illegal dances continued to be organized despite the opposition of the authorities. Gildea asks us to consider to what extent such practices were essentially recreational, the expression of a youth sub-culture, or to what extent they amounted to acts of symbolic resistance against the regime of Pétain.

The power of photographic images is examined by the late Claire Duchen. In her essay she asks whether it is adequate to consider the photographs of women collaborators with shorn heads simply as documentary. Why were women designated as symbolic targets for revenge after the Liberation? To what extent were they scapegoats for a collective sense of guilt? This conflation of the sexual and the political prompts a reassessment of the significance of gender in the post-Liberation purges. In a wide-ranging piece Roderick Kedward raises the issues of ruralism and regionalism in French political culture between the 1920s and the 1950s. At the centre of his argument lies an examination of the regressive ruralist ideology which was promoted by Vichy. However, Kedward invites us to envisage the possibility that alternative ruralisms existed, that alongside the anti-industrial, anti-urban ideology of the nationalist Right we can find examples of a republican, democratic vision of the rural world.

In his essay Alec Hargreaves asks why the contribution made by minority ethnic groups to French identity remains inadequately recognized. Hargreaves considers the place given to immigrants in republican constructions of French identity which take their source in the universalist rhetoric of 1789. Drawing on evidence provided by the Bicentenary of the French Revolution (including its archives), Hargreaves concludes that republicanism remains selective. The assimilationist model of the secular republic has intolerant edges and immigrants of Afro-Caribbean and especially of Algerian origin have been pushed into the margins of society. France is yet to come to terms with the postcolonial, post-imperialist dimension of immigration. In his contribution Maurice Larkin adopts a comparative approach and draws attention to both the similarities and differences between Spain and France in the twentieth century. He contrasts Vichy France with Spain under Franco and evaluates the decline in church attendance in both countries since the 1960s. In a thoughtful and reflective essay François Bédarida offers an overview of crucial recent debates on the competing claims of memory and history. His piece constitutes a penetrating meditation on the historian's mission to organize the past and present it faithfully to his/her contemporaries.

*

We should like, finally, to express our gratitude to Douglas Johnson himself, who, during several meetings at his former academic home in Birmingham, and on other snatched occasions, regaled us with his recollections. This made our task as editors all the more enjoyable. These recollections form the basis of the following account of his academic career.

Douglas Johnson was born in Edinburgh, Scotland, in 1925. As his parents moved to Lancaster, his schooling from 1934 was at Lancaster Grammar School. In 1942 he won an Open Scholarship to Worcester College, Oxford. He was called to military service during 1943 and 1944 (in the Northamptonshire Regiment), being invalided out in that year. He stayed on for a year in Oxford as a postgraduate student, writing a B.Litt. thesis on Sir James Graham, who was Home Secretary to Sir Robert Peel from 1841 to 1846. After the Liberation of France, in 1947 he was awarded a scholarship to the Ecole normale supérieure (ENS) in Paris, being the first British subject to have the title of 'élève à titre étranger', a status, or title, introduced after the war. He was the British student among a select group of pioneering 'foreign students', which included one Irish, one Egyptian, one Dutch, one Hungarian, one Czech, one Norwegian and one Canadian. These were effervescent times: in 1948, Douglas Johnson witnessed a return visit to the ENS (through a back entrance) of Jérôme Carcopino, the eminent historian of ancient Rome and sometime Minister of Education under the Vichy regime. This secret visit could have caused quite a scandal had news of it leaked out, so the witness kept quiet... Douglas Johnson's tutor at the Paris Ecole was René Rémond (latterly of the Académie française); his research on French history was supervised by Professor Charles Pouthas and his work on the beginnings of French colonization in Algeria by Professor Charles-André Julien. Fellow history students at the ENS included François Bédarida (who has contributed to this volume), Jacques Le Goff, Louis Bergeron, Alain Touraine, Maurice Agulhon and Pierre Ayçobéry. Apart from historians he became close friends with two philosophers who later became closely associated with Marxism: Louis Althusser and Maurice Caveing.

In 1949 Douglas Johnson was appointed Assistant Lecturer in History at the University of Birmingham. At that time and until 1960 the History Department was located in central Birmingham, in Edmund Street, although he was housed, along with Dr (later Professor) R. H. Hilton, in a distant part of the building situated in Easy Row. He had been expecting to teach French history and the history of the French conquest and colonization in Algeria after 1830, but he found himself teaching

Tudors and Stuarts. Eventually, however, he was able to take a Special Subject in French history, and although he published some of his work on Sir James Graham, his main preoccupation was with French history during the July Monarchy. His interest in Algerian history was maintained by invitations from Sir Keith Hancock to take part in his seminar on European overseas expansion at the Institute of Commonwealth Studies in the University of London, and by invitations to lecture at the School of Oriental and African Studies.

In 1963 he published *Guizot: Aspects of French History 1787–1874* and, in spite of assumptions that he would proceed by writing about Thiers, then published *France and the Dreyfus Affair* in 1966. A lecture on 'The Political Principles of General de Gaulle' to the Royal Institute of International Affairs (which appeared in the 1965 Chatham House publications) led to many opportunities to speak and to take part in discussions on contemporary French politics. There was much scope for this after he had succeeded Alfred Cobban as Professor of French History at University College, London, in 1968 (he had become Professor of Modern History in the University of Birmingham in 1963).

In London, apart from undergraduate teaching and postgraduate supervision (at one point he had nineteen students engaged on doctoral research), he wrote a number of general books (*A Concise History of France*; *France*, in the Thames and Hudson series on 'Nations and Peoples'; *The French Revolution*, for use in schools). But his main research preoccupation was now directed towards contemporary French history. He lectured in meetings organized by the Fondation Charles de Gaulle (and was elected British representative of the Comité scientifique of that body), the Association Maréchal Leclerc, the Association Georges Pompidou, the Société d'histoire moderne, the Institut d'histoire du temps présent and other organizations. Many of these lectures have been published. In 1972 the Franco-British Council was formed by President Pompidou and Sir Edward Heath and Professor Johnson was elected a member. The first publication of the Council was *France and Britain: Ten Centuries of History* (edited by François Bédarida, François Crouzet and Douglas Johnson) published in Britain and France in 1980. In 1982 Professor Johnson organized a series of public lectures on General de Gaulle, held in University College during the Christmas Term (these were published in *Espoir: Revue de l'Institut Charles de Gaulle*).

Douglas Johnson has published books with others, such as *The Idea of Europe* with Richard Hoggart (1987) and *The Age of Illusion* with Madeleine Johnson (1987), who is his wife. They met as students in Paris, when Madeleine Rébillard was studying at the Ecole normale supérieure de Sèvres, the then female part of the Ecole normale supérieure. (Richard Cobb wrote that the presence in London of a 'normalien' and a 'Sévrienne' was a coded message of excellence...)

Douglas Johnson has also contributed articles to many collective works, such as *Les Démocraties occidentales: craintes et espérances* (Fondation internationale des sciences humaines, 1977); *The Permanent Revolution* (edited by Geoffrey Best, 1988) and *De Gaulle and Twentieth-Century France* (edited by Hugh Gough and John Horne, 1994). Indeed it would be difficult to list here all the articles and book reviews which he has written for newspapers and periodicals. One should perhaps mention the weeklies, such as *New Society*, *The Times Literary Supplement*, *The Spectator*; daily newspapers, particularly *The Guardian*, *The Independent*, *The Birmingham Post* and *Ouest-France*; other reviews include *The London Review of Books*, *The New York Review of Books*, *The Los Angeles Times Review*, *Prospect*, *The Literary Review*, *History Today*, *Espoir*, *Histoire et Géographie*, *History*, *Revue d'histoire moderne et contemporaine*, *Raison présente* and *Modern and Contemporary France*. And of course, Douglas Johnson was a founder of the Association for the Study of Modern and Contemporary France which publishes this last-mentioned journal.

Michelet and the French Revolution, the Zaharoff Lecture at Oxford, *How European are the French?*, the Stenton Lecture at the University of Reading, are two examples of the formal lectures that Douglas Johnson has given, but like others who have worked in adult education, he likes to look back on the days when he taught for the Workers' Educational Association (WEA), for Extra-Mural Departments and for other bodies (like the Historical Association). The first course of lectures that he ever gave was for the WEA in Hanley, Stoke-on-Trent, in 1949. The recollections of those days do not affect the pleasurable memories of lecturing in French universities, notably as Visiting Professor in the University of Aix-Marseille in 1963 and in the University of Nancy in 1968.

On retiring from University College in 1990 and becoming Professor Emeritus in French History, he was appointed Visiting Professor to the French Department in King's College, London. He was the first President of the Association for the Study of Modern and Contemporary France on the creation of this society in 1980, and he remains very active in the Association as its Honorary Vice-President. He is a member of the National Commission for the publication of the works of Alexis de Tocqueville, a member of the jury for the Prix Guizot in history, an adviser to *Vingtième siècle*, the review of the Institut d'histoire du temps présent in Paris.

Editors' Note

We wish to thank the Executive Committee of the Association for the Study of Modern and Contemporary France for their support in this project. On behalf of Claire Duchen, the editors would like to acknowledge the assistance of Magnum and the Photographic Department of the Imperial War Museum.

Note

1   F. Braudel, *L'Identité de la France, espace et histoire* (Paris, 1986), book 1, chapter 1.

Select Bibliography

(including edited or co-authored works, in chronological order)

Guizot: Aspects of French History, 1787–1874 (London and Toronto, 1963).
France and the Dreyfus Affair (London, 1966).
France (London, 1969).
Editor of *The Making of the Modern World*, 2 vols (London, 1971, 1973).
Editor of *French Society and the Revolution* (Cambridge, 1976).
Editor, with F. Crouzet and F. Bédarida, of *Britain and France: Ten Centuries* (Folkestone, 1980). A French version was published under the title *De Guillaume le Conquérant au Marché commun: dix siècles d'histoire franco-britannique* (Paris, 1979).
Johnson, D. and Madeleine Johnson, *The Age of Illusion: Art and Politics in France, 1918–1940* (London, 1987).
Johnson, D. and Hoggart, R., *An Idea of Europe* (London, 1987).
Editor, with D. S. Bell and P. Morris, of *Biographical Dictionary of French Political Leaders since 1870* (London, 1990).
Michelet and the French Revolution (Oxford, 1990) (The Zaharoff lecture, 1989).
'L'Empire britannique en question', in G. Pedroncini and General P. Duplay (eds), *Leclerc et l'Indochine 1945–1947* (Paris, 1992).
'Historians, Nations and the Future of Europe', in M. Fulbrook (ed.), *National Histories and European History* (London, 1993).
'Introduction: Louis Althusser 1918–1990', in O. Corpet and Y.-M. Boutang (eds), *Louis Althusser: The Future Lasts a Long Time* (London, 1994).
'La Grande-Bretagne et la Libération de Paris', in C. Levisse-Touzé (ed.), *Paris 1944: les enjeux de la Libération* (Paris, 1994).
'De Gaulle and France's Role in the World', in H. Gough and J. Horne (eds), *De Gaulle and Twentieth-Century France* (London, 1994).
How European are the French? (Reading, 1996) (The Stenton lecture, 1995).
'La Politique européenne de Georges Pompidou', in *Georges Pompidou et l'Europe* (Paris, 1995).
'Témoignage étranger', in F. Bédarida, *L'Histoire et le métier d'historien en France 1945–1995* (Paris, 1995).

'Foreign Policy', in J. Gaffney and L. Milne (eds), *French Presidentialism and the Election of 1995* (Aldershot, 1997).

'France and Germany', in A. Pök (ed.), *Fabric of Modern Europe: Studies in Social and Diplomatic History – Essays in Honour of Eva Haraszti-Taylor* (London, 1999).

1

Catherine de' Medici: Saint or Sinner?

Robert J. Knecht

Propaganda feeds on history. In recent times we have seen how the reality of the Holocaust has been disputed by neo-Nazis. Historical biography, too, has been used to support or denigrate a political cause. Thus Joan of Arc has become a symbol of patriotic fervour and religious orthodoxy for the French National Front. Such a use of the past, however, has a long ancestry. In England, Richard III, the last of the Yorkist kings, became the target of a campaign of vilification by the Tudor monarchs who succeeded him. The unflattering portrait painted by Sir Thomas More in his *History of King Richard III* served as the model for the villainous hunchback of Shakespeare's play. One might assume that deliberate distortions of this kind would eventually be corrected by historians seriously interested in finding the truth. But sometimes distortion leads merely to counter-distortion. Thus the rehabilitation of Richard, which stems doubtless from a well-meaning desire to right an injustice rather than from any political stance, has become a veritable cult expressed in the activities of the Richard III Society, originally known as the Fellowship of the White Boar. These include the erection of a monument and the depositing of floral tributes on Bosworth field at the spot where the horseless Richard was allegedly slain. Yet the evidence that he did murder his royal nephews in the Tower of London seems overwhelming.[1]

Few historical reputations have fluctuated as wildly as that of Catherine de' Medici, queen of France from 1547 until 1559, and several times regent before her death in 1589. No historian would question the importance of her role in the civil wars, generally known as the Wars of Religion, which tore France apart in the second half of the sixteenth century. There is, however, a sharp difference of opinion as to her policies: did she consistently strive to bring peace to the kingdom by healing its religious divisions, or did she inflame the situation by playing one side

1

against the other and using violent means to get rid of political opponents? Traditionally, historians have been critical of Catherine, arguing that after she had failed to secure a religious compromise at the Colloquy of Poissy in 1561 and to impose a measure of religious toleration in a series of royal edicts, she threw in her lot with the Catholic extremists in 1572 and instigated the Massacre of St Bartholomew's Day in which thousands of Protestants or Huguenots were butchered in cold blood, including their leader, Admiral Coligny. Recently, there has been an attempt – initiated by Nicola Sutherland and continued by Jean-Louis Bourgeon – to whitewash Catherine.[2] They see her as the victim of a 'Black Legend'. Propaganda is seldom more effective than when it blames all the wrongs of this world on a single individual. In sixteenth-century France, Catherine offered the perfect target to a hate campaign fuelled by xenophobia, social snobbery and misogyny. As a Florentine, she was seen as deceitful by nature and a skilled poisoner; as the scion of a family deemed to be upstart (the Medici had risen socially through trade and banking), she was regarded as envious of France's ancient nobility; and, as a woman, she exemplified the 'monstrous regiment' denounced by John Knox and other male preachers and political theorists as unfit to rule.

Many Huguenot pamphlets published after the massacre extended to the royal family and to Catherine in particular attacks which had been previously directed only at the Guises. A notable example is the *Discours merveilleux de la vie, actions et déportements de Catherine de Médicis, Royne-mère* [Marvellous discourse on the life, actions and conduct of Catherine de' Medici, queen-mother] which purports to be a factual account of her life.[3] The anonymous author commends the wisdom of remembering good rulers and forgetting bad ones, but he thinks Catherine is an exception. She is 'a true picture and example of tyranny in its public manifestations and of kinds of vices in its private ones'. Fearful as he is of soiling his hands and making himself sick 'by stirring and smelling such a vile and stinking matter', he nevertheless feels bound 'to show to everyone the kind of woman who has us beneath her claws, and by looking at her past what can be expected of her in the future, unless we can find means of escape'. His purpose, he says, is not just to speak ill of Catherine but to prevent her from doing more harm. He reminds the reader that she is an Italian and a Florentine. 'Among the nations,' he writes, 'Italy takes the prize for cunning and guile; in Italy it is Tuscany, and in Tuscany, it is the city of Florence.' What is more, Catherine belongs to the house of Medici which for a long time lay 'unnoticed beneath the dregs of society'. Its rise was begun 'by a charcoal-burner (*charbonnier*) who acquired a little wealth'. His son, a doctor, adopted the family's badge of five pills 'as artisans today are wont to add the tools of their trade to their signs'. His family, which Florentine historians overlooked, began to draw notice to itself when one of its members 'put himself at the head of the common people against the patricians or nobles'. The Medici then enriched themselves 'through banking and usury' and gained control of Florence by corrupting the people. The author concludes that the French nobility can hope for nothing other than 'humiliation and extinction' from a woman of such lowly origins. After painting a lurid picture of the pontificates of Catherine's two papal uncles, Leo X and Clement VII, the *Discours* focuses on her father, Lorenzo, 'a man consumed by every kind of wickedness including adultery and incest; a man blinded by ambition who aspired to greatness simply to do more harm'.

As for Catherine herself, her birth was most inauspicious. According to the *Discours*, astrologers whom her parents consulted 'were unanimous in affirming that (if she lived) she would bring huge calamities and eventually total ruin to the house into which she would marry'. In 1533 Catherine married Francis I's younger son, Henry, and soon afterwards she arranged the poisoning of her brother-in-law, the Dauphin François, to clear the way for her husband to become king. Henry II, however, thought poorly of her. He told the Constable Montmorency, who backed her request for a greater share of government, 'that she was the biggest muddle-head in the world, and that, given the chance, she would spoil everything'. Following Henry's tragic death in a tournament in 1559, Catherine assumed a dominant role in government either as queen-mother or as regent for her three sons Francis II, Charles IX and Henry III in succession. Frustrated by Francis II's reliance on the Guises, she moved the Bourbons to attempt their overthrow in the Conspiracy of Amboise. When the coup failed, she disclaimed all responsibility and demanded that the plotters be punished. Under Charles IX, whom she soon corrupted, she courted the Protestants and promoted the tolerant edict of January 1562. Later, as the religious wars got underway, she switched to the Catholic side. After a vain attempt to poison the Huguenot leader, Condé, she had him murdered on the battlefield after he had surrendered. Another of her alleged victims was Odet de Châtillon, who died in England allegedly of poisoning. Catherine also rejoiced over the assassination of the second duc de Guise, saying that she had lost 'one of the men she hated most in all the world'. Her greatest coup, however, was the Massacre of St Bartholomew. Her intended target was not the Protestants, but the nobility. She had arranged the marriage of her daughter, Marguerite, to Henry of Navarre in order to lure Catholic and Huguenot nobles to their deaths in Paris. When the Catholics escaped, she tried to get them slaughtered at the siege of La Rochelle. Catherine then turned against her own sons. When Charles IX showed signs of independence, he was given poisoned fish to eat, while his brother, Henry of Anjou, was packed off to Poland in order to get him out of the way. When her youngest son, Alençon, tried to strike out on his own, Catherine set out to make him 'hateful to Catholics by spreading calumnies and false rumours'. He was accused of plotting with the Huguenots to set fire to Paris while Catholics were at mass. On the death of Charles IX, Catherine assumed the regency, despite the Salic law. Only the Estates-General, according to the *Discours*, had the right to appoint a regent. She could not be allowed to get away with her *coup d'état*, since she was 'a woman, a foreigner, an enemy hated by everyone'. At this point the *Discours* launches into a diatribe against government by women in France: 'no woman has ever ruled our kingdom without bringing to it misfortune.' Catherine is compared to Brunhild, queen of the Franks from 567 to 613, who was torn to pieces for her crimes by being dragged on the ground attached to a horse's tail. 'Now let everyone judge,' asks the *Discours*, 'what kind of sentence she deserves who has caused more men, women and children to be slaughtered in a single day than Brunhild caused men to die in all of her wars?' The author ends his book with a patriotic call to action.

> Let us recognize that whatever our differences of religion may be, we are all Frenchmen, the lawful offspring of one nation, born into the same kingdom, subject to the same king and that Brunhild should no longer make us believe, in order to make us kill each other, that our brothers are bastards or other than true Frenchmen. In the end, as you can see, she would make us all die. Let us therefore march together in step and with one heart of whatever class we are, nobles, bourgeois, peasants, and force her to liberate our princes and lords.

Internal evidence suggests that the *Discours* was written in mid-1574 after the death of Charles IX and before the return to France from Poland of his brother, Henry III. Catherine acted as regent during this period. Like many such works, the *Discours* was published without any indication of author, printer or place of publication. Many scholars have suggested that the author was a French Protestant exile, probably Henri Estienne, the classical scholar and printer of Greek texts. He was then living in Geneva and was soon to publish a polemic attacking the influence of Italians on the French language and French culture. The name of Innocent Gentillet, who wrote the *Anti-Machiavel*, has also been put forward as the author of the *Discours*. But Simon Goulart, who included it in his *Mémoires sur l'estat de la France*, claimed that it was the work of a 'politique', who was appalled by Catherine's seizure of power. The latest editor of the *Discours* favours someone who wished to promote an alliance between the Huguenots and malcontent Catholics against the house of Valois.[4] What is clear is that the author, whoever he was, had personal knowledge of the French court in the 1560s. The *Discours* was first published anonymously in 1575 and two years later in a revised edition, which shows signs of Huguenot tampering, possibly by Goulart. The *Discours* was an immediate success in France and abroad, running through several editions in French and other languages. As propaganda against female rulers, its impact was long-lasting: it was used against Marie de' Medici in the seventeenth century and against Marie-Antoinette in the eighteenth.

The *Discours* was used as a rich quarry by historical novelists in the nineteenth century. Perhaps as a reaction to the cult of Henry IV after 1815, a veritable vogue for the Wars of Religion and Catherine de' Medici in particular swept France in the 1820s. Charles d'Outrepont, a justly forgotten writer, wrote a play in 1826 on the Massacre of St Bartholomew in which Catherine is described as 'an execrable woman whose memory will remain in a bloody crape till the end of time'. Balzac, who at an early stage in his career hoped to become a French Walter Scott, wrote three essays, one of them a conversation between Catherine and Robespierre. They were subsequently published together under the title *Sur Catherine de Médicis*. In the author's judgement, her guiding stars were 'love of power and astrology'. Unlike most historians, he dismissed her predilection for her son, Henri. 'Her conduct,' he writes, 'proved the total hardness of her heart.' Historical fiction found its genial champion in Alexandre Dumas, who, as early as 1829, wrote a play, called *Henri III et sa cour*, in which Catherine is shown to be the real ruler of France.[5] Better known is his novel *La reine Margot*, in which Catherine is portrayed as a malevolent spirit presiding over a debauched court. In 1993 it was turned into an extremely gory film by Patrice Chéreau and Danièle Thompson, which will doubtless serve to extend the Black Legend.

Of course, the legend has not been the exclusive preserve of authors of fiction. It has been taken up by several generations of historians with varying degrees of enthusiasm. While admitting that it 'appears to have sprung from the extreme polemical writings of the Protestant pamphleteers, who laid on Catherine all the blame for the massacre of St Bartholomew', Nicola Sutherland argues that three serious histories of the age of Catherine published in the early seventeenth century, namely those of de Thou, Agrippa d'Aubigné and Davila, failed to contribute to the legend. 'It was only later,' she writes, 'that this began to seep into the mainstream of history.' Among the first to present it 'in all its ugly, incoherent vituperation' was Madame Thiroux d'Arconville's *Histoire de François II* (1783). The legend was endorsed by Henri Martin in his *Histoire de France* (1834–6), by Chateaubriand in numerous writings, and, above all by Jules Michelet in his *Histoire de France*, whom Sutherland charges with a 'staggering arrogance' for his claim to have seen the face of the sixteenth century. In her judgement, Michelet's version of the legend was quite as hysterical as that of Madame d'Arconville and 'descended to pure abuse of a remarkably vulgar kind'. Another of Catherine's detractors was Victor Duruy. 'Catherine,' he writes, 'decided then on a Machiavellian plan which was to have Coligny murdered by the Guises. The Huguenots would round on them to avenge their leader, then the royal troops would fall on both parties as violators of the public peace.'[6] Yet, as Sutherland indicates, Catherine has also had defenders among nineteenth-century historians, the first to denounce the legend being Capefigue, who published a life of the queen in 1856.[7] As for Jean-Louis Bourgeon, that other modern champion of Catherine, only the wildest speculation can account for the legend: 'Following the greatest names in French historiography, including Michelet, Lavisse, Hauser and Braudel, one has been satisfied for far too long with a history, both poorly "psychological" and haphazard, unconnected with the deep texture of the actual forces.'[8]

Michelet did not like Catherine, but, far from endorsing the 'Black Legend', he dismissed her as incapable of any initiative for good or evil. Henry II, he informs us, found her physically repulsive from the start of his marriage. He recoiled from her 'as from a maggot born in the tomb of Italy'. Her father, having died of syphilis, was soon followed to the grave by her mother; so that one may even wonder whether Catherine herself was born alive. 'Cold like the blood of the dead, she was able to bear children only at times forbidden by medical science' and aided by her physician. The result was an awful brood: 'a diseased king' (Francis II), a 'mad' one (Charles IX) and an 'unstable' one (Henry III). 'Thus purged,' writes Michelet, 'fertile with sick and dead children, she grew up, fat, jolly and laughing, amid our terrible disasters.'[9] He has nothing but contempt for her. 'She had little initiative,' he continues, 'and no boldness not even to do evil. She followed events from day to day, adapting her ethical indifference, her mendacity and her dexterity to whatever path seemed likely to succeed.'[10] Thus Michelet does not accuse Catherine of plotting Coligny's murder. Tavannes, he believes, exaggerated her role in this affair. Catherine may have agreed to the deed, 'but never would she have dared to commit such an act without foreign pressure and a great fear'.[11] Who, then, were responsible? Michelet points to the Guises: 'they took everything upon themselves: they provided the assassin, they provided the lodging from which the shot was to be fired; they provided the horse which was to save the assassin.'[12] According to Marguerite de Navarre, the duc de Retz informed the king that Catherine had plotted Coligny's death to avenge the killing of her servant, Charny. 'A big mistake,' writes Michelet, 'which Charles IX was unlikely to swallow. He knew his mother too well, she who had neither heart nor soul nor love nor hate, no *vendetta* for sure.'[13] This may be another Black Legend, but it is a far cry from that of the *Discours merveilleux*.

Catherine's unpopularity was late in coming. In the early stages of the Wars of Religion she was praised by the poet Ronsard for her peacemaking efforts. The turning point was provided by the Massacre of St Bartholomew in August 1572, a crime so heinous as to unleash a flood of pamphlets – most of them Huguenot – condemning the perpetrators.[14] Many Huguenots believed that Catherine had planned the massacre as long ago as 1565, when she had met the Duke of Alba, the chief minister of King Philip II of Spain, at Bayonne. Two years later, Alba was to make his mark on history by ruthlessly trying to crush the Dutch revolt against Spanish rule. He set up a special court, the Council of Blood, which sent many people to their deaths, including two noblemen, Count Egmont and Count Hoornes. No documentary evidence has yet come to light in support of a long-term plot hatched between Catherine and Alba at Bayonne, but this in itself is not enough to clear Catherine from some responsibility at least for the St Bartholomew's Massacre.[15] Let us now consider the evidence.

In April 1562 the Wars of Religion officially began when the Huguenot leader, Louis, prince of Condé, rebelled, seizing control of the town of Orléans. In a manifesto he claimed that he, not the king's ministers, was on the right side of the law.[16] The government of the young king, Charles IX, was then under the control of two members of the aristocratic family of Guise: Francis, second Duke of Guise, and his brother, Charles, Cardinal of Lorraine, both of them staunch Catholics. In February 1563 the duke, who had become a national hero following his recapture of Calais from the English and other military successes, was fatally shot in the back by a Huguenot nobleman outside Orléans. This was to prove a seminal event, for the duke's widow, Anne d'Este, and her family set themselves up as avenging angels. They believed, rightly or wrongly, that the duke's assassination had been instigated by Gaspard de Coligny, Admiral of France, who became the leader of the Huguenot party after the death of Condé at the battle of Jarnac in 1569. They refused to accept royal declarations which cleared Coligny of complicity in the crime, and employed 'hit-men' to murder him.[17] The vendetta, as we shall see, was to surface again on the eve of the St Bartholomew's Day Massacre.

Catherine, at this stage, was still hoping to cool the passions which were causing so much distress and violence in France. She certainly wanted her sons Charles IX and his successor, Henry III, to rule over a kingdom at peace. In 1561 at the Colloquy of Poissy she had tried unsuccessfully to reconcile the rival religious camps.[18] Profoundly ignorant of theology, she failed to realize the depths of passion which dogma is capable of arousing. She apparently believed that a reform of clerical abuses would suffice to reunite the churches.[19] But her initial efforts at peace-making foundered on the intransigence of the Calvinists and extreme Catholics. Catherine next tried to impose the edict of Amboise (19 March 1563), which conceded freedom of conscience to the Huguenots while regulating their rights of worship according to social status.[20] Although neither side liked the settlement, Catherine pressed for its acceptance by the nation in general. In August 1563 Charles IX's majority was declared and the Guises lost control of the government. Soon afterwards Catherine took the young king and his court on a 'grand tour' of France, which lasted two years.[21] She hoped that his authority would be strengthened by making him better known to his subjects. As the tour ended in 1566 the queen-mother seemed well satisfied; she felt that the recent pacification was holding.[22] But the situation in France was not self-contained. It could easily be upset by events elsewhere in Europe, notably in the Netherlands, where a serious revolt, partly Calvinist in character, erupted against Spanish overlordship in 1566. Philip II reacted by dispatching a large army under the Duke of Alba. It reached Flanders from the duchy of Milan by following the so-called 'Spanish Road', which skirted the eastern border of France.[23] At first no one knew exactly where it was heading for. The Huguenots, remembering Catherine's meeting with Alba at Bayonne in 1565, began to fear for their own safety. Even the French government took precautions: Charles IX raised a force of 6,000 Swiss mercenaries.[24] This move also upset the Huguenots, particularly when the Swiss remained in France after Alba had reached Flanders so that he no longer presented a threat to the kingdom. They wondered if the Swiss were going to be used against them. Fearing an attack, the Huguenot leaders decided to launch a pre-emptive strike, which has come to be known as the *Surprise de Meaux*. They tried to kidnap Catherine and her son, the king, as they were relaxing in the countryside at Montceaux. Warned in the nick of time, Catherine and Charles IX took shelter within the walls of Meaux, and, calling the Swiss to their aid, fled under their protection to Paris. The Huguenots, instead of admitting defeat, pursued them to the gates of the capital and proceeded to blockade it.[25] Catherine was understandably furious: her trust had been betrayed and the king's authority defied. In my judgement, neither she nor Charles ever forgave the Huguenot leaders. In June 1568 Catherine was informed that a large force of Huguenots under the Sieur de Cocqueville was marching towards the Flemish border in the hope of helping their hard-pressed co-religionists, but they were intercepted and routed by Marshal Cossé. Catherine instructed him to hand over his Flemish prisoners to Alba for punishment. As for his French prisoners, she wrote: 'I think some should be punished by execution and the rest sent to the galleys.'[26] Speaking to Alava, the Spanish ambassador, she described Alba's execution of Egmont and Hoornes as 'a holy decision', which she hoped soon to repeat in France.[27]

In August 1570 the third civil war was ended by the peace of Saint Germain, which was extraordinarily favourable to the Huguenots considering that they had been, in effect, defeated.[28] The peace may be interpreted in two ways: either as a genuine attempt to heal the religious division of France or as a trap designed to lull the Huguenots into a sense of false security. Arguments can be advanced in support of both readings, though historians generally believe that the French crown was sincere in its efforts to reconcile the Huguenots and Catholics. The peace enabled Catherine to do what she liked doing best: arranging marriages for her children. In particular, she worked for a marriage between the Huguenot prince, Henry of Navarre, and her flighty daughter Marguerite (the *Reine Margot* of Dumas's novel). For this, however, she needed a papal dispensation (which she failed to obtain) and the consent of Henry's mother, the redoubtable Jeanne d'Albret. The latter was an austere Calvinist who strongly disapproved of the French court and its morals. She was afraid that her son, once married, would be forced to abjure his Protestant faith in addition to picking up bad habits.[29]

On 12 September Admiral Coligny returned to the French court, remaining there for five weeks. He was admitted to the king's council and began to press Charles IX to intervene militarily in the Netherlands on the side of the Dutch rebels. Such a move was likely to provoke a hostile response from Spain, but Coligny argued that this would serve to unite Frenchmen against the traditional enemy and deter them from fighting each other.[30] Charles, it seems, was tempted by the Admiral's proposal, for he was madly jealous of the military successes scored by his younger brother, Henry Duke of Anjou, in the recent civil war. He was longing to emancipate himself from his mother's tutelage and to show his mettle by leading his armies into battle. Catherine, however, was horrified at the prospect of a war with Spain, which she knew France could not afford.[31]

Coligny was followed to the French court in March 1572 by Jeanne d'Albret, who had been under considerable pressure from Catherine for some time. Her stay at the court soon turned into an ordeal. The letters which she wrote to her son are filled with despair. Speaking of Catherine, she writes:

> She treats me so shamefully that you might say that my patience surpasses that of Griselda herself . . . I have come here on the sole understanding that the Queen and I would negotiate and be able to agree, but all she does is to mock me. She will not yield at all on the Mass, which she speaks of quite differently from formerly . . . Take note, my son, that they are making every effort to get you here, and watch it carefully . . . I am sure that if you knew the pain I feel you would pity me, for they treat me with all the harshness in the world and with empty and facetious remarks instead of behaving with the gravity the issue merits.[32]

In another letter Jeanne writes: 'I do not know how I can stand it: they scratch me, stick pins into me, flatter me, tear out my fingernails without letup . . . I am badly lodged, holes have been drilled in the walls of my apartment, and Madame d'Uzès spies on me.'[33] In June, Jeanne, who had been ill for some time, died. Catherine was later blamed for her death: her Florentine perfumer had allegedly sold Jeanne some lethal gloves. This charge, however, is nonsense: an autopsy clearly showed that Jeanne had died of natural causes.[34]

The wedding of Henry of Navarre and Marguerite de Valois took place in Paris on 18 August 1572 and brought to the capital a large number of Huguenot and Catholic nobles, including Coligny and Henry third Duke of Guise.[35] On the morning of 22 August, as Coligny was walking from the Louvre to his residence in the rue de Béthisy, he received an arquebus shot fired from an open window of a house close by. If he had not bent down to adjust a shoe at this instant, he would have been killed. Instead he suffered the loss of a finger and a fractured arm.[36] His companions rushed into the house: they found the arquebus still smoking, but the assailant had fled. He has since been identified as the Seigneur de Maurevert, who had tried to murder Coligny on an earlier occasion. An official enquiry revealed that the house from which he had fired the shot belonged to Anne d'Este and the horse on which Maurevert escaped had come from the Guise stables.[37] We can be reasonably sure that the attempt on Coligny's life was another instalment in the Guise vendetta against him, but Catherine has also been blamed. Her most recent biographer, Ivan Cloulas, is in no doubt. He follows a tradition which can be summed up as follows: Catherine was jealous of Coligny's influence over her son and afraid that he might drag him into a war with Spain. She, therefore, plotted with Anne d'Este to have him eliminated under cover of the Guise vendetta.[38] In 1973, however, Nicola Sutherland cast doubt on this explanation. Dismissing the 'maternal jealousy' theory as 'fatuous', she also suggested that Coligny had not had as much influence on Charles IX as historians had supposed.[39] But, as Marc Venard has shown, Sutherland has overlooked evidence contained in letters sent by the papal nuncio in Paris during the summer of 1572. These refer to the queen's jealousy of Coligny and refer to two very long conversations between the Admiral and the king. On 5 August they remained locked in a room with four secretaries from 11 pm until 2 am.[40] Building on Sutherland's research, another historian, Jean-Louis Bourgeon, has recently blamed the Spanish government for the attack on Coligny.[41] The truth is unlikely ever to be known. What is certain is that many people, including Catherine, had good reasons for wishing to be rid of Coligny. She may have been given more than her fair share of blame, but she was no saint and had dabbled in assassination before. Her complicity in the attack on Coligny remains an open question.

The botched attempt on Coligny's life was followed two days later by the Massacre of St Bartholomew's Day. Again the truth is not easily disentangled from the mass of partisan accounts. It is unlikely that the attack on the Admiral was intended to be the opening shot in a campaign against the Huguenots generally. Even if it had been successful, it would have been a tactical blunder, for it would have alerted the Huguenot leaders to the danger facing them; they would probably have left Paris and started a new civil war. The failure of the attack caused panic at court. According to Tavannes, the king and his council met on 23 August and decided that civil war had become inevitable. They thought 'it preferable to win a battle in Paris, where all the leaders were, than to risk one in the field'.[42] Tavannes' memoirs were written by his son long after the event. They may not be reliable, yet Catholics did fear a Huguenot uprising after the attack on their leader. Rumours of a plot to murder the king and his family were rife and Charles IX doubtless remembered the *Surprise de Meaux*. The idea of a pre-emptive strike may have commended itself to him. Be that as it may, we can be reasonably sure that on 23 August he ordered the Huguenot leaders to be wiped out. Catherine was almost certainly a party to that decision.[43] On 24 August members of the king's guard led by the Duke of Guise burst into the Hôtel de Béthisy and murdered Coligny. At the same moment, the great bell of Saint-Germain l'Auxerrois gave the signal for the mass slaughter of Parisian Huguenots. Although the king ordered the killing to stop, it lasted for almost a week. Massacres of Huguenots also took place in other towns.[44] Was the slaughter premeditated? Many contemporaries, both Catholic and Protestant, believed that it was. The Cardinal of Lorraine hinted that it had been planned by the Guises. The Genevan pastor, Simon Goulart, claimed that the planning had begun in 1570 with the peace of Saint-Germain.[45] The idea that the massacre was premeditated has its modern advocates. Bourgeon sees it as a campaign orchestrated by Spain and the Papacy to force a complete change of royal policies in France: the annulment of the Navarre marriage, the abrogation of the edict of Saint-Germain, the return to power of the Guises, the exclusion of the Huguenots from public office and the abandonment of French interference in the Netherlands.[46] However, Denis Crouzet has emphatically ruled out premeditation in his book on the massacre. He believes that royal policy following the pacification of 1570 was to seek to project the kingdom into a golden age of happiness through the fulfilment of a neo-Platonic 'dream of Love' instilled into the young king Charles IX by his humanistic tutors.[47] He does not deny the government's responsibility (and, therefore, Catherine's) for the massacre, but he suggests that the decision was taken precipitately in an ill-conceived attempt to rescue the dream from perdition.

Where does this leave Catherine? Outside France, Catholics greeted news of the massacre with jubilation. Gregory XIII held a *Te Deum*, which was followed by a celebration in the French church of St Louis under the direction of the Cardinal of Lorraine. A special commemorative plaque was struck showing an angel carrying a cross and superintending the killing of Coligny and his friends. Vasari painted a mural in praise of the massacre in one of the halls of the Vatican palace. Catherine was acclaimed as the Mother of the kingdom and the Conservator of the Christian name. Cardinal Orsini congratulated her on her Catholic zeal. Philip II 'praised the son for having such a mother . . . then the mother for having such a son'.[48] She was happy to let the Catholic powers believe that she had long planned the massacre. Her claim, however, was contradicted by the nuncio Salviati. Writing from Paris on 24 August, he said: 'if the Admiral had been killed by the arquebus which was fired at him, I cannot believe that there would have been such a great carnage.'[49] The Spanish ambassador, Zuñiga, gave a similar opinion. 'The Admiral's death,' he wrote, 'was a planned action; that of the Huguenots was the result of a sudden decision.'[50] Catherine hoped to gain something from the massacre. She tried to marry the duc d'Anjou in Spain, but Philip II failed to oblige; so she turned to the Protestant powers. On 13 September, she instructed Schomberg, who was going to Germany as ambassador, not to allow the princes to believe that the Admiral and his accomplices had been killed out of hatred for their religion, but only as a 'punishment for their wicked conspiracy'.[51] She also continued to negotiate with England over the Alençon marriage, and resumed relations with Nassau and the Prince of Orange.[52]

Catherine never showed any regret or remorse over the massacre. In fact, she seems to have rejoiced in its consequences: the Huguenot party had been left leaderless, and Henry of Navarre was now her son-in-law and a king in his own right. What is more, he and Henri de Condé had become Catholics (under duress). The queen-mother's joy became manifest during a ceremony held by the chapter of the Order of Saint-Michel on 29 September. As she watched Henri de Navarre kneel before the altar like any other good Catholic, she turned to the foreign ambassadors present and burst out laughing.[53] Such incidents make one wonder if the 'Black Legend' is as grossly unfair to the queen as her recent apologists have suggested.

It would be wrong to suggest that in France after 1572 Catherine was the object of universal opprobrium. Among contemporaries she had no greater admirer than Pierre de Bourdeille, seigneur de Brantôme, who, in his *Dames Galantes*, tried to refute the allegations contained in the *Discours Merveilleux*. Drawing upon the queen's funeral oration by Renaud de Beaune, Archbishop of Bourges, he glorifies her ancestry. From the Medici, who had been allowed by Louis XI to add the fleur de lys to their coat of arms, she had derived her taste for the arts, especially architecture. On her mother's side, her ancestry went back to Godefroy de Bouillon, leader of the first crusade, and to Saint Louis. Brantôme also praises Catherine's physique: her regal stature, her ample bosom, her fine legs and especially 'the most beautiful hand that was ever seen'. Good humoured and companionable, she liked to dance and go hunting. She was a fine horsewoman, liked archery and laughed heartily as she watched plays. She also excelled in such pursuits as silk embroidery which are associated with her sex. Politically, Brantôme bears witness to her efforts for peace in the early stages of the civil wars ('I know', he writes of one occasion, 'I was there'), but even he is hard put to justify Catherine's role in the massacre of St Bartholomew. He sidesteps the difficulty by saying that he was not in Paris on that 'bloody day' and must therefore rely on hearsay. He tries to diminish Catherine's responsibility by suggesting that she was not the prime mover behind the massacre nor its most zealous advocate. The threats uttered by Huguenot hotheads were, in his view, largely to blame for what followed. As for Coligny's murder, he sees in it an act of filial vengeance by Henri de Guise which no gentleman can possibly find unacceptable.[54]

So, even Brantôme, who knew and admired Catherine, leaves us in suspense. History is all too often like that. The truth about Catherine is unlikely ever to be established. That she was the object of a vile campaign of propaganda cannot be seriously doubted; yet this is no reason for exculpating her for some of the more sinister events of the French civil wars. We need to remember that, like Richard III, she was the product of a particularly ruthless and violent age.

Notes

1 C. Ross, *Richard III* (London, 1981), pp. xix–liii, 78–9.
2 N. M. Sutherland, 'Catherine de Medici: The Legend of the Wicked Italian Queen', in *Princes, Politics and Religion, 1547–1589* (London, 1984), pp. 237–48; J.-L. Bourgeon, *L'Assassinat de Coligny* (Geneva, 1992), pp. 14, 29–30, 48 n.9.
3 *Discours merveilleux de la vie, actions et déportements de Catherine de Médicis, Royne-mère*, ed. N. Cazauran (Geneva, 1995). See also R. M. Kingdon, *Myths about the St. Bartholomew's Day Massacres, 1572–1576* (Cambridge, Mass., 1988), pp. 200–11.
4 *Discours merveilleux*, ed. N. Cazauran, pp. 31–54.
5 *Henri III et son temps*, ed. R. Sauzet (Paris, 1992), pp. 16–19.
6 V. Duruy, *Histoire de France* (Paris, 1892), p. 417.
7 Sutherland, *Princes, Politics and Religion*, pp. 237–48.
8 Bourgeon, *L'Assassinat de Coligny*, p. 14.
9 J. Michelet, *Renaissance et réforme* (Paris, 1982), pp. 440–1.
10 *Ibid.*, p. 569.
11 *Ibid.*
12 *Ibid.*, p. 591.
13 *Ibid.*, p. 597.

14 Kingdon, *Myths about the St. Bartholomew's Day Massacres*, pp. 200–13.

15 *Ibid.*, pp. 42–6; L. Romier, 'La Saint-Barthélemy: les événements de Rome et la préméditation du massacre', *Revue du XVIe siècle* (1913), pp. 529–60; Camillo Capilupi, author of *Lo Stratagemma di Carlo Nono Re di Francia, contro i rebelli di Dio e suoi l'anno MDLXXII*, and the Venetian ambassador, Giovanni Michiel, believed that the massacre had been prepared over a long period. Capilupi, who was in Rome at the time, owed his information to the nuncio Salviati. He would also have had access to information sent to the cardinal of Lorraine, then in Rome, from Paris. See *The Massacre of St. Bartholomew*, ed. A. Soman (The Hague, 1974), p. 109.

16 Q. Skinner, *The Foundations of Modern Political Thought* (Cambridge, 1978), vol. 2, p. 302.

17 Sutherland, *Princes, Politics and Religion*, pp. 139–55.

18 D. Nugent, *Ecumenism in the Age of the Reformation: The Colloquy of Poissy* (Cambridge, Mass., 1974), pp. 42–3, 51–3, 58, 60–2, 64–7.

19 A. Talon, *La France et le Concile de Trente (1518–1563)* (Rome, 1997), p. 308.

20 A. Fontanon, *Les Édits et ordonnances des rois de France* (Paris, 1611), vol. 4, pp. 272–4; N. M. Sutherland, *The Huguenot Struggle for Recognition* (New Haven, Conn., 1980), pp. 346–57.

21 P. Champion, *Catherine de Médicis présente à Charles IX son royaume (1564–1566)* (Paris, 1937); J. Boutier, A. Dewerpe and D. Nordman, *Un Tour de France royal. Le voyage de Charles IX (1564–1566)* (Paris, 1984); V. E. Graham and W. McAllister Johnson, *The Royal Tour of France by Charles IX and Catherine de Medici. Festivals and Entries, 1564–6* (Toronto, 1979).

22 Catherine de Médicis, *Lettres*, H. de la Ferrière and G. Baguenault de Puchesse (eds), (Paris, 1880–1909), vol. 3, p. 59.

23 G. Parker, *The Army of Flanders and the Spanish Road, 1567–1659* (Cambridge, 1972), pp. 61–6.

24 Catherine de Médicis, *Lettres*, vol. 3, pp. 41–3.

25 F. de La Noue, *Discours politiques et militaires*, ed. F. E. Sutcliffe (Geneva, 1967), p. 682; B. B. Diefendorf, *Beneath the Cross: Catholics and Huguenots in Sixteenth-Century Paris* (Oxford, 1991), pp. 80–1.

26 Catherine de Médicis, *Lettres*, vol. 3, pp. 166–7.

27 I. Cloulas, *Catherine de Médicis* (Paris, 1979), p. 235.

28 Sutherland, *The Huguenot Struggle for Recognition*, pp. 358–60.

29 J. Shimizu, *Conflict of Loyalties: Politics and Religion in the Career of Gaspard de Coligny, Admiral of France, 1519–1572* (Geneva, 1970), pp. 150–2.

30 *Ibid.*, p. 154.

31 D. Crouzet, *La Nuit de la Saint-Barthélemy. Un rêve perdu de la Renaissance* (Paris, 1994), pp. 288–9; Shimizu, *Conflict of Loyalties*, p. 147.

32 N. L. Roelker, *Queen of Navarre. Jeanne d'Albret, 1528–1572* (Cambridge, Mass., 1968), pp. 372–4. The original letter is in *Mémoires de Michel de Castelnau*, ed. J. Le Laboureur (Paris, 1659), vol. 1, p. 858.

33 Roelker, *Queen of Navarre*, p. 376. The original letter is in Marquis de Rochambeau, *Lettres de Antoine de Bourbon et de Jehanne d'Albret* (Paris, 1877), no. CCXXXVI, pp. 345–53.

34 Roelker, *Queen of Navarre*, pp. 391–2. The accusation was first made in the *Discours merveilleux*, ed. cit., pp. 200–1.

35 Crouzet, *La Nuit de la Saint-Barthélemy*, pp. 357–61.

36   Crouzet, *La Nuit de la Saint-Barthélemy*, pp. 378–80. Accounts of the attempted assassination vary. Crouzet writes: 'History has revealed itself, suddenly and dramatically, bereft of all certainty.'

37   Bourgeon, *L'Assassinat de Coligny*, pp. 39–41.

38   Cloulas, *Catherine de Médicis*, p. 283.

39   Sutherland, *The Massacre of St. Bartholomew*, pp. 295–6.

40   M. Venard, 'Arrêtez le massacre!', *Bulletin d'histoire moderne et contemporaine*, vol. 39 (1992), pp. 645–61.

41   Bourgeon, *L'Assassinat de Coligny*, pp. 45, 51–4.

42   Cloulas, *Catherine de Médicis*, p. 289.

43   Venard, 'Arrêtez le massacre!', p. 661.

44   P. Benedict, 'The St. Bartholomew's massacres in the provinces', *Historical Journal*, vol. 21 (1978), pp. 201–25.

45   Kingdon, *Myths about the St. Bartholomew's Day Massacres*, pp. 42–3.

46   Bourgeon, *L'Assassinat de Coligny*, p. 57, n. 36; A. Baschet, *La diplomatie vénitienne. Les princes de l'Europe au XVIe siècle* (Paris, 1862), pp. 551–2.

47   Crouzet, *La nuit de la Saint-Barthélemy*, p. 183.

48   G. Van Prinsterer, *Archives de la maison de Nassau*. First series, Supplement (1847), pp. 125* and 127*.

49   A. Theiner, *Annales ecclesiastici*, I (Rome, 1856), p. 329.

50   F. Decrue, *Le parti des politiques* (Paris, 1892), p. 175; Mariéjol, *Catherine de Médicis*, p. 193.

51   Van Prinsterer, *Archives de la maison de Nassau*. First Series, vol. iv. Supplement, p. 12*.

52   M. P. Holt, *The Duke of Anjou and the Politique Struggle* (Cambridge, 1986), p. 25.

53   H. Forneron, *Histoire de Philippe II*, ii, p. 332, n. 1; Mariéjol, p. 195.

54   M. Lazard, *Pierre de Bourdeille seigneur de Brantôme* (Paris, 1995), pp. 271–7.

2

Protestantism and Jacobinism in the Department of the Aveyron, 1789–1815

Peter M. Jones

The purpose of this chapter is to cast some light into one of the darker corners of the French Revolutionary decade. It explores the ways in which the events of 1789 and beyond re-ignited and greatly embittered the confessional rivalries that had punctuated life in the small towns and villages of the Bas-Rouergue (subsequently the department of the Aveyron) since the late sixteenth century. While the recent celebrations marking the 400th anniversary of the promulgation of the Edict of Nantes quite understandably focused on the positive features of France's confessional history, this is a story of intolerance, of violence and of the curtailment of civil liberties. It does little credit to either Protestants or Catholics.

The year 1789 gave birth to the Declaration of the Rights of Man and of Citizens, as is well known. Those rights included an endorsement of the freedom of opinion, although the distaste with which some of the deputies extended this principle into the religious sphere (Article 10 asserted the right to freedom of thought even in matters religious) seemed to prefigure some of the difficulties that lay ahead. Within four years the attempt to live up to what Bentham notoriously described as the 'metaphysical nonsense' of natural and imprescriptible rights had been substantially abandoned. 1793, a year that began with the execution of the monarch, ended in quaking fear for many. The selective coercion of individuals (priests who refused to swear an oath of allegiance; *émigrés* who declined to return home, for example) gave place to a policy of state terror directed against whole categories of French citizens: aristocrats, royalists, moderates, and myriad denominations of suspects.

Historians may ask whether the language of freedom enunciated in 1789 did not contain a defective gene of intolerance, but by 1794 there were ample circumstantial reasons to explain, if not to justify, the Terror. Republican France was locked in a life or death struggle with enemies on land and at sea. Famine threatened the working population of the towns and cities, whilst royalists and a host of discarded Revolutionary politicians (Fayettists, Feuillants, Federalists, and so forth) appeared to be conspiring against the regime from within. In such conditions the 'popular front' of 1789 fell apart. One party after another learned the first lesson of revolution: that political legitimacy goes to the highest bidder.

By 1794, in consequence, France found herself in the grip of a Jacobin dictatorship. In nearly every town and city effective power had settled in the hands of small coteries of tried and trusted revolutionaries who were therefore able to determine the parameters of what was politically legitimate and what was not. The small towns and *bourgs* of the southern Aveyron were no exception to this general rule. From the autumn of 1793 they were all securely controlled by groups of Jacobins operating from the municipalities, the *comités de surveillance*, the *sociétés populaires* and the District administrations. For many of these militants the Year Two (1793–94) would be a defining moment in their lives; a moment of joyful Revolutionary communion when they served as the provincial eyes and ears of the all-powerful Committee of Public Safety. Such men would never forget the role which they had played in 1794. Indeed, they would never be *allowed* to forget, for the Aveyron was a land of incipient counter-Revolution. Reputations earned in 1793 and 1794 clung like a pall: 'ancien membre du comité de surveillance de l'an II' was a stigma that carried a powerful charge on the streets of Millau or Saint Affrique even as late as 1816.

Jacobinism in its visceral, real-life dimension is the theme of this chapter, then: its rise to power and its fall over a 25-year period stretching from the outbreak of the Revolution in 1789 to the aftermath of Napoleon's defeat at Waterloo in 1815. The argument which will be outlined is that the stamina and tenacity of Jacobinism in the small towns and villages of the southern Aveyron derived very substantially from the tortured confessional history of the region. By 1794, Jacobinism was Protestantism writ large. Under the pressures of Revolutionary politics Catholic support for the new status quo was squeezed to the point of destruction, leaving only hard nuclei of militants in towns such as Millau, Saint Affrique and Camarès which tended to be Calvinist almost to a man. And yet the exigencies of Revolutionary politics are not sufficient on their own to explain the tenacity and the longevity of these Jacobin cells. Their strength and impermeability owed much to the tribal organization of southern Protestantism; a structure that had developed as a reflex of defence during the long years of persecution and 'desert' worship. Such a pattern of organization and political commitment was not unique to the southern Aveyron, of course. It can be found in Montauban, in Nîmes and in most of the little textile towns dotted along the southern escarpments of the Massif Central where Catholics and Protestants lived in close and mutually suspicious proximity.

I

By the time Henri IV brought confessional strife to a close with an edict issued at Nantes in 1598, it is estimated that Calvinism had secured the allegiance of between 1 and 1.25 million of his subjects. In a population of some 15 million, the Protestants therefore represented a sizeable minority. However, the vicissitudes of the seventeenth century would exact a heavy toll; by the end of the *ancien régime* Protestant numbers had dwindled to around 700,000 in a population that had very nearly doubled in size.[1] But if 2 per cent or maybe 3 per cent of French men and women were Protestant, not all were of Huguenot ancestry. The Lutheran communities of the east of France have to be subtracted, with the result that adherents of the Reformed Church probably numbered no more than 500,000 on the eve of the Revolution. On the other hand, most of the Calvinists resided to the south of the river Loire and they could be locally quite dense. In fact, the majority lived in a fairly well-defined assortment of villages and towns distributed in the form of a crescent stretching from Poitou in the north to Gascony, Languedoc and the Dauphiné in the southeast. In the south and south-west, the two great citadels of the Reform remained Nîmes and Montauban, which mustered Calvinist populations of around 14,000 and 5,500 respectively.

By contrast, the towns of the upper reaches of the Tarn and of its tributaries were of mediocre size. On the evidence of the figures gathered for the 1790 census, which historians generally consider to have been inflated, Millau claimed a little over 6,000 inhabitants; Saint Affrique 4,262; Nant 2,636; Saint Jean-du-Bruel 2,388 and Pont de Camarès 2,271.[2] However, only the towns of Millau, Saint Affrique and Camarès contained sizeable concentrations of Calvinists. Protestantism by the eighteenth century was essentially a village-based religion. Precise figures are hard to come by, but roughly a third of the inhabitants of

Millau adhered to the Reformed Church at the end of the *ancien régime*, and a quarter of those of Saint Affrique. Only in Camarès do the Calvinists appear to have been in a majority, if some rather sketchy evidence of baptism rates can be trusted.[3] To these solid contingents need to be added relic Huguenot communities of uncertain size located in Saint Rome-de-Tarn, Saint Rome-de-Cernon, Saint Sever-du-Moustier, Saint Félix-de-Sorgues, Cornus, Montlaur, as well as in the town of Saint Jean-du-Bruel.

The demographic health of Bas-Rouergue Protestantism in the decades before the Revolution will probably never be known in detail. After the Revocation of the Edict of Nantes in 1685, the Reformed Church largely ceased to exist in an institutional sense and accurate record keeping became the least of its worries. Not until 1787, in the so-called Edict of Toleration, were Calvinists granted legal recognition once more. Before that date official persecution remained a real, if infrequent, possibility. Every time Catholic and Bourbon France went to war with the Protestant powers, the Calvinists of Guyenne and Languedoc knew to expect the worst. In the 1730s and 1740s illicit open-air assemblies for the purpose of worship 'in the desert' were repeatedly broken up by troops. After one such 'dragonnade' in the neighbourhood of Saint Affrique in 1734, five Calvinists were condemned to the galleys for life and their property seized by the Crown. More punitive measures followed in 1745, not least an order from the *intendant* in Montauban imposing discriminatory taxes on the Protestant inhabitants of Millau, Saint Affrique and Camarès. The Calvinist bourgeoisie of Saint Affrique riposted by hauling the mayor, the *greffier* and the *consul* of the venal Catholic municipality before the Cour des Aides of Montauban. Needless to say, they lost their case and the *syndics* of the Protestant community – Pierre Girbal *père*, Etienne Sarrus *aîné* and Pierre-Guillaume Bourgougnan *père* – were ordered to pay substantial damages. Finally, in 1762, at the height of the Seven Years War, two celebrated Protestant victims were executed on the Place du Salin in Toulouse: Pastor François Rochette[4] who had been captured by the *guet* of Caussade, and Jean Calas.

Unsurprisingly, such experiences left a deep impression on the Protestant communities of the Montalbanais and the Bas-Rouergue. They riposted by organizing themselves for defence and mutual support; and in place of the formal structures of a church there developed what Louis Dermigny has described as 'une sorte de rigueur morale', rooted in an exaggerated 'esprit de clan'.[5] These characteristics, and perhaps also the spiritual barrenness of eighteenth-century Calvinism, would provide an excellent context for the vocation of Jacobinism.

Nevertheless, it should be acknowledged that the judicial extermination of Rochette and Calas by the Parlement of Toulouse served to close the chapter of institutionalized persecution. In the 1770s and 1780s the Protestant merchants and artisans of Millau and Saint Affrique breathed more easily, for toleration was in the air. From the time of Controller-General Turgot onwards a succession of government ministers agitated for the cause of religious freedom. Finally, in 1787, Louis XVI was induced to depart from the policy laid down by his royal great-grandfather and to begin the task of rehabilitating his non-Catholic subjects. In fact, the Edict of 1787 did not grant to Protestants a right of public worship, much to their disappointment; it merely restored some of their entitlements, such as the right to live legally in France, to exercise a profession and to marry legally before a magistrate. But with the Revolution beckoning, unconditional toleration was not far off. In December 1789 a decree of the National Assembly accorded full religious and political rights to Protestants on a par with those of Catholics. The following year families that had emigrated rather than forswear their faith were invited to return to France with a promise that property confiscated by the state would be restored.

II

Let us now turn to the early years of the Revolution in the towns of Millau and Saint Affrique. The train of events in Camarès is less thoroughly documented which serves to underline an important truth about provincial Jacobinism: it tended to display its most brutal and domineering characteristics in fractured communities. What Protestants in the southernmost towns of the Aveyron lacked in numbers, they more than made up for in terms of economic weight. Also, they displayed a capacity for staying united that their Catholic opponents were quite unable to match. The elastic patriotism of Calvinist merchants and craftworkers readily adjusted to each new political development as the Revolution slipped remorselessly from constitutional monarchism to republicanism, and from republicanism to Montagnard Jacobinism. In this respect the 1790s provide an early demonstration of the 'toujours à gauche' principle which characterized Protestant politics for much of the nineteenth century.

The economic strength of the Calvinists stemmed, of course, from the fact that they dominated the commercial and manufacturing life of the Bas-Rouergue and the Bas-Languedoc. According to an early nineteenth-century source, five-sixths of the population of Saint Affrique were involved in the woollen industry in some capacity or other.[6] But cloth production and marketing were firmly controlled by the Protestant bourgeoisie. In 1780 these *marchands-fabricants* claimed that they paid out 500,000 *livres* each year in wages to some 4,000 families of wool spinners. Many, probably most, of these outworkers would have been Catholics, living in the villages of the nearby Causse du Larzac. Camarès, likewise, constituted a major textile centre whose output was chiefly destined for markets in the Levant and in the Spanish colonies as Dermigny's study of the *réseau* Solier has demonstrated.[7] By contrast, the economy of Millau was more diversified. A *rentier* class of nobles and robe bourgeois had brought wealth and a veneer of urban civilization to the town, but leather-working and textiles (cotton as well as wool) provided the major source of employment. In 1799 the town claimed 60 premises for the preparing and tanning of hides; 20 glove-sewing workshops; 15 hatters and approximately 2,000 individuals employed in wool and cotton spinning.[8] It goes without saying that the bulk of this industry was controlled by a tight mesh of Calvinist households: the Montets, the Caldesaigues, the Cabantous; the Carrières, the Buscarlets, the Nazons, and so on.

How did these Protestant dynasties react to the irruption of Revolution in their midst? With open arms, it appears, and for three main reasons. For a start, the empty formalism of eighteenth-century Calvinist religion readily adjusted to the prevailing spirit of the age, that is to say to Enlightenment utilitarianism and the notion of a remote and impersonal God. In consequence the Calvinists found the secularizing policies of successive Revolutionary legislatures far easier to accept than did their Catholic contemporaries. Indeed, this contrast was nowhere more sharply drawn than in the villages of the southern Aveyron where the Catholic population remained deeply attached to the pre-Tridentine version of their faith. A second and more obvious reason is that the Revolution emancipated the Protestant community in the fullest sense of that term. Calvinists were at last permitted to leave the constitutional ghetto, to become fully participating members of the polity. Furthermore, the authority to restore the institutional structures of their church and to indulge in public acts of worship (withheld in the 1787 Edict) was now explicitly, if somewhat reluctantly, acknowledged in the Declaration of the Rights of Man. Scarcely less important in the eyes of the Calvinist bourgeoisie of Millau and Saint Affrique was the recognition of their economic contribution as industrious citizens whose skill and energy had brought employment and material well-being to their respective communities. The decision of the National Assembly to open public office to election on the basis of a property franchise seemed to them only right and proper, therefore. And besides, the restriction of the vote to the better-off members of the community helped to compensate for their numerical inferiority *vis-à-vis* the Catholic population. The third reason why the Protestant dynasties welcomed the Revolution, of course, is that it provided an opportunity to settle old scores. But this vengeful reflex only came into play later, in 1793 and 1794.

III

Let us consider first the case of Millau. Even before the Revolution the journeymen and apprentices of Millau had a reputation for excitability, even violence, and it was chiefly in order to contain this restless population of craftworkers that the *intendant* confirmed Louis de Bonald in the ancestral office of mayor in 1785. The town was also noted for being dominated (as a later petition put it) 'par une caste de privilégiés'[9] who were, by definition, Catholics. In February 1790 de Bonald's mandate was renewed by popular election, but he served only a few months before being called to higher office in Rodez. In any case his support for the Revolution, which had never been more than lukewarm, was cooling swiftly. Later that year he resigned from the post of president of the Department administration, and by 1792 we find the future theoretician of counter-Revolution living the life of an *émigré* in Koblenz.

Instead, the town switched to a joint Catholic and Protestant administration under the patriotic leadership of Louis de Bourzès. However, this did not prevent a sharp deterioration in law and order from taking place towards the end of 1790 and early in 1791. Most of the other local authorities (and notably the District administration) were malevolently hostile to the Revolution, and in no mood to enforce the National Assembly's decrees. In fact it was abundantly obvious before the end of 1790 that two parties were jockeying for power and busily recruiting supporters – both inside and outside the town – in readiness for a trial of strength. Symptomatic of this polarization was the establishment of rival political clubs. The Amis de la Constitution in which the patriots congregated seems to have been founded in the summer of 1790 at the behest of an emissary of the Société des Amis de la Constitution of Rodez. But that same autumn a club calling itself Les Amis de l'Ordre et de la Paix started to meet in the *hôtel* of the Marquis de Pégayrolles. This was the headquarters of those who were becoming known, both socially and politically, as the 'aristocrates'. The club refused admission to non-Catholics and channelled its energies into a campaign against the Civil Constitution of the Clergy.

In the event it was the attempt to administer the oath of loyalty which the National Assembly expected all members of the parish clergy to take that triggered the first call-to-arms in the commune. This took place in January 1791 and it resulted in a short-lived victory for the royalists and their Catholic street supporters. However, in May the patriots used similar tactics against the monarchist club which was systematically pillaged and destroyed. Under pressure from the crowd, the municipality was obliged to order the disarmament of nine of the royalist leaders. The attempt to divert the course of the Revolution and to impede the application of the Civil Constitution of the Clergy proved futile in any case, for in July the monastic churches of the Jacobins, the Carmelites and the Capuchins in Millau were closed down willy-nilly. Whereupon, as if to add insult to injury, the church of the Jacobins was commandeered by the Calvinists for use as their temple. Further clashes occurred in the town later in the year and again in April 1792, but by that date it was apparent to everyone that the balance of power had shifted decisively in favour of the Calvinists. The local authorities were no longer in full control of the situation, and effective power was fast devolving upon the Jacobin Club (that is to say the Amis de la Constitution) and its street-level gangs of Protestant workmen.

What of Saint Affrique? Here a parallel descent into sectarian politics took place, albeit on a slightly longer time-scale. Moreover, when the narrative history of Jacobinism is pieced together, we find a familiar set of denominators: deep-seated antagonisms carried over from the *ancien régime*; a moderate and overwhelmingly Catholic municipal body; informal institutions of power devised by the Protestants in order to compensate for their inability to win over the local authorities; and finally repeated trials of strength on the issue of the Civil Constitution of the Clergy.

As in Millau, the creation of rival political clubs produced the first clear signal that public opinion was polarizing along sectarian lines. In May 1791 an intensely patriotic Club des Amis de la Constitution was set up, and within a month we find traces of its polar opposite, referred to as 'un club noirâtre'[10] in the sessional minutes. In fact, the Saint Affrique royalists and would-be counter-revolutionaries tended to rely more heavily on the municipality which was overwhelmingly moderate and Catholic in outlook and headed by a former canon, named Le Rat. Be that as it may, the first serious clashes occurred in August when the new 'juring' cleric, Gabriel Constans, was installed in the place of the outgoing parish priest who had objected to swearing the oath. A significant feature of the ensuing riot was that it triggered a general call to arms, with Catholics from nearby villages rushing into the town, while Protestant national guardsmen mobilized from as far afield as Millau and Camarès in order to offer protection to the constitutional priest. More clashes took place in September: the minutes of the Club des Amis de la Constitution record that shouts of 'les démocrates à la lanterne'[11] could be heard echoing through the alleyways. However, the climax occurred early in February of the following year when seasonal shortages caused bread prices to spiral sharply. On receipt of the news (inaccurate as it turned out) that an army of Millau Protestants with murderous intentions was approaching, a full-scale 'fear' swept through the Catholic population of the town and surrounding villages. In the clashes that followed, the Protestants emerged the losers.

Far worse was to come, though. The municipality had been deeply implicated in the mobilization of Catholic 'defenders', and as a result it was suspended by the Department authorities in Rodez. In fact the whole affair was referred to the Minister of the Interior for a definitive judgement. There followed a period of frantic petitioning as each party manoeuvred for advantage in Paris; and, in July 1792, the Minister Roland ruled in favour of the Catholic municipality. This decision had fatal consequences for the rule of law in Saint Affrique, for it prompted the ringleaders of the Calvinist community to resort to blatantly illegal and violent tactics in order to win control of the commune. Between July and November, they organized and directed a strike force known as 'la bande noire' or 'le pouvoir exécutif' whose function was to beat up or otherwise intimidate their opponents into submission.[12] The activities of this unsavoury gang of Protestant workmen and apprentices laid the foundations for the subsequent Jacobin dictatorship in the town.

When in October 1792 the Catholic-dominated municipality of Saint Affrique fell due for renewal, few voters were prepared to risk attending the electoral assembly and the Protestant leaders were hoisted into power virtually unopposed. Daniel Bourgougnon *fils aîné* de La Grave was chosen as mayor on a turn-out of 41 voters (the commune contained 600 'active' citizens alone), while Etienne Sarrus *fils aîné* dit Parfait was appointed procurator. One final attempt to rescue the town from the grip of sectarian politics was made, it is true. In November the higher authorities in Rodez appointed commissioners to investigate the activities of the 'la bande noire', and it is thanks to the weighty dossier which they compiled that we are able to penetrate the schemes of the Protestant leaders.[13] But this eleventh-hour attempt to uphold the standards of pre-Terror legality was blocked in Paris by deputy François Chabot, the evil genius of Aveyronnais Jacobinism. In January 1793 a decree of the National Convention exonerated Bourgougnon and Sarrus and their street bully boys. Henceforth, nothing further would stand in the way of a Protestant takeover of the town in the name of the Republic One and Indivisible.

A blow-by-blow account of the functioning of factional politics throughout 1793 and 1794 scarcely seems necessary. It is sufficient to say that the tactics extemporized by the Saint Affrique Jacobins were widely copied. Shadowy squads of militants, also calling themselves 'le pouvoir exécutif', emerged in Saint Rome-de-Tarn, in Roquefort, in Brusque, and in several other localities. By the autumn of 1793 and the advent of the Terror proper, these Jacobin cells were in an unassailable position in so far as local rivals for power were concerned. Rather, the risk to their authority came from outside the region; that is to say from the vagaries of Revolutionary politics in Paris which could pose difficulties of interpretation at 600 kilometres distance. Another source of vulnerability lay in the roving *représentants-en-mission* whose political instincts and moral scruples when confronted with palpable acts of terrorism could be difficult to anticipate.

Représentants Bo and Chabot who arrived in the Aveyron in the early spring of 1793 were both natives of the department and could be relied upon to condone the unsavoury politics of factional domination, but *représentant* Paganel who appeared in the region in February 1794 was less accommodating. Disturbed by the serious allegations levelled at the Saint Affrique Jacobins, he dismissed Bourgougnon and Sarrus from their posts in the municipality. After a stand-off, the pair decamped to Paris in order to petition the Committee of Public Safety for their reinstatement. On the intervention of Couthon and a stalwart Robespierrist from Millau named Joseph Lagarde, they obtained their rehabilitation, largely, it seems, on the evidence of their resolutely Anti-Federalist stance during the previous summer. But a fortnight later the coup of 9 Thermidor took place and swiftly erased their friends in the Great Committees. In a matter of days the Terror began to totter, and in due course the entire edifice of provincial Jacobinism followed suit. After attempting, vainly, to contain the flowing tide of moderation in Saint Affrique, Bourgougnon and Sarrus prudently removed themselves from the scenes. They were pursued by arrest warrants issued by Goupilleau, the Thermidorian *représentant-en-mission* to the Aveyron.

IV

What explains the extraordinary stamina of these bands of provincial militants? They were bound together by a common attachment to the Jacobin Republic of the Year Two, of course, but in small towns such as Millau, Saint Affrique and Camarès they were bound together even more tightly by ties of kinship, and a collective memory of persecutions at the hands of Catholics. When, following the events of February 1792, the patriots of Saint Affrique tried to settle the blame for the violence on their Catholic adversaries, the municipality mounted a spirited and revealing counter-attack. With little difficulty, it was able to discredit and disqualify every one of the 17 potential Protestant witnesses on grounds of kinship or master–servant relationships.[14]

As for the long memories and bitter resentments with which the Calvinist bourgeoisie embarked upon their struggle for power, what better example than the behaviour of the sons of Bourgougnon, Sarrus and Girbal? In the summer of 1795, when the nefarious activities of 'la bande noire' of Saint Affrique finally caught up with Bourgougnon and Sarrus *fils* and they were committed for trial before the Criminal Court of the Aveyron, a witness named Anne Tourtoulou was called. She was the widow of Jean Milhau, the mayor who had sued Pierre Girbal *père*, Etienne Sarrus *aîné* and Pierre-Guillaume Bourgougnon *père* for damages in 1767. Widow Milhau recounted how Bourgougnon *fils*, Girbal *fils* and several others had ransacked her house during the Terror in order to discover the original of the *arrêt* of the Cour des Aides of Montauban which had condemned their fathers and kinsmen.[15]

Although the coup of Thermidor removed from the Jacobin elites of Millau and Saint Affrique their political immunity, the process of retreat from the Terror was not as swift and as definitive as historians have often supposed. Detained suspects inundated the authorities with petitions for their release, but few of the men holding local office still in late 1794 and even in 1795 could afford to make a clean breast of the past. After five years of incessant Revolutionary turmoil, nearly everyone had something to hide or to reproach himself for. Consequently when Bourgougnon and Sarrus came to trial in July 1795, they were acquitted on 21 of 27 counts.[16] A jury made up of former revolutionaries refused to take notice of anything that might be construed as a political offence. Nevertheless, punishment could come from unexpected quarters. Joseph Lagarde, one of the main architects of the Terror in Millau and the man who had vouched for Bourgougnon and Sarrus before the Committee of Public Safety, was assassinated by a royalist murder gang in May 1796.[17]

Most of the key players of 1793 and 1794 survived, however, and they would emerge as senior cadres of the neo-Jacobin party under the Directory and the Consulate. Such was the politico-religious makeup of Jacobinism in the towns of the southern Aveyron that it was all but impossible to switch loyalties in any case. The elections of the Year Six (March 1798) triggered further outbreaks of neo-Jacobin activity in Millau and Saint Affrique, not to mention other localities in the department. Among the militants of Saint Affrique who drew attention to themselves in that year, we can identify 11 former members of the *comité de surveillance*.[18] Millau also succumbed to a recrudescence of party in-fighting in 1802 as the old Jacobin faction mobilized its supporters to secure the election of its candidate as *juge de paix*.

Then silence, as though the Empire really did succeed in putting the political passions of the 1790s to sleep. Appearances were a little deceptive, however, as the topsy-turvy events of 1814–15 would demonstrate. Predictably, Napoleon's brief return to power following his flight from Elba enabled the Calvinist bourgeoisie of Millau and Saint Affrique to get their hands on the levers of local government once more. As a result the first reports of a reverse or even a defeat in the Low Countries arrived in an atmosphere of simmering party-political animosities. Millau remained tense, but contained, as the definitive news of the battle at Waterloo came through. Nevertheless, the new Legitimist subprefect thought it expedient to seek permission to carry out a precautionary disarmament of 'tous les Montet, Caldesaigues, Carrière (excepté Jean Carrière père et fils), les Buscarlet, les Belluch, les Guy Faillet, les Nazon etc.'.[19] All were Calvinist families whose guilt was presumed to be collective.

The town of Saint Affrique had been living on tenterhooks ever since the First Restoration, and when Napoleon returned from Elba the neo-Jacobins took it as a signal to restore the Terror dictatorship of the Year Two. Indeed, if the subsequent judicial enquiry can be relied upon, Daniel Bourgougnon had been forewarned of Bonaparte's intentions and had even announced his imminent disembarkation.[20] Thus, when the defeat at Waterloo prematurely terminated the revived Empire, the triumphalism of Saint Affrique's royalists knew no bounds. As rejoicing turned to anger and reports of events in Nîmes and Avignon began to circulate, the Catholic population resolved to avenge all the humiliations and insults they had suffered since 1789. On 19 August 1815 a large crowd smashed down the doors of the Protestant temple with a battering ram. The building had been dedicated just ten years earlier, having been constructed on the site of the former chapel of the Cordeliers, but it was pillaged from top to bottom. Furnishings, psalm books, bibles, vestments were all thrown into the square. That night and for several nights following, the town and its surrounding hillsides were brightly lit as fires steadily consumed more than 100 3-metre long benches dragged from the shell of the temple.

This festive pyre was intended to signal a final settling of accounts with the Jacobins of the southern Aveyron, and so it was taken. After all, these men would never rally to the restored Bourbon monarchy. As the *procureur du roi* attached to the Civil Court of Saint Affrique put it in a report written in October 1815:

> Their principles are too deeply rooted; these principles are a legacy from their forebears, they draw on them in their education and religion; and history teaches us that Protestants are the irreconcilable enemies of the throne and the clergy.[21]

Notes

1 S. Mours, *Les Eglises réformées en France: tableaux et cartes* (Paris, 1958).

2 Archives Départementales de l'Aveyron (hereafter ADA) L664, Etats de population des municipalités, 1790–91.

3 See D. Robert, *Les Eglises réformées en France, 1800–1830* (Paris, 1961), pp. 139–40; H. Guilhamon (ed.), *Journal des voyages en Haute-Guienne de J.-F. Henry de Richeprey* (2 vols, Rodez, 1952–67), I, pp. 243–52.

4 In June 1794, on receipt of the news that the parish priest and a number of royalists of Caussade had been executed in Paris, the local Jacobins poured onto the streets shouting 'Rochette est vengé' [Rochette is avenged]; see F. Galabert and L. Boscus, *La Ville de Caussade (Tarn-et-Garonne): ses vicomtes et ses barons* (Montauban, 1908), p. 303. Also D. D. Bien, *The Calas Affair: Persecution, Toleration and Heresy in Eighteenth-Century Toulouse* (Westport, Conn., 1979).

5 L. Dermigny, *Cargaisons indiennes, Solier et Cie, 1781–1793* (2 vols., Paris, 1959), i, p. 28.

6 Archives Nationales (AN) F⁷ 8522, subprefect of St Affrique to prefect of the Aveyron, St Affrique, March 1807.

7 Dermigny, *Cargaisons indiennes*.

8 J. Artières, *Annales de Millau* (Millau, 1894–99), p. 267.

9 AN F⁷ 3657I, Rapport sur les troubles du district de Millau adressé au Ministre de l'Intérieur, Rodez, 20 April 1792.

10 See L. Raylet, *Procès-verbaux de la Société des Amis de la Constitution de Saint Affrique* (Rodez, 1942), session of 10 June 1791.

11 *Ibid.*, session of 30 September 1791.

12 To judge from the enquiry conducted by commissioners Daudé and Ser in November 1792, the force numbered around 30 individuals, among whom David Sarrus, *bourgeois*; Jacques Sarrus, *grenadier*; Etienne Sarrus, *négociant*,

procurator of the commune; Daniel Bourgougnon *fils aîné* of La Grave, *marchand-tanneur*, mayor of the commune; Pierre Bourgougnon *cadet* of La Grave, *marchand-tanneur*; Jean Bourgougnon *fils aîné* of Le Griffoul Haut, *mégissier*; Pierre Bourgougnon *cadet* of Le Griffoul Haut, *mégissier*; Pierre Grand, *maréchal à forge*; Paul Grand; François Peyre, *mégissier*; Jean-Pierre Peyre *fils*, of the Lion d'Or, *négociant*; Denis Rachou, *négociant*; Hyacinthe Rachou, *négociant*; David Valès *père*, *trafficant*; David Pierre Valès, *perruquier*; Antoine Martin, *perruquier*; Joseph Brugière dit Firmy, *charpentier*; Jean Roulandes, *huissier* at the District Court; Jean Roumagnac, *huissier* at the District Court; Jacques Connac, *menuisier*; Jean Anglade, *brassier*; Joseph Cuillé, *menuisier*; Thomas Hugonenc, *caporal*; Jean Millau, *valet de ville*; Etienne Arlès *fils* of Les Albarèdes; Jacques Cambon dit la Perle; Antoine Malet, *valet de ville*; Girbals *fils* dit l'Anglais; Jean du Gras *fils aîné*; Pierre Cadenat, *tanneur*; Jacques Céré, *commis marchand*; Boulogne *fils aîné* of La Grave (ADA 90L57, Tribunal criminel de l'Aveyron: affaire Bourgougnon, Messidor, an III).

13   *Ibid.*
14   AN F7 3657[1].
15   ADA 90L57, Tribunal criminel de l'Aveyron: affaire Bourgougnon, Messidor, an III, witness declarations.
16   *Ibid.*, décisions du jury.
17   ADA L700, Dumas, captain of the National Guard to Rouvelet, commissaire du Directoire-Exécutif près le département de l'Aveyron, Millau, 4 Prairial IV.
18   AN AFIII 216 Aveyron: petition of the republicans of St Affrique addressed to the Directoire-Exécutif, St Affrique, 10 Germinal VI.
19   AN F7 8998, Dulac, subprefect of Millau to prefect of the Aveyron, Millau, 23 March 1816.
20   *Ibid.*, procureur du roi près le Tribunal de St Affrique to prefect of the Aveyron, St Affrique, 24 August 1815.
21   AN BB[3] 151, copy of a letter addressed to procureur-général en la Cour Royale de Montpellier, St Affrique, 8 October 1815.

3

Chalk and Cheese or Bread and Butter: Political Culture and Ecology in the Seine-Inférieure during the Napoleonic Period

John Dunne

> The arrondissement of Neufchatel presents an entirely different picture. It is made up, for the most part, of pastures and is given over to the rearing of livestock. The local inhabitant is unadventurous, good-natured, even slow. He prefers above all else to rest in the midst of his enclosures, his milk and his cows. There is little crime because people live simply. Administration is easy there, because people put up with a lot before taking the trouble to complain or protest, and, for that reason, this arrondissement deserves to be treated more scrupulously by the administrator.... This arrondissement is then peaceful, and offers in several cantons the touching and astonishing picture of a patriarchal society twenty five leagues from Paris.
>
> Beugnot, *Tableau de situation du département pendant l'an X*[1]

When, in 1802, on the eve of Bonaparte's official visit, the first prefect of the Seine-Inférieure wrote the briefing paper in which this passage appears, there was nothing new in the idea of the distinctiveness of the arrondissement of Neufchatel on its eastern borders – except, that is, the arrondissement itself. Long before the territorial divisions of modern France were invented, writers commented on the striking contrasts in the natural and man-made landscape between the open arable plain of the Pays de Caux, which constituted the heartland of the future department, and the verdant valleys and increasingly pastoral economy of the

31

Pays de Bray, from which the arrondissement was later to be created.[2] Where Beugnot was original was in affirming the existence of the area's *political* peculiarity. The pastoral scene he depicts features in a report on *esprit public* in the department. In it, the unruffled calm of the arrondissement of Neufchatel stands in contrast to the preceding accounts of political tensions and discord still evident elsewhere, despite the new regime's successes in bringing down the political temperature. If the area was an exception to the region's rural political rule, Beugnot tells us, it was because of the peculiar political mentality of the local population.

In this Beugnot was anticipating by more than a century ideas expressed in André Siegfried's celebrated *Tableau politique de l'Ouest*, in which the inhabitants of the Pays de Bray are attributed with a different political temperament from their neighbours on the Caux plateau, who represent the authentic 'Norman spirit'. Equally, Beugnot's implicit association of the mentality of the population of the arrondissement with its distinctive rural habitat may also be seen as an early anticipation of Siegfried's general thesis and its application to this area. Introducing his list of determinants of the regional temperament, he writes, 'The character of the peasants of the Bray is to be explained by the nature of their *pays* . . . '[3]

Taking Beugnot's assertions as hypotheses to be explored, this chapter asks, first, whether the difference in *esprit public* which in his view marked off this particular arrondissement from the other four in his charge, is borne out by a study of political behaviour in the department in the Napoleonic period. And second, it considers whether any differences in political culture that are apparent between different parts of the department can be linked to contrasts in the prevailing socio-economic structures. In so doing, it draws on a long established but currently neglected analytical tradition which runs from Siegfried to Le Bras and Todd.[4]

*

This study has its origins in doctoral research I carried out in the 1970s, under Douglas Johnson's supervision, on the department's notables.[5] Yet its appearance now is timely since it raises problems and issues which for the last couple of decades have found little place on the research agenda of specialists in late eighteenth- or early nineteenth-century France, shaped as it has been by the 'culturalist' paradigm. In the first place, the general preoccupation with the creation of democratic politics in France in the late 1780s has not been matched by any great interest in their apparent demise 10 or 20 years later. Second, while numerous historians have gone off in search of Revolutionary political culture, few have attempted to delimit its appeal either socially or geographically, or even posed the question of the relationship between political culture and socio-economic structures.[6]

When, on beginning archival work in France, I came across Beugnot's remarks about the peculiarity of the arrondissement of Neufchatel, my inclination was not to take them seriously. Surely Beugnot was dealing in well-worn literary images rather than empirical observation. (Perhaps, to that extent, he deserved Bonaparte's dismissal of him, after their first meeting in Rouen, as 'a pure ideologue'.[7]) When I first encountered signs of difference, they concerned the failings of its administration rather than the attitudes or behaviour of the citizenry. On several occasions Beugnot's long-suffering successors had cause to reprimand the sub-prefect, Pocholle, for overdue returns to the prefecture; occasionally to no avail, it seems, from the odd lacuna in the archives in Rouen.[8] Eventually, though, when looking at the results of elections to the electoral colleges, I came across clear evidence of the political distinctiveness of this part of the department.

In the only detailed study of Napoleonic elections to date, Coppolani succeeds in dispelling various misconceptions surrounding the complicated electoral system which lasted from its introduction in 1802 until the fall of the Empire in 1814, and shows that the practice was not far removed from the constitutional theory.[9] This does not prevent him endorsing Taine's view of the process as an elaborate 'electoral comedy'.[10] Even in theory voters had little leverage over parliament, let alone government. Primary voters in each canton elected to the departmental and arrondissement electoral colleges their quota of life members, who went on to designate candidates for the national legislature, from whom Napoleon selected the eventual incumbents.

It is perhaps not surprising that Coppolani makes little attempt to use the voluminous statistics he gathers, to shed light on the state of public opinion. Whatever meanings may have attached to the electoral process under that authoritarian regime are peculiarly hard to recover; partly because of technicalities in the obscure electoral process but more as a result of official constraints on political life. In the absence of candidates, parties, political pamphlets and newspapers other than official ones, it is virtually impossible to ascribe votes to particular political tendencies. However, in relation to the problem at hand even the raw statistics of voting in the elections of 1809 provide interesting food for thought.

In September 1809, for the first time since the elections of 1792 to the Convention, the Seine-Inférieure underwent the experience of legislative-type elections under a system of virtually universal male suffrage. (This franchise was envisaged in the Constitution of 1802 but not actually introduced in time for the 1803 elections in the department.) The polls opened with the usual ceremony: the sound of the drum called voters to the polling stations, where they were to be met by guards of honour. The only thing missing was the voters: in half the 50 cantons, fewer than one in 10 electors appear to have registered a vote at any one of the several ballots – 3 were often required to complete the business. The average turnout in the department (based on the highest poll in each canton for which figures are available) was 14 per cent. Altogether, around 18,000 people, out of an electorate of 120,000, were involved in electing notables to the electoral colleges.[11]

The interest comes from the variation in turnout across the department, ranging from 3 per cent in the canton of Valmont to 73 per cent in Londinières. The latter figure, though, was truly exceptional. In only 13 of the 50 cantons did more than one in five voters turn out; against that, 11 cantons failed to muster even one in ten. Crucial for our purpose is the geography of participation. With a single exception, the areas of lowest turnout either comprised the large urban constituencies – Le Havre, Dieppe and Rouen – or formed a bloc of cantons occupying the western promontory running from Fécamp to the mouth of the Seine. By contrast, 12 out of the 13 'high polling' cantons were either strung out along the Seine or formed a solid phalanx entirely within the arrondissement of Neufchatel. With only one of its cantons securing a less than average turnout, this arrondissement achieved an overall participation rate of 31 per cent, which was virtually double that achieved by three of its counterparts, and nearly 50 per cent higher than Rouen.

The electorate of the arrondissement of Neufchatel distinguished itself even more during Napoleon's Hundred Days by its response to the plebiscite on the Additional Act, which liberalized the constitutional arrangements prevailing before the fall of the Empire in 1814. The Seine-Inférieure as a whole was one of those departments which witnessed, in Bluche's words, a 'spectacular drop' in participation in 1815 as compared to the previous plebiscite of 1804 on the founding of the Empire: turnout plummeted from 42 per cent to 7 per cent.[12] Under these circumstances, with virtually one in five of its electors registering a vote, the arrondissement of Neufchatel appears an island of participation in a sea of abstentionism. Its nearest rival was neighbouring Dieppe with only 8 per cent, which itself was well over twice the figure for Le Havre.[13]

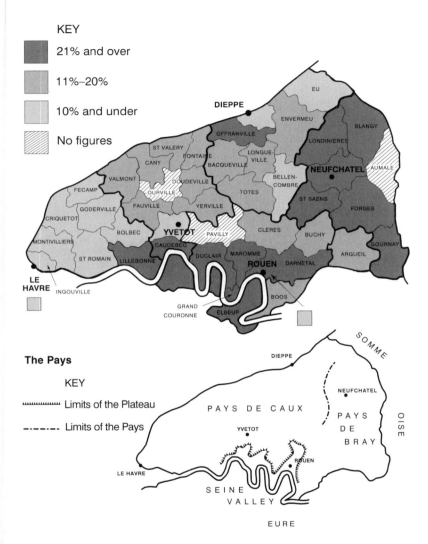

KEY

21% and over

11%–20%

10% and under

No figures

The Pays

KEY

ᴡᴡᴡᴡᴡᴡ Limits of the Plateau

ᴰ.ᴰ.ᴰ.ᴰ. Limits of the Pays

Map 1 Turnout in the 1809 Elections.

Turnout in the elections of 1809 and the plebiscite of 1815 then provide strong retrospective support for Beugnot's claim that the Napoleonic arrondissement of Neufchatel was somehow politically as well as geographically distinct from the greater part of the department he administered until his appointment to the Council of State in 1806. Whether it was different in a way that accorded with his characterization of its inhabitants is another matter. At first sight the higher turnout here seem to suggest a more vibrant political culture at odds with their reputed liking for the quiet life. However, the common-sense notion that electoral participation equates with political awareness finds little favour with the academic specialists, who prefer two other interpretations. For Coppolani, exceptions to the norm of low polling at Napoleonic elections, where they were not a consequence of vote-rigging, were the result of the manipulation of voters by rival notables.[14] High turnouts may indicate the presence not of a more aware electorate but of a more than usually gullible or deferential one. The other interpretation centres on people's attitude to the government of the day as the crucial factor in their decision whether to vote. Thus, writing of the plebiscite of the Hundred Days, Bluche concludes that 'the highest rates of abstention correspond unfailingly with a strong royalist presence'. Whereas the 'deferential voter' thesis has not been applied to plebiscites, the effect of attitudes to the regime on participation in elections has often been noted for the Revolutionary period.[15]

There can be little doubt that with regard to the plebiscite of 1815, for the Seine-Inférieure as elsewhere, variations in the level of participation are best explained in terms of attitudes to the precarious regime. Even more than previous plebiscites, this one, with another standing in the wings, was, as Bluche has remarked, effectively a referendum. With 99 per cent of voters voting 'yes', rejection of the regime was expressed essentially through abstention.[16] Certainly, the highest rates of abstention occur in areas evidently hostile to Napoleon's return, some even in more or less open revolt. In the port of Le Havre the plebiscite followed weeks of disturbances between garrison and citizenry and a major royalist demonstration, which resulted in a state of siege being declared.[17] Unsurprisingly the turnout here was minuscule, and nearly a quarter of voters explicitly registered their opposition. Equally, many of the small rural communes on the Caux plateau which returned their registers completely blank were scenes of escalating royalist unrest.[18] Whether, at the other end of the scale, high participation rates in the arrondissement of Neufchatel were an expression of genuine support for the regime or just conformism, they clearly signify a willingness to work with the regime. A similar attitude was evident on the part of the arrondissement's notables who, in far greater numbers than their peers elsewhere, carried on business as usual in their local electoral college.[19]

It is tempting to interpret the variations in turnout in the elections of 1809 in the same way. In addition to the resemblance in the pattern of participation on the two occasions, there is reason to believe that the respective positions taken up in the Hundred Days, far from the product of circumstances, reflected more fundamental and enduring preferences. It was on the basis of election results under the early Third Republic that Siegfried concluded that the population of the department's Cauchois and Brayon cantons were governed by opposing political temperaments – conservative and republican or 'governmental', respectively. Some of the maps of voting in 1848–9 suggest a similar regional division of opinion.[20] Furthermore, there is some evidence for suggesting the political divide was already in place within a couple of years of the start of the Revolution. It consists of local clergy's responses to the requirement, by a law of 27 November 1790, to take an oath to the new Civil Constitution of the Clergy. Whereas 67 and 70 per cent respectively of priests in the two districts which were later to make up the arrondissement of Neufchatel took the oath, non-jurors were in a clear majority in all the other five districts.[21] Can we assume some connection between clergy's response to the oath and their communities' acceptance or rejection of the Revolution? If so, can we see the voting in the arrondissement of Neufchatel not only in the 'referendum' of 1815, but also in the elections of 1809, as yet another expression of the region's commitment to the 'new' France?

As attractive as the notion of such fixed, clear-cut 'geopolitical' divisions may be, there is no shortage of anomalous indications. On one side of the imaginary internal political frontier, the districts of the Caux plateaux, although undeniably a centre of militant royalist activity after Thermidor Year III, contained a much denser implantation of Revolutionary political societies than did the Brayon districts of Neufchatel and Gournay. Nor were the latter uniformly 'pro-revolutionary' or conformist: the forests around Blangy were a refuge for royalist paramilitaries, and in the elections of the year VI, which were generally favourable to the Republic, the assemblies of Gournay, Forges and Gaillefontaine have all been classified as royalist by a recent researcher. Whether such facts simply complicate the original picture or require it to be entirely redrawn cannot be answered here.[22]

Even accepting the reality of two fixed ideological blocs, it is doubtful whether in the end it holds the key to regional variations in polling in 1809. Part of the reason derives from the difference in character between Napoleonic 'legislative' elections and plebiscites. Whereas the latter were explicitly a vote of confidence in the regime, the job of the primary elections was simply to fill the electoral colleges. Provided they succeeded in this, did it really matter how few people voted? True, Beugnot himself had felt it necessary to excuse the low attendance at the previous primary elections in 1803 to his superiors, but abstentions had assumed much greater significance on that occasion, mainly because, unlike in 1809, ballots were invalid unless an absolute majority of the electorate voted. As a result large numbers of seats in the colleges had remained unfilled. It is also true that a few officials, after the introduction of universal male suffrage, expressed their concern at the general lack of interest shown by the electorate. However, the men concerned were, it seems, speaking for themselves and not the regime, which some suspected of wishing to end elections altogether.[23] In general the concern of the local authorities seems to have been not how many voted but who was elected. Doubtless the low poll made it easier for officials and notables alike to exercise influence over voters' choices.

More positively, there may have been more to voters' readiness to turn out in high polling areas than either enthusiasm for the regime or sheer conformism. These areas tended not to exhibit the remarkable convergence of opinion evident in the voting elsewhere. Generally speaking, the higher the poll, the lower the share of the votes obtained by the most popular candidates. The extreme case is again Londinières where less than half of the 73 per cent of the electorate who voted opted for the most popular choice, Pernet, a lawyer from the *chef-lieu* of Neufchatel. In adjoining Blangy, it may have been the extent of divisions among the voters which caused the presiding official to close the polls after one ballot with most of its quota of college members still to be elected.[24] In other words, the high-polling cantons of the arrondissement of Neufchatel may have experienced real electoral contests.

Moreover there is evidence from outside the electoral arena that some form of political life did indeed survive in the arrondissement of Neufchatel well into the Empire. Judicial records provide a detailed account of two cases of factional conflict: one in the canton of Blangy (where the 1809 primaries may have been particularly acrimonious) and the other in Argueil (where in all probability they were not). At first sight both disputes seem classic instances of the sort of 'individual struggles' between rival notables, which Coppolani holds responsible for unusual flurries of activity at the polls. Both were examples of 'place politics': according to Chapais, the *procureur-général* at Rouen, one or two troublesome individuals in each area were waging determined campaigns to unseat the local justice of the peace, in the hope of getting their own man appointed.[25]

In fact, there was more to the disputes than first meets the eye. First, however self-interested the motives of the leading protagonists, the contests were not devoid of 'party-political' content. As a former member of the Constituent Assembly, the leading spokesman for the 'opposition' group in Blangy, Simon, could claim to represent the cause of 'good citizens' against the now ascendant 'counter-Revolutionaries' under the leadership of a local noble who had earlier been suspected of involvement in Cadoudal's plot to assassinate Bonaparte.[26] Second, these were not intra-notable conflicts in the strict sense. No doubt the justices of the peace had powerful backers: one denunciation of the justice of Argueil accuses him of letting his 'obligations' towards the canton's leading notable affect the way he dispensed justice.[27] The 'anti-establishment' parties, though, were led, in Argueil, by an innkeeper-cum-attorney, and in Blangy by a bailiff and the former deputy Simon, who had clearly fallen on hard times.[28] Lacking the economic power and social influence of genuine notables, they relied on gaining support through more democratic or demagogic means. In Blangy this included one-way pamphlet warfare, 'publishing libels' aimed at the local magistrate. In the other case, whatever the medium, the populist message seems to have hit home; early in 1807 the campaign against the magistrate was said to be 'stirring up the *pays*'.[29]

Finally, whereas the trivial personal struggles Coppolani refers to were isolated, localized affairs, factional conflicts in the arrondissement of Neufchatel in the early years of the Empire were neither. Reverberations from the two conflicts reached well beyond the limits of the jurisdictions concerned. Moreover, it seems they were far from the only instances of their kind in the area: Chapais explained to the minister that he reported in such detail on events in Argueil not simply for their own importance but also to 'illustrate the agitation which torments the arrondissement'.[30] Ultimately, we can only speculate on whether the waves from such disputes were still being felt at the polls in 1809. What is beyond question is that the area, which a few years earlier was heralded as a haven of calm, now appeared peculiarly resistant to Napoleonic depoliticization. The implicit contrast with the 'return to normal' elsewhere is evident in Chapais' complaint: 'there reigns in the arrondissement of Neufchatel a violence of faction which is compressed only in appearance but which continues to simmer fuelled by recollections of the past.'[31]

*

Why was the arrondissement of Neufchatel, compared to the department as a whole, so supportive of the Napoleonic regime and at the same time prone to factionalism? No doubt, individual personalities, such as the former regicide sub-prefect, Pocholle, played their part. Yet, the existence of such marked, if as yet inadequately defined, differences in political culture between adjoining areas, commonly regarded as comprising distinct 'natural' regions – the Pays de Bray and the Pays de Caux – suggests the need ultimately for a more structural approach. But if political contrasts had their roots in socio-economic differences between the two regional communities, the question is 'which'?

Clearly the problem cries out for the kind of rigorous comparative ecological and sociological analysis that Paul Bois applied to 'white' and 'republican' districts of the Sarthe in his *Paysans de l'Ouest*.[32] Here I can only offer some brief reflections on the similarities and differences between the probably 'off-white' Caux and 'pinkish' Bray based largely on secondary material. A less obvious difference from Bois's work is that whereas in his argument with the ghost of Siegfried he took on a single, dominant intellectual tradition; now one is faced with an embarrassing choice of 'geopolitical' models which from their different disciplinary perspectives all claim to hold the key to the structural linkages between politics and society.

The most recent and ambitious of these models may be despatched quickly. In a number of works, Emmanuel Todd and Hervé Le Bras have between them expounded the thesis that not merely throughout France but universally the political choices of different regions are to be explained in terms of the ideological values inculcated by different family systems.[33] In relation to the problem at hand, we are not obliged to respond to what Vovelle has called 'the provocation of the anthropologists'. The main reason is that their maps rely overwhelmingly on data aggregated to the department level. As a result they are unable to reflect or identify the sort of internal contrasts with which we are here concerned. It is true that if what they say about the connection *between* prevailing inheritance rules and family structure were correct, then there would be an *a priori* basis for suggesting that family and household arrangements may have been different in our two regions. For up to the Revolution the Pays de Caux had its own customary law which, in contrast to the more 'egalitarian' Custom of Normandy in force elsewhere, featured strict primogeniture.[34] However, historians are generally sceptical of the notion that family systems vary with inheritance customs, and no demographic study of upper Normandy has yet to point to significant differences in family structure.

Perhaps Beugnot's original 'untheorized' insight connecting Neufchatel man and his pastoral surroundings opens up a more promising line of enquiry. What most struck contemporary visitors as they went down from the plateaux into the sunken valley system of the Bray was the way cornfields gave way to pastures. Was this contrast in the regional agrarian economies the key to the political differences we are looking for? Certainly there is a sufficient fit between the geography of turnout in 1809 and the major agricultural zones as defined by contemporaries. All but one of the high polling cantons lay either within the largely pastoral valleys of the Pays de Bray or had substantial territory in the more mixed but largely pastoral farming zone in the Seine valley, while the lowest turnouts occurred within the two predominantly arable zones.[35]

But if there was a link between the regional farming systems and their respective political cultures, what precisely was it? Following on from Beugnot's reference to the unexacting nature of pastoral work, it is tempting to speculate about the impact on social relations of the differing extent and pattern of labour demands in arable and pastoral agriculture. Unfortunately, current disagreements among the specialists about the respective requirements of the two systems would make this a particularly hazardous undertaking.

There is, too, the awkward fact of economic differences within the pastoral region. For all the contemporary talk of pastures, diversity of land use was the hallmark of the arrondissement of Neufchatel at the time. As well as the great forests, which survive to this day, the arrondissement also contained very extensive wasteland. Of course, through the common grazing they afforded, both made a considerable contribution to the pastoral economy. Despite the progress of agricultural specialization, a lot of land was still under the plough. As late as 1830, whereas well under two-fifths of the core of the Bray was given over to cereals, in the rest of the arrondissement the share was often in excess of two-thirds.[36]

Help with these difficulties may be available from across the Channel. Originally devised by agrarian historians as a means of distinguishing regional farming systems, the distinction between arable, frequently downland, 'fielden' areas and 'wood-pasture' regions has more recently come to be seen by some as 'the key social and cultural divide' in early modern England.[37] Wood-pasture villages, it has been variously suggested, were more prone to popular disorder and crime, proved fertile ground for the spread of Puritanism, and (in the southwest) preferred 'bat and ball games' to football. Finally, and most controversially, the model has been applied to the problem of regional political differences. David Underdown has argued that the contrasting popular cultures of wood pasture and arable regions inclined their respective communities to opposite sides in the English Civil War with pastoral or 'cheese' districts opting for the parliamentarian cause and arable villages on the chalk downlands siding with the king.[38]

Are the differences in political culture between the Caux plateau and the arrondissement of Neufchatel another case of 'chalk and cheese'? Or, perhaps, as butter production for the Paris market seems to have been the major concern of the regional 'dairy industry', it should be 'bread and butter'. Certainly the model answers the problems already encountered. The wood-pasture category makes a feature of the kind of relative economic diversity evident on the east of the department's internal political frontier. It also supplies the missing explanatory element: according to Underdown, in the English context, the differing degrees of social cohesion within the communities of the two farming regions was at the root of their other cultural and political differences.

On closer inspection, however, the model does not travel well. This is largely because Underdown's chalk/cheese dichotomy involves not only farming systems, but also takes in other ecological or sociological contrasts, which are crucial to his argument. The supposed organic nature of communities in arable zones results from their nucleated character and collective regulation of agriculture required by the open field system. The celebrated individualism of wood-pasture settlements stems primarily from the physical and social isolation that comes with scattered settlement and the absence of any collective economic endeavour, but is greatly amplified by their association with the rural textile industry, which, it is argued, has a solvent effect on traditional social ties. In Napoleonic Seine-Inférieure, however, these various socio-economic characteristics do not align with the two farming systems in the same way. In the chalk-arable part of the Seine-Inférieure, despite greater population density, the settlement pattern was no less dispersed than in the neighbouring Bray, and here it was in the arable region that the rural cotton industry was concentrated, making it 'one of the most remarkable sites of proto-industrialization'.[39]

More strikingly, unlike in other open-field areas, in northern France as well as sixteenth-century England, the Caux was characterized by agrarian individualism: collective grazing rights had been successfully curtailed, and production was carried on in large-scale capitalist farms – cut off from their neighbours, symbolically, by screens of tall beech trees.[40] In fact, if one were to place the respective communities of the two areas on an organic–individualist continuum, Brayon ones would surely be at the organic end. For the population living on the edges of its large forest tracts or surrounded by wasteland, the exercise of customary rights still constituted a crucial resource. Predictably, attempts to liquidate those rights both on the eve of the Revolution and again during the Napoleonic 'restoration' met with the combined resistance of a dozen or so villages.[41] In so far as the English model has drawn attention to fundamental socio-economic contrasts between our regions, it has served a valuable purpose, even if in Normandy their relationship to the respective farming systems was the reverse of that prescribed.

Siegfried's model, which began the whole 'geopolitical' tradition, and closes our discussion, is in some respects complementary to the one above. His methodological approach involved a consideration of almost every possible component of the rural environment except that relating directly to the cultivation of the soil: the distribution of settlements, patterns of land ownership and tenurial arrangements, along with two 'non-ecological' factors: state penetration and clerical influence. However, in explaining why different regional peasantries vote differently he ultimately comes down in favour of the structure of landholding, combined with 'the religious factor'. Where, thanks to their ownership of the soil, peasants were sheltered from the influence of the landlord, and his able assistant the parish priest, they naturally voted for the Republic.

Over the last 30 years or so historians have become increasingly impatient of Siegfried's arguments and conclusions, even while admiring his methodological line. Tony Judt expresses the general view: 'Siegfried's own work is now largely discredited, its ecological and territorial determinism a barrier to more subtle accounts of social change.'[42] Other charges include intrusive political sympathies, dubious statistical procedures, flirtation with racial theory and *lèse-histoire*. In our case, however, there are two compelling reasons for giving him a careful hearing. First, the problem at issue is very much his own. The Seine-Inférieure was one of 14 western departments which form the subject of the *Tableau politique de l'Ouest*. But not only did Siegfried study our two regions; most of what he has to say about the department depends upon the contrasts between the Pays de Caux, the epitome of Norman conservatism, and the Pays de Bray, one of a number of Republican enclaves within western France. These pages no doubt have done more than anything else to fix in the general consciousness the notion of the Pays de Caux and the Pays de Bray as eternal opposites.

Second, there is the matter of the fit between his map of landholding for the 1880s and our Napoleonic political data. With regard to the elections of 1809, the lowest polling cantons fall almost entirely within his area of 'large-scale property', which is more or less coterminous with the Caux. Conversely, the six high polling cantons of the arrondissement of Neufchatel, and two more adjoining Rouen, fall within his two areas of middling and mixed property ownership, which together form a cordon on the department's eastern border. However, the general inverse correlation of high participation and large-scale property is by no means absolute: the four high polling cantons, with territory both in the Seine valley and on the Caux plateau, are included in this category.[43]

To counter the charge of anachronism, it is important to emphasize that, in outline, Siegfried's account of landholding in the 1880s holds good for our period. Even without a detailed or complete picture of property distribution within the department for the late eighteenth or early nineteenth century, there is no doubt that the Caux plateau, which covered so much of it, was an area of *grande propriété* as well as *grande culture*.[44] Siegfried's observation on the region in the 1880s – 'The essential fact is there is no peasant ownership whatsoever' – is not much of an exaggeration for the earlier period.[45] A recent study of a score of *terriers* of the late *ancien régime* puts peasants' share of the land at no more than 10–15 per cent.[46] Nor do Revolutionary land sales in the area seem to have done much to reverse the process. Peasant purchases were comparatively limited, and, according to one Napoleonic official in the area, many of the smaller plots peasants had gained, had disappeared ten years later.[47] Data specific to the arrondissement of Neufchatel is more scarce, but what there is conforms to Siegfried's picture for the later period. On the basis of limited samplings from the *ancien régime*, the historical geographer Sion, writing shortly before Siegfried, also classified the core of the Bray region as one of middle-sized ownership, while a recent study of the sale of national lands in the area refers to its 'lack of big landowners'.[48]

There remains a very real problem with Siegfried's argument in relation to this area, for both his and our own period. It is not that the distribution of ownership is *not* as he claimed, but whether the differences in distribution between the two regions are sufficient to explain the *marked political* contrasts with which we are concerned. After all, the comparison is not between regions at opposing ends of the property-owning scale, but large-scale versus middling and 'mixed' categories. Doubtless Siegfried's more empirically minded critics are right to be concerned by the apparent disparity here between modest cause and great effect. This, though, is not to say that such differences are not a critical part of the explanation. Surely, as de Planhol remarks in relation to enduring differences in religious practice between the two Pays, it is not a matter of rejecting the basic schema out of hand, but rather of adding to it further explanatory 'local factors'.[49] None of the additional factors I am about to propose would come as any great surprise to Siegfried, and two already have some place in his analysis.

For Siegfried the normal influence of the large landed estate was considerably increased when in noble hands. In the Caux of his day he believed 'feudal' influence, though surviving, was 'half dismantled', as the bourgeoisie of Rouen and Le Havre acquired an increasing share of the countryside.[50] In the Napoleonic period, however, it is now clear the economic basis of that influence remained largely intact. In Seine-Inférieure as a whole, even in the aftermath of the Revolution, the really big estates were overwhelmingly in former noble hands. The full extent of noble dominance in parts of the Caux is confirmed in a document, in which the sub-prefect of Yvetot, when submitting his recommendations for the new post of *président de canton*, identified the one or two biggest landowners in each of the ten cantons he administered. All but a couple were of noble origin, and several were former *émigrés*.[51]

By contrast, Siegfried's characteristic observation that the Neufchatel area lacked an 'essentially feudal, landed infrastructure' may well apply with almost equal force to the 1800s as the 1880s.[52] This is not based solely on the inference that where land ownership was somewhat less concentrated, noble estates would also be commensurately reduced. The distribution of land confiscated from *émigrés* leads Bodinier to consider a large part of the Bray as a 'desert' as far as noble fortunes were concerned.[53] In addition, noble estates in this area were disproportionately made up of forest and woodland, which, not only in Siegfried's view, did little to create social dependence. Finally, it is worth noting, as an indication of the lower profile of the nobility in the area, that the 'superstructure' of the great noble estate, the château, was a comparatively unusual feature on the rural landscape in this part of the department.[54]

Another factor concerns landholding rather than ownership. This occupies an important part in Siegfried's interpretative schema generally, with different modes of tenure enhancing or undercutting landowners' influence. However, his explicit discussion of the subject with regard to the Seine-Inférieure is more or less limited to the important fact that everywhere the same 'tenurial regime' operated: land farmed by tenants on a nine-year lease was the norm. However, within the same prevailing system there are still regional differences to be highlighted.

First, there were more numerous exceptions to the rule of indirect farm tenure in the arrondissement of Neufchatel owing to the fact that forest and pastoral land lent themselves more readily to exploitation by their owners. More important for its likely social effects was the difference in farm size either side of the internal frontier. In Sion's sample parishes for the late 1770s, in the Caux farms of over 40 hectares occupied most of the land area; in the Bray holdings in the 10–40 hectares bracket had the largest share.[55] As a result, inside Cauchois villages economic power was frequently monopolized by a few great tenants of the large estate, whereas in the Bray, with access to land not general but widespread, there was something closer to a property-holding, if not property-owning, democracy.

Adding to this series of modest but cumulatively significant important structural differences the two elements of differentiation highlighted by the fielden–forest model earlier, the enjoyment of collective rights and, more crucially, the location of the rural cotton industry, we can begin to build up a picture of two widely divergent societies. Furthermore, to a large extent they diverge in ways originally signposted by Siegfried. By and large, all the various factors of differentiation have the effect on the Caux plateau of increasing the concentration of economic power and social disparities, and of reducing or containing them in the arrondissement of Neufchatel. It would, of course, be reductionist to sum up the two regional societies as, respectively, 'a land fit for notables' and a 'social democracy'. Nevertheless such labels do convey much of what was different about them.

Such social differences do not actually explain, far less did they determine, the kind of political and ideological differences with which we started – attitudes to the regime and its political process – but they do make them more understandable. This is obvious, but it is something which in recent years, with the widespread disenchantment with social history, has frequently been overlooked. In case this sounds simply like a call to turn back the clock, I should add that if we are to establish properly, for our bread and butter regions, the links between landholding patterns and prevailing economic activities and their associated social structures, on the one hand, and political behaviour on the other, we shall need to examine much more closely social relations and cultural practice within the two regional communities. Following the example of Agulhon and Underdown, we could start by looking into the 'rowdy drinking holes' of Gournay and Forges, and the village bars of the Caux, which, unlike in the Bray, respectfully closed during mass.[56]

Acknowledgement

My thanks to Jan Boldt, Duncan Alexander, Michael Zell and Malcolm Crook for reading and commenting on drafts of this chapter. Professor Clout and Daniel Fauvel kindly made their works available to me.

Notes

All translations from the French are my own.

1 Archives nationales (hereafter AN) F1c III Seine-Inférieure 8.
2 For the problem of delimiting the Pays de Bray, see Professor Clout's unpublished M.Phil. thesis, 'The Pays de Bray: A Study in Land Use Change 1750–1965' (University of London, 1967). Here I take the shortcut solution generally adopted of treating it as roughly equivalent to the arrondissement.
3 A. Siegfried, *Tableau politique de la France de l'Ouest sous la Troisième République* (Paris, 1913), pp. 248 and 250.
4 For example, E. Todd, *The Explanation of Ideology: Family Structures and Social Systems* (Oxford, 1985) and *The Making of Modern France: Politics, Ideology and Culture* (Oxford, 1991); and H. Le Bras, *Les Trois France*, new edition (Paris, 1995).
5 Unpublished PhD thesis 'Notables and Society in Napoleonic France: the Seine-Inférieure, 1799–1815' (University of London, 1988).
6 Notable exceptions, outside of France, are Lynn Hunt's *Politics, Class and Culture in the French Revolution* (London, 1986), which attempts to bridge the gap between discourse analysis and social history, and Timothy Tackett's *Religion, Revolution and Regional Culture in Eighteenth-Century France* (Princeton, 1986). Michel Vovelle's important work *La Découverte de la politique: géopolitique de la Révolution française* (Paris, 1993) may signal a climate change.
7 E. Dejean, *Un Préfet du Consulat. Jacques-Claude Beugnot* (Paris, 1907), p. 326.
8 Archives départementales de la Seine-Maritime (hereafter ADSM) 6 MP 5114–5 (registers of heads of leading families).
9 J.-Y. Coppolani, *Les Elections en France à l'époque napoléonienne* (Paris, 1980).
10 *Ibid.*, p. 4.
11 Figures are calculated from college membership lists: AN F1c III Seine-Inférieure 3.
12 F. Bluche, *Le Plébiscite des Cent-Jours* (Geneva, 1974), p. 46.
13 ADSM 3M149 (results of voting dated 23 May 1815). On the basis of Malcolm Crook's research in progress, I have calculated the electorate at 20 per cent of population. Figures for the latter are from the *Annuaire statistique pour l'année 1809*.
14 Coppolani, *Les Elections en France*, p. 234.
15 Bluche, *Le Plébiscite des Cent-Jours*, p. 96; Lynn Hunt, *Politics, Class and Culture*, p. 127.
16 Bluche, *Le Plébiscite des Cent-Jours*, p. 89.
17 AN F1c III Seine-Inférieure 4 (prefect's letter dated 15 May 1815).
18 AN BII 942A and C (registers for the plebiscite, Le Havre and Yvetot), BB3 163 (letter of procurator-general dated 8 May 1815).

19   On attendance in the colleges, see my thesis, 'Notables and Society in Napoleonic France', pp. 261–72.

20   M. Boivin, *L'Opinion publique en Seine-Inférieure: élections et plébiscites (1848–1914)* (Rouen, 1971).

21   Tackett, *Religion, Revolution and Regional Culture*, p. 355, figures up to April 1791.

22   *Atlas de la Révolution française 6: Les sociétés politiques* (Paris, 1992), map pp. 18–19; D. Pingue, 'Elections et politique religieuse sous le Directoire en Seine-Inférieure', in Comité Regional d'Histoire de la Révolution française, *A Travers la Haute Normandie en Révolution 1789–1815* (Comité Regional d'Histoire de la Révolution française, 1992), p. 65.

23   AN F1c III Seine-Inférieure 2 (letter dated 7 Vendémiaire Year XII. ADSM 3MP 3014 (Gueroult's letter dated 15 September 1809).

24   AN F1c III Seine-Inférieure 3 (departmental college list, 1809).

25   AN BB18 818 (Chapais's letter, 25 June 1807).

26   AN BB18 816 B (Thomas's letter, 27 Ventôse Year XII).

27   AN BB18 818 (petition dated 1 June 1807).

28   Notes 23 and 24 above, and L. Soublin, *Le premier vote des normands (1789)* (Fécamp, 1981), p. 98.

29   See note 23.

30   AN BB18 818 (Chapais to *Grand juge*, 2 February 1807).

31   AN BB18 818 (Chapais to *Grand juge*, 27 January 1807).

32   P. Bois, *Paysans de l'Ouest. Des structures économiques et sociales aux options politiques depuis l'époque révolutionnaire* (Paris, 1960).

33   See note 4 above.

34   E. Todd, *The Explanation of Ideology*, map, p. 63.

35   *Annuaire statistique du département de la Seine-Inférieure pour l'an XIII de l'ère française*, p. 258.

36   Clout, *The Pays de Bray*, map, p. 186.

37   N. Davie, 'Chalk and Cheese? "Fielden" and "Forest" Communities in Early Modern England', *Journal of Historical Sociology*, vol. 4 (March 1991), pp. 1–31.

38   D. Underdown, 'The Chalk and the Cheese: Contrasts among the English Clubmen', *Past and Present*, no. 85 (November 1979), pp. 25–48, and *Revel, Riot and Rebellion: Popular Politics and Culture in England 1603–1660*, (Oxford, 1987).

39   His preface to G. Lemarchand, *La Fin du féodalisme dans le Pays de Caux. Conjoncture économique et démographique et structure sociale dans une région de grande culture de la crise du XVIIe siècle à la stabilisation de la Révolution (1640–1795)* (Paris, 1989), p. ix.

40   *Ibid.*, pp. 60–2.

41   Clout, *The Pays de Bray*, pp. 156 and 166.

42   T. Judt, *Socialism in Provence, 1871–1914* (Cambridge, 1979), p. 239.

43   Siegfried, *Tableau politique de la France de l'Ouest*, map opposite, p. 536.

44   J. Bottin, *Seigneurs et paysans* (Paris, 1983), p. 320.

45   Siegfried, *Tableau politique de la France de l'Ouest*, p. 235.

46   G. Lemarchand, 'L'abolition de la féodalité', in *A Travers la Haute-Normandie*, p. 184.

47   On sales in the department, see *ibid.* (note 22), pp. 179–251.

48   J. Sion, *Les Paysans de la Normandie orientale* (Paris, 1908), pp. 259 and 271; Hélène Pierreuse, 'La vente des biens nationaux de première origine dans le Pays de Bray (1791–1832)' in *A Travers la Haute Normandie*, p. 221.

49 X. de Planhol, *An Historical Geography of France* (Cambridge, 1994), p. 330.
50 Siegfried, *Tableau politique de la France de l'Ouest*, p. 236.
51 ADSM M Elections: Nominations de président de canton an XI–XII (Legrand's letter to prefect, undated).
52 Siegfried, *Tableau politique de la France de l'Ouest*, p. 250.
53 B. Bodinier in *A Travers la Haute Normandie*, p. 244.
54 P. Seydoux, *Châteaux du Pays de Caux et du Pays de Bray* (Editions de Morand, 1983), map opposite title page.
55 Sion, *Les Paysans de la Normandie orientale*, p. 271.
56 Siegfried, *Tableau politique de la France de l'Ouest*, p. 251; de Planhol, *An Historical Geography of France*, p. 330 and note p. 498.

4
Getting out the Vote: Electoral Participation in France, 1789–1851

Malcolm Crook

Just over a decade ago a stimulating article by Peter McPhee drew attention to a striking contrast between rather low rates of electoral participation in the French Revolution of 1789 and the extremely high turnout recorded in 1848.[1] It was a splendidly argued, provocative piece which sought to contest existing explanations for political modernization. Yet it seemed to confirm the widely held belief that the tradition of democratic elections was not really established in France before the mid-nineteenth century; what went before merely constituted, in the words of one historian, an 'anticipation' or, according to another, 'the prehistory of modern democracy'.[2] However, save for his own research in the Mediterranean department of the Pyrénées-Orientales, McPhee's argument was based on partial and sometimes defective sources, which distorted the contrast between the two experiments with democracy that he sought to compare and contrast.

Much more information is now available on the role of elections in the development of French political culture during the 50 years that followed the Revolution of 1789 and it is possible to reappraise McPhee's contentions in three respects: first, to demonstrate that turnout during the revolutionary decade from 1789 to 1799 was not always as lamentable as he suggested; second, to show that, conversely, the level of electoral participation during the Second Republic was sometimes lower than he believed; and, lastly, to reopen the debate on the meaning of this relative contrast in rates of voting between the Revolution of 1789 and that of 1848.

Various explanations have been advanced for increased electoral turnout in mid-nineteenth-century France. These include: the effects of economic and social development; the impact of a uniform administrative framework; the role of the elites as culture-brokers in the countryside;

**THE DEPARTMENTS OF FRANCE
IN 1790 AND 1848**

Map 2 The Departments of France in 1790 and 1848.

and the political apprenticeship facilitated by local elections under the July Monarchy. Yet it is worth questioning the significance of getting out the vote. Should it be equated automatically with a modern culture of elections? There is good reason to believe that despite the rising numbers of participants, attitudes towards the electoral process evolved only slowly and that strong polling under the Second Republic did not necessarily amount to the advent of political maturity. Eugen Weber's *Peasants into Frenchmen* has been at the centre of a controversy over when and how the politicization of the French people took place. Electoral behaviour is a vital aspect of this debate, though not the only indicator, and it must be studied over a long period of time and not simply confined to the exceptional circumstances of the mid-nineteenth century.[3]

I

McPhee highlighted the paradox of weak electoral participation in the 1790s, set in a context of unprecedented political mobilization via Jacobin clubs, collective petitions and popular action. Yet when he wrote his challenging article, turnout in revolutionary elections had attracted little interest and only one attempt at a general analysis.[4] The question of who could vote and whom they voted for in legislative elections during the 1790s had been explored, but not *how* they did so. The astonishing neglect of a subject so central to the politics of the French Revolution has now been reversed and during the last decade work has been proceeding apace on various aspects of its political culture.[5] However, the investigation of voter turnout continues to be inhibited by the absence of any reliable, national statistics for the 1790s. Recourse to local archives is essential, but the cumbersome, two-tier procedure for elections beyond municipal and cantonal level, renders the interpretation of data a difficult business.

McPhee, for example, asserts that in 1791 in Paris, the hub of the political nation, elections to send deputies to the Legislative Assembly involved a mere 200 out of 946 members of the capital's electoral college.[6] He cites this as evidence of a lack of commitment to the electoral process on the part of the metropolitan elite, which controlled this second stage of the two-step electoral system. What he fails to recognize, however, is that this particular ballot was taken towards the end of a gruelling series of votes. The session had already lasted more than a fortnight and attendance was inevitably waning.[7] When the proceedings had commenced, by contrast, turnout was far higher: the earlier election of Parisian deputies to the Legislature had attracted 85 per cent of the electors. Nor was high turnout to choose national legislators confined to Paris. In second-tier assemblies throughout the country the major ballots in 1791 mobilized over 80 per cent of the membership. A year earlier attendance had been higher still and it would remain elevated for the rest of the decade.

The departmental electoral colleges of the 1790s comprised an elite of some 40,000 wealthy notables, seasoned campaigners who had been chosen as local representatives at primary assemblies in the cantons, where the mass electorate had participated. Even during the early years of the Revolution when the suffrage was more restricted, roughly 60 per cent of adult males were able to participate at this level. The real novelty under the Second Republic of 1848 was not the granting of the vote to all adult males in France, but rather the fact that they were now *directly* electing their national deputies (something that was envisaged in the stillborn Constitution of 1793, but never implemented). None the less, it should be emphasized that local elections in the 1790s, for municipal councils, national guard officers and justices of the peace, were *direct*.[8]

To be sure, attendance at the primary assemblies left much to be desired during the Revolution. Among ordinary voters turnout was extremely variable, from one department to another and even from one canton to the next. It was unusual for the degree of participation to top 30 per cent and it might fall to single figures. The proceedings of these primary assemblies were, like their 'secondary' counterparts, extremely complicated and rarely completed in a single day; once again, attendance inevitably diminished the longer the session went on. At Saint-Vallier in the Var in 1791, for example, 249 voters turned up at the cantonal assembly for the morning session, but only 91 of them returned after lunch. Or again, at Chahaignes in the Sarthe, in the same year, 157 citizens were present at the beginning of the proceedings, but fewer than 30 remained at the finish.[9]

It was not simply a question of submitting a ballot paper and retiring gracefully, but of choosing assembly officials at the outset, before going on to conduct a series of exhaustive (and exhausting) ballots in pursuit of absolute majorities for the posts to be filled. For the most part, the revolutionaries had retained the electoral system in local use at the end of the *ancien régime*, which was employed for elections to the Estates General of 1789. Yet the tradition of voting in assemblies was ill-adapted to the larger electorate of the revolutionary decade and to the cohabitation of rival communities which often vied for control. It is difficult to determine whether parochial or political tensions were responsible for the ensuing disorder.

The complex electoral process, which was radically simplified in 1848, was clearly a deterrent to high turnout in the 1790s, but there were other inhibiting factors at work too. Since the authorities produced voting lists on the basis of tax or census returns, they were actually listing the *potential* electorate. Those who wanted to vote had to fulfil additional formalities, which included paying their taxes promptly and registering for service in the National Guard. They also needed to swear a civic oath which some opponents of the Revolution were unwilling to take. Lists of fully registered voters rarely come to light, but they suggest that such requirements could remove up to half the electorate from further contention.

No such drawbacks operated under the Second Republic, when the authorities drew up the registers and most elections took place in an orderly atmosphere. The Revolution of 1789, by contrast, was accompanied by severe disruption from 1791 onwards. The state of civil war, actual or latent, which afflicted many parts of France obviously had a significant impact on electoral activity: in areas of insurrection, elections were simply not held or involved only the hardiest of citizens. In the Morbihan in 1799, for example, guerilla action prevented the polls from opening in over a third of the cantons, while elsewhere in the department voting was a declaration of support for the Republic liable to provoke reprisals. Moreover, the outbreak of war in 1792, which contributed to internal unrest, necessitated the enlistment of hundreds of thousands of young, adult male voters, who continued to be included on the electoral rolls, regardless of their inability to vote.

The 'quiet' year of 1790 has often been called the 'golden age of voting', because turnout in excess of 50 per cent was not unusual. Yet later in the revolutionary decade, especially in urban areas, similar enthusiasm was not entirely lacking: in Toulouse in 1797, for instance, primary elections attracted 70 per cent of the eligible voters. It would, therefore, be wrong to posit a consistent decline in participation during the 1790s; instead there were fluctuations from one year to another. The Directory has often been derided as a time of acute electoral apathy, but polling was relatively heavy in 1797 and 1798. The generally low levels of voting recorded in 1799 must be set in the context of two successive years when the government had restricted political liberties and interfered with the results of elections.

Indeed, it can be argued that the very intensity of the electoral process, not to mention its frequency, meant that for the minority of ordinary Frenchmen who did participate regularly, the degree of politicization was truly profound. Neither declared candidates nor recognized parties existed during the Revolution, yet canvassing was evidently conducted outside and inside the meeting halls. The assembly method may have been undisciplined, but it facilitated a good deal of debate and the passage of resolutions as well as voting. In 1793, for example, amendments to franchise provisions in the Constitution were proposed by citizens who assembled to approve the document; it was much more than a simple referendum.[10] These aspects of the revolutionary electoral process should not be dismissed as archaic survivals from the past. They continue to characterize participation in contemporary associations like trade unions and clubs. Today's elections are rather austere occasions, whereas in the 1790s assemblies offered a much richer experience than the bald figures for turnout would suggest.

Table 1 Approximate percentage turnout at primary assemblies, 1790–99*

1790	1791	1792	1793	1795	1797	1798	1799
45	20	15	33	20	25	20	10

*These figures are rough calculations from incomplete returns which offer no more than an order of magnitude.

II

In the light of what has been said above, comparisons with the elections that followed the reintroduction of universal male suffrage in 1848 should not be undertaken lightly. Global figures are certainly easier to come by and, at first sight, they point to a huge contrast: in elections to the Constituent Assembly of the Second Republic, held in April 1848, it appears that 84 per cent of a vast electorate of some 9 million adult Frenchmen cast their votes.[11] As under the Revolution, they voted at the *chef-lieu* of each canton. This meant a journey of at least a few kilometres for most rural inhabitants, yet the overall level of participation has been bettered on only one occasion since then. Not until the interwar years of the twentieth century did abstention of less than 20 per cent become the norm.[12]

As with all statistics, some caution must be exercised, for careful scrutiny suggests that the figures for 1848 are open to question. How many Frenchmen were entitled to vote? Local authorities had only a month to compile registers according to the instructions issued on 8 March 1848. Some 9 million voters were apparently enrolled, yet it is generally accepted that by the end of the year this total had risen to almost 10 million. How many actually voted? Returns to Paris in April 1848 often indicated approximate numbers of participants.[13] Verification reveals a good deal of inaccuracy: in the Var, for example, the turnout of 91 per cent must be reduced to a relatively modest 74 per cent, because officials confused the number of registered voters with those who took part.[14] There is good reason to suppose that the national average might be revised downwards in the light of more detailed study.

Yet there seems no doubt that over 7 million individuals, representing at least 70 per cent of the potential electorate, did vote for departmental slates of deputies in April 1848. What is equally striking is the uniform manner in which they did so, from one department to another and even from canton to canton; in the Var, for instance, only three cantons recorded a rate of abstention 10 per cent above the average. Citizens in northwestern France and the Massif Central, who had a long history of electoral absenteeism, participated as strongly as their counterparts in traditionally high polling departments like the Marne or the Aube to the north-east. Thus the Vendée recorded over 80 per cent, while even the Cantal managed more than 70 per cent. In 1848 Paris returned the lowest proportion, with just 68 per cent, and many urban areas failed to match the attendance at rural polls.

Was this heavy turnout sustained in the elections that followed? The presidential contest of December 1848 attracted at least as many individuals to the polls as six months earlier, though since there were more registered voters it is usually reckoned that participation dipped below 75 per cent. Despite the fact that abstentions rose to almost 50 per cent in one or two wintry upland departments such as the Haute-Loire, this was an outstanding performance.[15] National turnout subsequently fell slightly, to an average of 68 per cent in the legislative elections of May 1849, with marked variations from one department to another, though this was the third national poll in little more than 12 months.[16]

The overall trend from 1848 to 1849 was one of gentle decline after a strong start, but rather less interest was evident at by-elections, perhaps because voters were unwilling to travel to the cantonal *chef-lieu* to elect a single deputy. The choice of the same deputies in several departments (the simultaneous candidature was a feature of electoral practice in France which helped facilitate the emergence of Louis Napoleon in 1848), as well as deaths and expulsions, necessitated numerous by-elections during the Second Republic. There were a clutch of these contests in June 1848 which saw turnout at least one-third lower than in April. In the Côte-d'Or, for instance, this election attracted only 43 per cent of the electorate, as opposed to nearly 80 per cent six weeks earlier. When a further round of by-elections was held in September, attendance fell to as little as 20 per cent in the Charente-Inférieure and the Nord, though turnout in similar polls at Paris remained above 60 per cent.[17]

It was the same in the wake of the legislative elections of 1849. In July of that year a fresh crop of partial elections took place and few contests attracted more than 40 per cent of the electorate. There was talk of compulsory voting, but in the spring of 1850, another round of by-elections (caused by exile or imprisonment for those involved in the insurrection of 13 June 1849 at Paris) produced a renewed wave of enthusiasm, and turnout in excess of 70 per cent was once again the norm. This surge of participation, which was accompanied by the return of numerous radicals, prompted the Right in the Assembly to introduce fresh restrictions on the franchise: the law of 31 May 1850 removed roughly one-third of voters from the registers. The reduced electorate had few opportunities to name deputies after this date, before Louis Napoleon's *coup d'état* in December 1851 brought the experiment with competitive elections to a halt.

Not all national elections in the Second Republic attracted the huge turnout for which they are renowned, but participation was clearly much heavier than at the primary assemblies of the 1790s. The comparison in terms of local elections is, however, rather less flattering to the Second Republic. A generally low level of participation greeted the series of municipal and departmental elections that were held after the revolutions of both 1789 and 1848. To be sure, turnout in 1790 was often high, but thereafter it fell and fluctuated in the same manner as primary elections. In 1848, on the other hand, local polls rarely evoked the same degree of enthusiasm as national ones, perhaps because the novelty factor soon wore off. Municipal elections in the Var, for instance, produced an average turnout of less than 40 per cent.[18] Figures for 1849 suggest declining attendance, while the election of *arrondissement* and departmental councils in 1848 had attracted fewer than 30 per cent of the voters.[19]

III

Closer inspection usually reveals greater complexity than a single glance. A detailed comparison of turnout in 1789 and 1848 suggests that getting out the vote under the Second Republic was much easier than during the 1790s, though the contrast is less pronounced than Peter McPhee claimed in his pioneering article. Perhaps, as he suggests, an explanation is to be found in economic development which, by the middle of the nineteenth century, was enabling local communities to grasp the significance of national politics to a greater extent than before.[20] However, the significance of this factor is not easily established and the unevenness of economic change is not reflected in the consistently high levels of turnout recorded everywhere in 1848.

Universal electoral participation was doubtless assisted by a uniform and centralized framework of administrative and political institutions which had lasted for half a century.[21] The political elite had been involved in national electoral politics for a similar period. After 1830, the July Monarchy doubled the electorate so that some 240,000 Frenchmen were eligible to vote. Although this represented only 3 per cent of males over 21 years old, legislative elections were held regularly and a rising proportion of these wealthy electors were participating: over 80 per cent cast a ballot in the general election of 1846.[22] It should also be stressed that widespread electioneering involved a substantial number of non-voters in the accompanying rituals.

Maurice Agulhon has suggested that political activities of republican notables paved the way for the advent of popular sovereignty in 1848.[23] In fact, as he acknowledges, a large segment of the French male electorate had already served an electoral apprenticeship under the July Monarchy, if not earlier under the Revolution. The law of 1831 on municipal elections enfranchised some 3 million Frenchmen, who were also entitled to take part in choosing officers for the national guard.[24] Approximately one-third of the electorate of 1848 was given a limited opportunity to vote under the July Monarchy. According to a national survey, over half of them participated in the local polls of 1834.[25] It seems that a similar proportion continued to do so in the triennial contests which followed and there is evidence to suggest that turnout rose in the polls of 1846.

Electoral participation in 1848 was thus built upon solid foundations. The degree of mobilization was also far greater than during the 1790s, when press, 'parties' and declared candidates had played only a tentative role.[26] Under the Second Republic newspapers were involved in the process in a far more conspicuous fashion, while electoral committees operated overtly and lists of candidates were freely circulated, together with countless manifestoes. The much vaunted election of 23 April 1848 began on Easter Day (then, as now, Sunday was the habitual day for the start of polling in France) and concluded on the Monday. Despite a number of commentators who thought this coincidence with the religious holiday would diminish attendance, it undoubtedly worked in the opposite direction.[27] In more 'faithful' areas, where Easter communion was *de rigueur*, citizens set out to vote together after an early mass. Clerical influence over the elections was accordingly facilitated as priests distributed or made out their parishioners' ballot slips.[28]

The actual system of voting, to which far too little attention has been paid, also played a role in encouraging turnout. Voting was conducted at the cantonal centre, or *chef lieu*, as it had been in the 1790s, at what was still called an electoral *assembly* (though any discussion inside the polling station was now strictly forbidden).[29] Despite a barrage of demands for voting in each *commune*, the government felt that voting for a list of deputies at cantonal level would diminish the influence of local notables.[30] It was recommended that countryfolk travel together to the polling stations, where the electors from one village after another were called upon to vote in turn. This naturally encouraged rural dwellers to transact their civic duty as a *community*. Indeed, they sometimes clashed violently with contingents from other villages, as they had in the past.[31] Pierre Rosanvallon and others have accordingly suggested that voting in 1848 reflected local solidarities, rather than the exercise of a partisan political choice.[32] Only a handful of voters cast their ballots independently of their fellow villagers and even in towns there are examples of traditional processions to the polls, which continued to characterize the national elections that were subsequently held.[33]

The famous passage from Alexis de Tocqueville's memoirs bears eloquent testimony to the collective act of voting:

> We had to go in a body to vote at the town of Saint-Pierre, a league away from our village. On the morning of election day all the electors, that is to say the whole male population over 20 years old, assembled in front of the church. They formed themselves into a double column in alphabetical order; I preferred to take the place my name warranted, for I knew that in democratic times and countries one must allow oneself to be put at the head of the people, but must not put oneself there. The crippled and sick who wished to follow us came on pack horses or in carts at the end of this long procession. Only the women and children were left behind. We were in all 170 persons ... [34]

De Tocqueville continues by stating, somewhat modestly, that 'All the votes were given at the same time, and I have reason to believe that almost all were for the same candidate.' He was in fact elected as a deputy for the Manche, his home department, 'with 110,704 votes out of about 120,000'.[35] This degree of unanimity was apparent elsewhere in the elections of April 1848 and one candidate frequently swept the board at the cantonal level, notwithstanding a more evenly split departmental vote. Even in the more keenly contested legislative polls of 1849, candidates like de Tocqueville continued to attract almost 90 per cent of the votes cast.[36] Nor was such influence over voters confined to conservative or liberal forces; radicals also drew on deference. In the Corrèze in 1849, one democrat wrote to the candidate: 'Six hundred voters bow to my command, so any untoward ballot papers will immediately be torn up or changed.'[37] The intervention of these 'hidden persuaders', as Peter Jones calls them, was not always successful (as the presidential elections of 1848 revealed), but their influence was none the less paramount.

During the Revolution voters wrote down, or had written for them, the names they wished to select when they were asked to do so by assembly officials. Under the Second Republic, by contrast, they brought completed ballot papers with them, which were usually supplied by the candidates themselves. Since these were often distinctive in shape or colour, it was clear whom voters had nominated when their slips were inserted into the urn, or ballot box.[38] There were many allegations of electoral agents or local officials foisting these handwritten or printed lists on to poorly informed or illiterate voters, who found it difficult to exercise a considered individual choice. In April 1848, for example, the mayor from one *commune* in the canton of Tournay, in the Hautes-Pyrénées, distributed voters' cards when villagers arrived together at the *chef-lieu*. He gave everyone a ballot paper with six names on it and told them to use it.[39]

Though the Provisional Government of 1848 declared that it was abandoning the promotion of official candidates practised by the July Monarchy, old habits died hard and it was not averse to offering 'guidance' to the electorate via the administrative machinery. The advice was not always heeded, but nor was it always unsolicited. One mayor in the Loir-et-Cher wrote to the *commissaire de la République* in March 1848 demanding to be told whom he should recommend to his fellow inhabitants; he was certain of delivering their votes.[40] Meanwhile, a mayor in the Doubs stated of his villagers: 'Every day they ask me whom they should vote for, seeing that we are not in the centre of the department and that we do not know any truly republican men capable of upholding our rights in the National Assembly.'[41] The following year, the Prefect of the Yonne reported that he was being pestered for similar information during the campaign that preceded the legislative elections of May 1849.[42]

Historians need to be much more sensitive to the culture of voting in order to tease out the real meaning of the practice. The personal and private approach, which became established in the twentieth century, was hardly apparent in 1848, either in town or countryside. The high turnout actually militated against 'modern' electoral behaviour of today's sort and, paradoxically, the lower turnout of 1849 might be seen as reflecting the development of a more individualistic approach. Abstention should not automatically be regarded as tantamount to indifference any more than participation can be seen as proof of politicization, though both were essentially products of collective pressures. In 1848, despite some procedural changes, many voters were simply re-enacting the communal rituals they had inherited from the past.

In sum, it would be wrong to see 1789 as the beginning and 1848 as the end of a process of electoral 'modernization' purely on account of rising turnout. There were undoubtedly improvements in getting out the vote in 1848, which stemmed from changes in the electoral system itself and the apprenticeship offered by the July Monarchy, besides developments in the wider political and social context. Traditions that stretched back beyond 1789 were modified by regulations which simplified the electoral process, removed the deliberative element and rendered the assemblies more perfunctory. In the legislative elections of 1849, in particular, there was an undeniable degree of pluralistic competition: voters were more frequently invited to choose between party lists rather than to nominate well-known local personalities.

Yet the political apprenticeship of French men (no mention has been made here of women, though they were not entirely absent) would prove an extremely protracted process. It advanced in an uneven and non-linear fashion, in both geographical and chronological terms: the initial impact of universal suffrage in 1848, for example, was quite exceptional. If 'modern' electoral behaviour is construed as a private transaction, which involves casting of a secret ballot for a variety of candidates in response to national issues, then its development involved a lengthy process of change in France. It requires much broader and deeper exploration from historians than it has so far been given.

Acknowledgement

I would like to thank John Dunne, Sharif Gemie, Peter Jones and Peter McPhee for their invaluable comments on an earlier draft of this chapter.

Notes

All translations from the French are my own, except where otherwise indicated.

1 P. McPhee, 'Electoral Democracy and Direct Democracy in France, 1789–1851', *European History Quarterly*, vol. 16 (1986), pp. 77–96.

2 R. Huard, *Le suffrage universel en France (1848–1946)* (Paris, 1991), p. 14 and M. Agulhon, *La République au village. Les populations du Var de la Révolution à la Seconde République* (Paris, 1970), p. 289.

3 E. Weber, *Peasants into Frenchmen. The Modernization of Rural France 1870–1914* (London, 1977). See P. M. Jones, *Politics and Rural Society. The Southern Massif Central c.1750–1880* (Cambridge, 1985), for an outstanding attempt to set the issue in the long-term context it demands.

4   M. Edelstein, 'Vers une "sociologie électorale" de la Révolution française: la participation des citadins et campagnards (1789–1793)', *Revue d'histoire moderne et contemporaine*, vol. 22 (1975), pp. 508–29.

5   Apart from my own work cited below, see P. Gueniffey, *Le Nombre et la raison. La Révolution française et les élections* (Paris, 1993) and numerous articles by M. Edelstein, most notably 'La participation électorale des Français (1789–1870)', *Revue d'histoire moderne et contemporaine*, vol. 40 (1993), pp. 629–42. A study group meets regularly in Paris and will shortly be publishing a *Guide des élections de la Révolution*.

6   McPhee, 'Electoral and Direct Democracy', p. 81.

7   The secondary source on which McPhee draws is based on erroneous information in J. M. Thompson, *The French Revolution*, rev. edn (Oxford, 1985), p. 227. See instead M. Crook, *Elections in the French Revolution: An Apprenticeship in Democracy, 1789–1799* (Cambridge, 1996), pp. 165–6 and *idem.*, 'Masses de granit ou grains de sable? Les électeurs des Assemblées départementales sous la Révolution française, 1790–1799', in D. Turrel (ed.), *Mélanges offerts à Claude Petitfrère. Regards sur les sociétés modernes (XVIe–XVIIIe siècle)* (Tours, 1997), pp. 203–10.

8   P. Rosanvallon, in an otherwise excellent survey entitled *Le Sacre du citoyen: Histoire du suffrage universel en France* (Paris, 1992), p. 269, is reluctant to acknowledge this fact.

9   Crook, *Elections in the French Revolution*, p. 68.

10 Archives nationales (referred to hereafter as AN) BII 1–33, Procès-verbaux, 1793 and R. Baticle, 'Le plébiscite sur la Constitution de 1793', *La Révolution française*, vol. LVII (1909), pp. 496–524.

11 Huard, *Le suffrage universel*, p. 38, for example.

12 A. Lancelot, *L'abstentionnisme électoral en France* (Paris, 1968), pp. 14–17.

13 AN C*II 384, Liste des représentants du peuple, 1848.

14 Archives départementales du Var (referred to hereafter as AD Var) 2M3–13, Elections à l'Assemblée nationale constituante, 23 April 1848.

15 AN C*II 384, Procès-verbal de la commission électorale, December 1848.

16 AN C1330–5, Procès-verbal de la commission électorale, 1849.

17 AN C1325, Elections partielles, 1848.

18 AD Var 2M7 7, Election des conseils municipaux, 1848–52.

19 As Huard comments, in 'Comment apprivoiser le suffrage universel', in D. Gaxie (ed.), *L'explication du vote* (Paris, 1985), p. 147, there are few comparative studies of local elections after 1848.

20 McPhee, 'Electoral Democracy', pp. 90–1.

21 M. Edelstein, 'Integrating French Peasants into the Nation-State: The Transformation of Electoral Participation (1789–1870)', *History of European Ideas*, vol. 15 (1992), p. 323.

22 AN C*II 383, Les élections d'août 1846 et les élections précédentes.

23 Agulhon, *La République au village*, pp. 259 *et seq.*

24 C. Guionnet, *L'Apprentissage de la politique moderne. Les élections municipales sous la monarchie de Juillet* (Paris, 1997).

25 AN F1BI 257, Compte rendu au roi. Les élections municipales de 1834. For a local example see: AD Haute-Garonne 3M4, Municipal elections, *arrondissement* of Toulouse, 1831–46.

26 M. Crook, 'La plume et l'urne: la presse et les élections sous le Directoire', in P. Bourdin and B. Gainot (eds), *La République directoriale* (2 vols., Paris, 1998), 1, pp. 295–310.

27 AN F1cII 56, Correspondance concernant les élections, March 1848.

28 A. Corbin, *Archaïsme et modernité en Limousin au XIXe siècle, 1845–1880* (2 vols., Paris, 1975), tome 2, p. 796 and A. Cobban, 'The Influence of the Clergy and the "Instituteurs Primaires" in the Election of the French Constituent Assembly April 1848', *English Historical Review*, vol. LVII (1942), pp. 334–9.

29 S. Aberdam, 'Elire, voter, s'assembler. Trois logiques à l'œuvre dans l'élargissement du droit de suffrage: 1789–1849'; unpublished paper, delivered to a conference at Macerata (Italy) in 1995.

30 AN F1cII 56, Protestations, March–April 1848.

31 P. M. Jones, 'An Improbable Democracy: Nineteenth-Century Elections in the Massif Central', *English Historical Review*, vol. XCVII (1982), p. 551.

32 Rosanvallon, *Le Sacre du citoyen*, p. 292.

33 A. Garrigou, *Le Vote et la vertu. Comment les Français sont devenus électeurs* (Paris, 1992), pp. 58–9.

34 A. de Tocqueville, *Recollections*, ed. J. P. Mayer (New York, 1971), pp. 119–20.

35 *Ibid.*, p. 121.

36 A. Jardin, *Tocqueville. A Biography* (New York, 1988), pp. 425–6.

37 Corbin, *Archaïsme et modernité*, tome 2, p. 794.

38 P. McPhee, *The Politics of Rural Life. Political Mobilization in the French Countryside 1846–1852* (Oxford, 1992), p. 93.

39 AN C1328, Protestation des citoyens du canton de Tournay, 25 April 1848.

40 G. Dupeux, *Aspects de l'histoire sociale et politique du Loir-et-Cher (1848–1914)* (Paris, The Hague, 1962), p. 332.

41 A. Cobban, 'Administrative Pressure in the Election of the French Constituent Assembly, April, 1848', *Bulletin of the Institute of Historical Research*, vol. 25 (1952), p. 144.

42 T. Zeldin, 'Government Policy in the French General Election of 1849', *English Historical Review*, vol. LXXIV (1959), p. 242.

5
A Forgotten Socialist and Feminist: Ange Guépin

Pamela Pilbeam

Ange Guépin was a socialist and a feminist, revered as 'the advocate and protector of the disinherited,'[1] who had a leading role in western France in 1830, 1848 and 1870, yet he has been almost forgotten since his death in 1873. This chapter, which is based on a rich and largely untapped archive offers a unique opportunity to allow him to explain his ideas and his actions. His manuscript writings, including a diary for 1833–39 and 1848, autobiographical fragments and many letters written and received by him, his family and friends survive.[2] Guépin wrote half a dozen major books, a substantial number of pamphlets and countless newspaper articles. In a single month in 1833 he dispatched at least 20 articles on topics from fifteenth-century Nantes to the colonization of Algeria. Typical of his generation he was a vigorous intellectual omnivore, lecturing on science and history. When he became a freemason in 1867, he was soon the local expert. For a man who never stopped writing, few have chosen to write about him. A biography appeared months after his death and another was published in 1964.[3] A commemorative exhibition was organized in Nantes in 1997.

Ange Guépin (1805–73) was born into radical, anti-clerical republicanism. His father, Victor, was one of a notable family of lawyers in Pontivy (Morbihan), who supported the Girondins in the early 1790s, survived Jacobin imprisonment, resumed his official career and was elected to the Assembly of the Hundred Days. In 1823 Guépin wrote to a close friend: 'My mind is made up. I am a materialist, I can disprove the immortality of the soul, yet I believe in the existence of a God.'[4] Later he asserted that the socialism he shared with many others was 'derived from our fathers, part of an intellectual tradition which began with Robespierre and Condorcet'.[5] Apparently his father's radicalism denied Ange a place at the Ecole polytechnique. Instead he studied medicine in Paris from 1824 and was enlisted into the last *charbonnerie* group created. He experimented with radical ideas, political, social and scientific, encountered in lectures and in informal gatherings in his room and elsewhere. He joined the Saint-Simonians and got to know Trélat, Buchez and Leroux.

Like other radical doctors trained in Paris in the 1820s,[6] Guépin was preoccupied with social medicine, especially the links between disease and poverty. His doctoral dissertation examined the types and causes of cancer. He stressed the significance of damp, dark living conditions and a poor diet in proliferating the disease.[7] In 1828 he graduated with the first prize for chemistry and became a lecturer in Chemistry and Industrial Economics in the Nantes medical school.

Throughout his life Guépin looked back to the 1830 Revolution as a turning point when workers and bourgeoisie found a common cause in the social question. Just before his death he spoke at the annual commemoration of those who died in the July Days in Nantes.[8] On his arrival in Nantes, Guépin quickly became a leading figure in the liberal opposition to Polignac and in the popular demonstrations which followed the news of the four ordinances in July 1830. Guépin was keen to point out afterwards that these demonstrations against the Restoration officials were not lawless; the leaders promised to pay the naval gunsmith from whom they requisitioned weapons. During their successful march to the *château* to release the prisoners incarcerated after a previous demonstration, shots were exchanged with the troops and nine citizens were killed.[9] Guépin published an account of the local revolution and used the profits to provide for those injured in the fighting and the dependants of those killed. Guépin was proud of his role in 1830 and organized a monument and annual memorial gatherings.

Guépin soon established himself in a successful medical career, in which his social reforming ideas found expression. He was made responsible for women's medicine, surgeon at the local hospital and medical officer for the customs' service when the incumbent succumbed to cholera in 1832. In 1841 he also became chief surgeon at the *Hôtel-Dieu*. He gave his services free to the nursery school[10] and was for many years a member of the public health committee.

Guépin was constantly harried by administrators to cut the cost of medical services. His solution was to develop out-patients' clinics, which the sick could afford in place of quacks and charlatans. He ran a clinic for syphilitics. He pointed out that 50 out-patients were treated for a mere 200 francs in his skin complaints clinic while his eye clinic offered attention for 20 centimes each, a tenth of the cost of in-hospital treatment. He urged the case for clinics in rural areas and dispensaries, modelled on Italian prototypes. He observed that if local services were not developed rapidly, when the new rail network was completed, treatment would be centred in Paris, depriving provincial towns of local access to skilled professionals.[11]

In 1830 he set up the first specialized, and free, eye clinic in France, which became famous throughout Europe and America. He became a leading expert in cataract surgery and colour blindness. A large proportion of his patients were workers, especially in the building trades. In 1840 he opened a free dispensary for the poor. He was involved in a number of cooperative projects in addition to his own research. He was also fascinated by economic innovation, invented an artificial fertilizer and set up several joint ventures.

He became well integrated into the intellectual life of the area in a variety of learned societies. Unforgettable was a series of controversial weekly lectures on ethology (*ethropologie*), which the Academy of Nantes, of which he was secretary, invited him to give in 1835. There were vociferous complaints that Guépin was 'an atheist, a materialist, even a madman, that I explained the harmonies of Nature through periodic fractions'.[12]

His early Saint-Simonianism matured into successful practical socialist initiatives. Disappointed with the failure of the Orleanists to address the problem of poverty, in November 1830 Guépin founded the Société industrielle de Nantes to provide work for the unemployed, mainly navvying. It was based on the mutual aid principles that were becoming increasingly popular as a means of cushioning workers against unemployment.[13] Plans to provide retirement pensions for workers were included. Along with other radicals of his generation, Guépin was convinced that human progress depended on education: the society also ran evening classes and, in 1832, secured a grant of 6,000 francs from the government for an apprenticeship scheme. Guépin persuaded the duc d'Orléans to give 2,000 francs. Generous donations allowed the society to acquire its own building, with a library and a clinic, which Guépin ran. In 1832 he also helped to set up a short-lived tailors' cooperative.

At first Guépin did not see that his socialism made him an enemy of the Orleanists.[14] His first wife, Adélaïde Le Sant (1810–31), was the daughter of a pharmacist, who was deputy mayor of Nantes. Their lives were closely interconnected with leading figures in the local judiciary, administration and education. Philosophically and experimentally, he gradually became an eclectic socialist, drawing his concepts from a wide range of his contemporaries, particularly Saint-Simonians and Fourierists.

He maintained links with most contemporary socialists from Cabet to Buchez, Leroux, Perdiguier and especially Lamennais. He had no contact with revolutionary socialists like Blanqui and disagreed both with Blanc's hopes that the state would fund artisan cooperatives and Cabet's vision of a society where everything was held in common. Although, like Proudhon, he favoured producer and retail cooperatives, he ridiculed Proudhon's plans for a People's Bank where notes of exchange would be the basis for a system of barter.[15] Guépin always hoped that mutual aid, producer and retail cooperatives and a universal system of education would solve the social question, convinced that the real enemy of society was ignorance.

Although Guépin always claimed to be a democrat, he was a patriarchal socialist, keen to help the poor, but sceptical of their morality, dirt and shiftlessness. In the mid-1840s he defined socialism as 'a duty of protection owed by the rich to the poor', but he distinguished this from charity, which he argued was medieval and only led to idleness.[16] Not untypically, his views were more radical in youth than maturity, to the extent that he is often referred to locally as a liberal, rather than a socialist. By 1869 he was calling himself a radical republican rather than a socialist.

In early 1833 Guépin was still closely involved with the Saint-Simonians through his friendship with Michel Chevalier, then editor of their newspaper, *Le Globe*, despite the imprisonment of Chevalier and Enfantin accused of corrupting morals by advocating trial marriage. Guépin invited Chevalier and Barrault, another leading Saint-Simonian, to send 'scientific missionaries' to Nantes to give lectures to the local population, including the dockworkers, on 'geology, geography, the industrial future of the planet...theology...and human progress', arguing that they would be 'very useful and would thus earn some money'.[17] A group of eight duly arrived from Lyon, but working men rather than teachers. The doctor helped them to produce a publicity brochure, although apparently he did not agree with all their ideas. The dockers, whom they had hoped to convert, were unsympathetic and stones as well as insults were flung by both parties. The workers were antagonized by the Saint-Simonian costume of white trousers, a blue tunic and red waistcoat, buttoned at the back as a constant reminder of men's interdependence. Guépin protected them and paid their bills. The sect was deeply in debt; three years on Chevalier appealed to Guépin for advice on how to sort out their finances.[18] Although Enfantin's sexual politics shocked Guépin,[19] the two kept in touch; Guépin referred to Enfantin as 'cet adorable Satan' in his later history of socialism. Saint-Simonianism had a permanent influence on Guépin's socialism, although Chevalier's defection to Orleanism led to a prolonged break in their friendship.[20] In January 1833 Guépin started a Saint-Simonist society, the Réunion de l'Ouest, after taking preliminary soundings among friends and relatives. He aimed to create an intellectual focus for the half-dozen surrounding departments and make Nantes the 'Edinburgh of the west'.[21] At first there were 50 members, including women. In his opening speech Guépin proposed that they write a new Declaration of the Rights of Man, or rather, of Destinies and Rights (Guépin thought the original phrase might scare some listeners). It began: '[Man] is born sociable and free. Equality is a consequence of liberty. Liberty is the right to do what the law does not forbid.' Guépin emphasized that their ideas were drawn from Gall (his old teacher), Owen and Saint-Simon, in particular their stress on the role of education in moulding human nature. The Réunion was divided into three sections: Science, Industry and the Arts. The scientific group was to consist of experts, who would also be popularizers. Elected president, Guépin summarized their objectives: free evening classes in a wide range of subjects, the extension of voting to a wider range of *capacités* (that is, local worthies who paid less than the 200 francs tax needed to qualify for a vote after 1830; Guépin at that time barely qualified as an elector) and more opportunities for women.

Guépin believed that in the long term France would become a republic, but in the early 1830s he tried to distance himself from the republican clubs, affiliated to the Society of the Rights of Man in Paris. He claimed to be only vaguely aware that a workers' republican club existed in Nantes in November 1833, although it has been asserted that he ran the Cercle national, composed of *avocats* (lawyers), some members of the local garrison and workers, especially tailors and wigmakers. The 'very bloodthirsty' doctrines of 1793 were supposed to be aired twice weekly.[22] Guépin admitted that in early 1834 both the organizers of Aide-toi, le ciel t'aidera, a radical movement for electoral reform, and the Society of the Rights of Man tried unsuccessfully to recruit him. He explained that to declare a republic prematurely would bring war and that without adequate funds, republicans would be forced to institute a new Terror. 'I believe that a republic will emerge, but the longer the delay, the better the chance that its leaders will be radicals who understand political economy and the better its chance of surviving.'[23]

In common with other radicals, Guépin admired Napoleon. In March 1833 he ordered a wax bust of the Emperor, used later in the year with his own reconstruction of the battle of Austerlitz, in a procession to celebrate the July Days, which he was chosen to organize. Guépin criticized

Napoleon's dictatorial rule, his dependence on a narrow aristocratic elite, his failure to realize that the potential existed to base his regime on popular support and his hostility to women.[24] What did he admire? Did he see Napoleon as a patriot? Guépin always described himself and his friends as 'patriotes' and was proud to be elected captain of the Nantes National Guard in 1834.

Government legislation in April 1834 banning all clubs and subsequent censorship which forbade the press even to use the term republican made 'patriots' like Guépin more circumspect. In April 1834 he confided in his diary that if his notes fell into the wrong hands they would be burned by those 'who want to destroy the last vestiges of *the revolutionary spirit*'.[25] His diary entries became very brief and his journalism tailed away as censorship forced radical newspapers to close. In September 1834 Guépin organized a scientific conference in Poitiers at which he spoke on social reform, but advised by friends left out the rights of women, an issue considered explosive after Enfantin's imprisonment. He invited assorted Orleanist critics including Carnot, son of the republican regicide, and Arlès Dufour, a Saint-Simonian from Lyon.

His response to press censorship was to bury himself in his professional work and to write books. Guépin's writing gave him a national reputation. He wrote on 'social economy' in a series of tiny volumes, 'The Popular Library', launched in Paris in 1832 'to make knowledge accessible to people of all classes and intellectual levels'. The committee included the great and good of all political persuasions, such as Chateaubriand, Cauchois-Lemaire, Charles Dupin, Victor Hugo, Arago, Villermé, Sainte-Beuve, and even a woman, Mme Waldor.

His volume started with an ancient Chinese proverb: 'Whoever wishes to enjoy the pleasure of wealth, must accept the burden of work.'[26] Guépin went on to describe a systematic approach to economic planning, summarizing the radical schemes with which he and others had been experimenting.[27] He recognized that the biggest problem was the conflict between master and man and was convinced that the solution lay in mutual aid associations. Weekly subscriptions of about 10 per cent of earnings produced enough to pay out for accidents, unemployment and retirement. Wage levels would be negotiated by mutual agreements. Sons would be guaranteed employment and the society would loan cash to workers to start their own firm.[28] Conflict would thus be avoided and everyone would be provided for, no one luxuriously. Guépin disapproved of excessive wealth.

In 1835 Guépin co-authored a study of contemporary Nantes whose shocking statistics on poverty, diet, income and the health of poor people were much copied and quoted by Villermé, Blanc and others.[29] Two years later he brought out his own history of the city from the earliest times, an abridged version of which appeared in an 1844 historical dictionary of Brittany.[30] In a now familiar format, Guépin's views on how to solve the social problems of the city were interwoven with an account of its architecture and history.

Unlike many of his contemporaries, Guépin had a very practical, unsentimental approach to poverty. Working people, he argued, lived in damp, dark cellars, running with water, surrounded only by a make-shift bed and their spinning-wheel. They worked for up to 14 hours a day for 15–20 sous. On his day off, Sunday, the worker had no money to go anywhere, no clubs, no entertainment. In the summer he might go out to the country. His children lived in the gutter and three-quarters of them died. 'At 20, people thrived, or they were dead.'

Guépin printed out a detailed family budget to prove his point and went on to recite the illnesses to which the poor were prey. Their children were killed by chest infections in the winter, diarrhoea in the summer.[31] One illness which was no respecter of income and living conditions was syphilis. After years of battling with the disease in his clinics, in 1846 Guépin petitioned the Chamber of Deputies to follow the example of Belgium, legalize soliciting, oblige whores to undergo regular medical checks and provide free hospital treatment.[32]

Fourierism attracted Guépin along with others disillusioned with Enfantin. Fourier's ideas, interpreted by Considérant, offered a decentralized, practical solution to the social question, which did not question private property, did not insist on total equality and could, in principle, subsist within a capitalist system.[33] In 1832 the Fourierists launched their first experimental community at Condé-sur-Vesgre, near Paris, on land owned by Baudet-Delary, a member of parliament and doctor friend of Guépin.[34] Guépin took a lively interest in this disastrous project and wrote an approving newspaper article in January 1833. Another friend of his, Leroux, also founded an experimental commune at Citeaux.[35]

In 1840 Guépin supported the first Banquet campaign for electoral reform and also backed the strikes of local workers for better wages and conditions. In 1846 he was elected to the municipal council of Nantes and chosen as its secretary, despite the attempt of Achille Chaper, Guizot's loyal prefect, to claim that Guépin's professional work and municipal office were incompatible.[36] In 1847 Guépin presided over a banquet to commemorate the 1830 Revolution. In his speech he again urged cooperation between workers and the middle classes. He scorned the main Banquet campaign for its narrow concentration on suffrage, suggesting a more comprehensive 'cahier de doléances' on the lines of 1789 which would include social reform. He helped to set up nine charity workshops when national shortages of both work and bread led to great deprivation and demonstrations in Nantes against the *octroi* and the price of bread.[37] In 1847 he secured 6,800 signatures in Nantes for his petition to Parliament demanding that they take steps towards 'the organization of work', which had become a key issue for Fourierists[38] and other socialists. The petition criticized the Orleanists for treating the poor as barbarians. For 17 years they had divided the French into 'majors and minors', depriving the 'minors', the poor, of rights and encouraging the rich, the 'majors', to ignore their responsibilities.[39] Guépin tried to persuade Chaper, who was not insensitive, to organize the local notables to provide practical training for rural workers and education for artisans.[40]

Guépin took the lead in declaring a republic in Nantes on 26 February 1848 and two days later was named *commissaire* (the term briefly preferred to prefect in 1848).[41] Nantes was in the throes of a financial as well as a political crisis. Since the news of the February revolution a few days earlier, the local bank had been drained of money and its director was desperately trying to find cash to pay anxious customers. Guépin set up a committee of experts to secure credit. He gave up his salary for March and used his own savings, some of which the government later refunded, to set up workshops for the growing number of unemployed. On a single day in early March 20 businesses closed.[42] He toured the radical workers' clubs, such as the Club de l'Oratoire in Nantes lecturing them on socialism, starting with St Bernard and ending with Enfantin. He also spoke to more bourgeois clubs demanding work, security and schooling for the children of workers.[43] Less than three weeks later, after much criticism from former friends, he was moved to Morbihan, where he proved too radical for Parisian republicans, particularly in his hostility to the supercilious bossiness of local legitimists. He accused them of riding roughshod over the poorer voters in the way their publicity was presented; money and free drink were pushed around. The electoral traditions of the past had to go. 'The sovereign people are now free; yesterday slaves, today they are equals.'[44] Guépin underestimated the notables; their candidates were elected to the National Assembly rather than Guépin and his friends.

His blunt, rather naive, approach was not suited to the duplicity of politics.[45] In early April he escaped to his old haunts, his surgery and local cafés in Nantes, boring anyone who would listen to his claims to be only a 'soldier of 1830'.[46] In mid-June, harried by the new conservative deputies, he agreed to leave as soon as a replacement could be found. Meanwhile his second wife, Clotide Maussion (1800–1849), whom he had married a few months earlier and who had vigorously opposed his appointment, was exchanging frantic letters with his relatives in Nantes which ranged from anxiety about her husband's health to concern that their precipitous move to Vannes left her without the necessary table linen and so on vital to entertain and appease local dignitaries.

If he found politics exasperating, Guépin used his position to initiate valuable social reform. In May 1848 he started La Fraternelle universelle, a Proudhonian-style mutual aid society. A membership fee of 1fr.50 a month provided sickness benefit, pensions and 2 francs a day unemployment pay for its 700 members until the prefect forced it to close in November 1850. Like many such plans it was rooted in *compagnonnage* structures and only those holding a *livret* could join. Similar schemes operated in Lorient, Rennes, Brest, Angers, Tours, Lyon and Paris. Guépin was proud of the cooperative bakery he organized. Eventually it forced private bakeries to reduce their prices so much for the 400 families who joined that the bakers banded together to get the prefect to close it down. A later project survived until a fire in 1862.[47]

The most controversial decision Guépin made at Vannes was his much-publicized dispatch of National Guardsmen to Paris to fight against the workers in June. Given Guépin's role in 1830 and 1848 and his life-long preoccupation with the problem of poverty, at first sight his active support for the government in the June Days was contradictory. At the time he was genuinely convinced that the rebellion was an attack on the republic itself and that it was the Christian duty of all democrats to help defend it. He organized a detachment of the Morbihan National Guard to go to Paris to do just that and publicly thanked them for their participation. He was convinced, he insisted, that the Parisian workers had been led astray by 'dishonest advice and provocative posters'.[48] However, in a letter to a close friend less than a month later, Guépin wrote that although on 24 June he would willingly have fought with the National Guard against the rebels, now he would refuse, because he realized that the episode had been used to undermine 'the small amount of freedom gained in February'.[49] Guépin was not the only socialist to oppose the Paris workers in June, Louis Blanc followed the same route, Guépin perhaps more naively. No wonder another old friend Michel Chevalier referred to him some years earlier in a letter as 'goodhearted, but not very calculating'.[50]

Guépin was opposed to Louis-Napoleon's election as president. He helped to organize the radical-democrats in the legislative elections of 1849, but was not himself elected. After the coup of December 1851 he resigned from the departmental general council, unwilling to take the oath to a regime which had 'interned, exiled and transported so many of my friends. The new order calls for new men and I am and always will be, the same; with faith in liberty for all, openness, goodwilled tolerance. To my mind that is the real national tradition.'[51] Guépin, however, retained the friendship and protection of the Prince Napoleon.

Guépin returned once more to his own medical practice and his writing, but continued to produce the sort of radical polemic which was bound to attract criticism from conservatives. He completed a study of socialism begun during the July Monarchy,[52] followed by a more popular version[53] and a study of nineteenth-century philosophy,[54] which were all basically in defence of the socialists of 1848. The first began, in typical Guépin style, with the geological origins of the earth, developing themes he had raised in his course of ethology in 1835. Guépin made no claim to be an original thinker, he was a moralist and socialist popularizer, still attracted to Fourierist ideas on the primacy of the commune.[55] When the book came out in 1850, a witch-hunt arose to deprive him of his Chair at the medical school. He was accused of attacking Christianity by his praise of other faiths and immorality because he said men and women should be equal partners in marriage. Guépin was also reviled for praising the Convention of 1793 and lending people copies of its debates. Although he insisted that he had never condoned the cruelty, violence or dictatorship of 1793, but recognized that the revolutionaries had been striving for 'the intellectual and moral well-being of all', he was dismissed.[56] Irrepressible, his account of nineteenth-century philosophy revisited the same themes: praise for all religions and all races, urging racial inter-marriage[57] and concluding that to understand God, one had to understand science.[58]

Was Guépin an unworldly intellectual, or was he out to shock? In an attempt to explain the contribution of socialists to 1848, Guépin came out with the idea that he was really a communist, defining communism as the expansion of the role of the commune, not communal ownership. This version of communism, he explained, was an historical belief, dating back to the tenth century.[59] Fortunately, this article does not seem to have found a publisher, for even the Fourierist would have objected to being called communist. Although Guépin's active socialism faded after 1850, he retained an interest in Fourierism and years later was one of a tiny number who subscribed to a new Fourierist newspaper, *La Solidarité*,

Journal des principes et du socialisme scientifique. The venture was short-lived, almost no copies of the paper survive, but a handful have been preserved among Guépin's papers.[60] Guépin always described himself and his friends as 'patriotes' or 'co-religionnaires'. He seemed to include in this term historic anti-clerical resistance to what he thought of as 'Chouan terrorism' as well as the open-minded tolerant patriotism typical of contemporary radicals. Later in his life it became more all-embracing: 'before I was French, I was a European and a Gallo-Roman.'[61] He took an interest in the cause of Italian nationalism, and in the late 1860s joined an international organization for European peace.

Guépin was a life-long feminist outraged that the influence of Rousseau, the revolutionaries and Napoleon had left women with the legal status of children.[62] He claimed that married women should be recognized as adults, not servants, should be equal in law with their partners and should have the same rights as husbands to protest against flagrantly adulterous spouses.[63] He was particularly interested in girls' education and vocational training. He was convinced that successful marriages and contented families depended on educated women, and was always outspoken in his scorn for clerical influence in the education of girls in Brittany, which, he asserted, left women dominated by saints and superstition. He asserted that women should also have a public role, as teachers and decently paid workers, as members of learned societies. He was sympathetic to American plans to give women the vote.

Guépin developed and maintained contact with a wide range of organizations committed to women's education. He admired Michelet's *La Femme*, commenting that workers in Nantes saved up to buy copies. He commented too on Pauline Roland's contribution to the creation of producer cooperatives in Paris in 1848[64] and was full of praise for women writers, including Madame de Staël, Madame de Girardin, George Sand and American writers like Harriet Beecher Stowe.[65] He surrounded himself with three dominant, articulate wives, and bred daughters with similar attitudes. Possibly in reaction, his only son was at first a rebel spendthrift, although he came into line, marrying the daughter of a Protestant pastor.

Guépin's views on women and marriage were succinctly expressed in his family letters, especially during his battle to end his career as a 'vache à lait' for his son. He advised Ange junior that a man could not hope to succeed without the right wife. If his son found a woman who was 'straightforward, modest, distinguished and orderly', Ange senior would continue to bankroll him (by this stage the renowned doctor was a wealthy property-owner in his own right). He warned the young man that he must not make the mistake his cousin had just made, marrying a seductively charming and agreeable lass, but who was given to 'elegant outfits, expensive train journeys, no sense of economy and headstrong. Look for a woman who is organized and principled; without this last quality you will not find intelligence'.[66] When Ange junior married, his father reminded his son that he must make a marriage settlement which would ensure his future wife's financial security, in return for which the father would make a proper financial settlement on any future children.[67]

Guépin's emphasis on the importance of the woman's role in the family was explored in *Marie de Beauval*, a 'philosophical novel', about a Breton girl born in 1785. Several hundred pages of the partial manuscript of this novel survive among his papers. Guépin was obviously inspired by Rousseau's *Emile*, in both the structure of the novel and the approach of Marialla, Marie's tutor. Preparation began with Marie's mother; Guépin argued that what a pregnant woman saw and read, as well as what she ate, influenced the baby. Marie's mother observed a routine of a healthy and substantial diet, exercise and uplifting reading. She surrounded herself with beautiful paintings. Guépin whimsically observed that in one commune he knew well almost everyone resembled a particularly repulsive figure of Christ in the local church.

Bringing up Marie was a balance of heredity and environment. Guépin did not deny the importance of heredity, seen in the breeding of Arab horses. Over-emphasis would lead to fatalism, too much stress on tradition and on the inheritance of characteristics from the male (who could never be guaranteed to be the father). Education was the counter-weight to heredity. It was a family responsibility and included environment, food, friends and animals.

Marie received a rational, scientific education, rooted in the empirical and experimental study of the natural world and an investigative approach to history. Marialla, a disenchanted former soldier of the Revolutionary wars, adopted a Socratic method of questioning. He was very conscious that his mission was to educate a soul. Observation and questioning needed the support of books. Marie read Descartes, Rabelais and Voltaire. Her tutor was careful to select books which explained the role of women, past and present.

Guépin noted that most current female education emphasized materialism, furniture, servants, strutting around in fancy clothes. The contemporary image of the feminine was rooted in the unhelpful perception of Mary, mother of Jesus. A more positive model would be the way Robinson Crusoe taught Man Friday. Religion was important. Marialla taught his pupil two religions: the perfect religion, made by God, the music of the spheres, an awareness of the stars and planets; secondly imperfect religion, the work of man, education. The life experiences of men and women were different. Women were bound to be wives and mothers, and as mothers they educated their children. Yet girls received no education themselves. Girls were equally unprepared for marriage, which was a moral and intellectual, as well as a physical experience. A husband should choose a wife for her intellect, her mastery of domestic economy, her courage and her prudence. He admired the freedom of girls in Britain and America to associate with boys and argued that hard work would protect their virtue.[68]

Guépin emphasized that the best teacher was the mother. In breast-feeding her child and keeping it clean she taught the infant regular habits, and in answering its simple questions the child would learn about the world. Ideally, he thought a child should be educated by its own mother until 10, but he recognized that this was rarely possible. He thus advocated free, non-denominational state education from nursery age. Guépin urged that nursery schools distance themselves from the idea of a charity workhouse. He envisaged them well equipped with huge globes and magic lanterns to teach children about the world. He considered the role of the woman teacher 'a saintly mission'.[69] He demanded that a fund be set aside to provide food for the 400 small children who turned up each day to the schools in Nantes with no packed lunch.[70] At the end of primary schooling, all pupils should have the chance to go on to secondary education and university at no cost to their parents, while the majority should be provided with vocational education, specializing in agriculture or a trade, coupled with an apprenticeship.[71]

Vocational training for girls was Guépin's pet project. He recognized, however, that although men might provide the cash, it was important that the scheme was visibly run by women, because they provided the pupils with role-models. Guépin's third wife, Floreska Leconte, a militant, highly educated feminist, took a leading role in bringing his ideas to fruition. Inspired by Jean Macé's prototype educational society for girls, in 1864 Floreska started a vocational school for girls in association with four local women, including a retired teacher. Guépin drew up their 'mission statement'. They had two objectives. The first, and less urgent, was to create a sort of higher primary school for girls from bourgeois families, with scholarships for able poor girls, where they would learn appropriate applied science, book-keeping, how to write business as well as personal letters, and industrial design for the clothing and furniture trades. More important, the association would offer apprenticeship and classes in industrial design, book-keeping and arithmetic for young girls of respectable families, bourgeois or workers.

The founding members' aims were to inspire love and respect for work and teach moral behaviour in an atmosphere of total religious toleration. The society would be run by women with money behind them, husbands prepared to put in funds but not interfere, and particularly women whose husbands ran the sort of industrial and commercial enterprises in which pupils might find employment afterwards.[72] The school, which could take up to 80 girls aged 12 and above, consisted of school and workshop, separated by a garden. Girls learned useful trades, not lace-making and embroidery, which were judged 'good for nothing', but dressmaking, tailoring, corsetry and underwear-making. Each day was divided between school and work. School pupils were encouraged to join the workshop, but were not obliged to do so. Designed to help girls from poor, or unemployed families, the school was so successful that plans for a second were soon drafted.

Guépin's first objective in his unsuccessful attempt to get into parliament in 1869 was free schooling and vocational training for children of both sexes.[73] When he was briefly prefect in 1870 he actively promoted the opening of a second vocational school for girls. He also repeatedly campaigned for a decent salary for women teachers. As a member of the departmental education committee until his death, he demanded an *école municipale professionnelle* in Nantes, which would admit both sexes.[74]

This attempt to understand Guépin's career draws out inconsistencies which Guépin shared with other contemporary middle-class socialists. His political career was disastrous. Three times in charge of a department, he left within a few weeks; he was never elected to parliament. Guépin's didactic, emotional and uncompromising personality did not bend with the political wind. In addition in 1848, Guépin faced the prospect of losing a second wife after a very brief marriage. He returned to Vannes shortly after his departure as *commissaire*, not to sort out the administration, but to visit the graves of his first wife, who died in 1831, and that of his father.

However, Guépin's failure as a politician in 1848 must be seen in the context of the dismissal of other radical socialists as the traditional conservative elite re-established its control. Nor was Guépin the only middle-class radical, then and now, to swing between intellectual socialism and distaste for insurrection. He thought of himself as a republican soldier, but he kept himself apart from the republican movement of the early 1830s, opposed the workers' rebellion in June 1848 and the Paris Commune of 1871. Indeed, after a small demonstration outside his home in the summer of 1849, Guépin wrote to the mayor of Nantes threatening to withhold his taxes if the municipality could not protect his person and property. For all his genuine concern about health care, poverty and unemployment, and insistence that in 1870 there were no more 'misters and workmen' – everyone was a worker at work and a mister otherwise – Guépin always saw the poor as fairly passive recipients. He was on cordial terms with artisan socialists like Greppo and Perdiguier, but his friends were nearly all, like him, patrician intellectuals. He always thought of himself as a democrat, but disappointment with the election of 1871 pushed him to conclude: 'We have given the vote to *political minors*, to illiterate peasants who only think of money.'[75]

Although he never became a national political figure, Guépin had an international professional reputation, his books had a national impact and he shaped much of the intellectual and medical life of western France. The cooperatives and mutual aid societies he created made a real contribution to social problems and his forceful action on girls' education set a formative example locally. Guépin had a favourite aphorism, which encapsulated his fundamental optimism. 'Aimer, c'est vivre! être aimé, c'est vivre encore!' [To love is to live! To be loved, is to live still!][76] He would have been thrilled to know that his civil funeral was attended by 50,000 mourners.

Notes

1 Vote of thanks to Guépin, *commissaire* in Nantes, on his departure to Vannes, from the Club national électoral de l'Oratoire. Archives Départmentales Loire-Atlantique 19J6 (hereafter ADLA). I am grateful to the British Academy and to Douglas Johnson for their support for a grant which enabled me to work on the Guépin papers in Nantes.
2 They were acquired in 1958–9 and a typed list of contents exists. ADLA 19J1–19J26; Microfilm of family letters, ADLA 1MI8.
3 O. Monprofit, *Le docteur Guépin, sa vie, ses œuvres, son caractère* (Nantes, 1873); G. Frambourg, *Le docteur Guépin 1805–73* (Nantes, 1964).
4 Guépin to Emile Souvestre, a school friend; ADLA 19J12 [undated, but clearly 1823].
5 Guépin, 'Dix huit jours de l'administration d'un communiste dans l'une des grandes villes de la France'; [nd 1852?] ADLA 19J6.
6 A. B. Spitzer, *The French Generation of 1820* (Princeton, 1987).
7 Guépin, *Quelques considérations sur le cancer* (Paris, 1828), pp. 15–17.
8 Guépin, 'Vie politique 1870–73'; ADLA 19J9.
9 Monprofit, *Le docteur Guépin*, pp. 12–14.
10 Mayor of Nantes to Guépin, 4 April 1842; ADLA 1MI8.

11 A. Guépin and C. E. Bonamy, *Nantes au XIXe siècle* (Nantes, 1835), p. 225.
12 Guépin, notes in defence of his Chair to Orfila, 1850; ADLA 19J1.
13 M. D. Sibalis, 'The Mutual-Aid Societies of Paris, 1789–1848', *French History*, 3 (1989), pp. 1–30.
14 Guépin, 'Faits divers'; ADLA 19J4. (He kept this diary from 1 January 1833 to 12 May 1839.) Typical of his initial support for Orleanism was his *De l'abolition de l'hérédité de la Pairie* (Nantes, 1830).
15 Guépin, *La Philosophie du socialisme, ou étude sur les transformations dans le monde et l'humanité* (Nantes, 1850).
16 Guépin, 'Nantes', in Ogee, *Dictionnaire historique de la Bretagne*, p. 224.
17 Guépin, 'Faits divers', 13 January 1833; ADLA 19J4.
18 ADLA 19J12.
19 Guépin, 'Nantes', in Ogee, *Dictionnaire historique*, p. 224.
20 13 December 1843; ADLA 19J12.
21 Guépin, 'Faits divers'; ADLA 19J4.
22 Frambourg, *Guépin*, p. 110.
23 Guépin, 'Faits divers', February 1834; ADLA 19J4.
24 Guépin, *Philosophie du XIXe siècle. Etude encyclopédique sur le monde et l'humanité* (Nantes, 1854), pp. 827–9.
25 Guépin, 'Faits divers', April 1834; ADLA 19J4.
26 A. Guépin, *Traité d'économie sociale* (Paris, 1835), p. 12.
27 *Ibid.*, p. 53.
28 *Ibid.*, pp. 81–3.
29 Guépin and Bonamy, *Nantes au XIXe siècle*.
30 Ogee, *Dictionnaire historique de la Bretagne*.
31 *Ibid.*, p. 228.
32 Guépin, *Suppression de la syphilis. Pétition à la Chambre des Députés* (Paris [printed in Nantes], 1846), p. 61.
33 V. Considérant, *Destinée sociale* (Paris, 1837).
34 *Le Phalanstère. Journal pour la fondation d'une Phalange agricole et manufacturière associée en travaux et en ménage*, founded by Considérant in 1832 to promote an experimental 'phalange'.
35 Guépin, 'Faits divers', November 1833; ADLA 2J4.
36 Chaper, prefect Loire-Inférieure to Minister of Interior, 18 September 1846; ADLA 55J1, bound volumes of correspondence, p. 279.
37 Prefect to minister of interior, 18 and 22 January 1847; ADLA 55J1.
38 V. Considérant, *Principes du socialisme* (Paris, 1847).
39 ADLA 19J18.
40 ADLA 19J18.
41 Guépin, 'Dix huit jours de l'administration d'un communiste dans l'une des grandes villes de la France'; [nd 1852?] ADLA 19J6.
42 Guépin, 'Notes et souvenirs du commissaire Loire-Inférieure 1848', ADLA 19J6.
43 Guépin, 'Notes et souvenirs', ADLA 19J6.
44 Commissaire Morbihan to Juges de paix, 1 May 1848; ADLA 19J7.
45 Mme Clotide Guépin to M. Le Sant, Guépin's first wife's father, from the prefecture in Vannes, 1 June 1848; ADLA 1MI8.
46 Secretary of *commissaire*, Nantes, Luminais to Menard-Blanchard, *commissaire*, 7 April 1848; ADLA 13J6.

47 Frambourg, *Guépin*, p. 236.
48 Guépin, 'Gardes Nationaux et Volontaires du Morbihan', 28 June 1848; ADLA 19J7.
49 Guépin to Richard de Rennes, *professeur*, lycée Louis-le-Grand, 17 July 1848; ADLA 19J5.
50 Chevalier to Guépin, trying to rebuild their friendship, 13 December 1843; ADLA 19J12.
51 Guépin to prefect, 29 April 1852; ADLA 19J8.
52 Guépin, *La Philosophie du socialisme, ou étude sur les transformations dans le monde et l'humanité* (Nantes, 1850).
53 Guépin, *Le Socialisme expliqué aux enfants du peuple* (Nantes, 1851).
54 Guépin, *Philosophie du XIXe siècle. Etude encyclopédique sur le monde et l'humanité* (Nantes, 1854).
55 Guépin, *Philosophie du socialisme*, p. 717.
56 Guépin to Orfila, rector at Rennes (who had originally appointed him to the Chair of Medicine in 1830), 24 February 1850; ADLA 19J8.
57 Guépin, *Philosophie du XIXe siècle*, p. 357.
58 *Ibid.*, p. 34.
59 Guépin, 'Dix huit jours de l'administration d'un communiste', ADLA 19J6.
60 In ADLA 19J18.
61 9 December 1859; ADLA 19J8.
62 Guépin, *Philosophie du XIXe siècle*, p. 817.
63 Guépin, *Marie de Beauval. La mission de la femme*, part I, conclusion; ADLA 19J14.
64 Guépin, *Philosophie du socialisme*, p. 680.
65 Guépin, *Philosophie du XIXe siècle*, p. 918.
66 Father to son, 26 April 1860; ADLA 1MI8.
67 3 March 1863.
68 Guépin, *Marie de Beauval. La mission de la femme*, ADLA 19J14.
69 Guépin, 'Réflexions dressées à une jeune femme qui se destine à l'enseignement des enfants de son sexe', 17 août 1853; ADLA 19J18.
70 Guépin, 'Réforme municipal à faire dans la ville de Nantes' [1863]; ADLA 19J5.
71 Guépin, 'Education' [nd]; ADLA 19J18.
72 Guépin, 'Education professionnelle' [nd]; ADLA 19J18.
73 Guépin to Evariste Mangin, 8 April 1869; ADLA 19J8.
74 ADLA 19J18.
75 October 1870; ADLA 19J9.
76 Guépin, *Philosophie du XIXe siècle*.

6

Attitudes towards Animals and Vegetarianism in Nineteenth-century France

Ceri Crossley

Since the 1970s the issues surrounding the moral status of animals have received a remarkable amount of attention from English-speaking philosophers, anthropologists and cultural historians. The best-known work has been done by Stephen Clark, Mary Douglas and Peter Singer.[1] Much attention has been paid to the central question of animal rights and to related matters of experimentation and animal welfare. The counter-case against animal rights has recently eloquently been put by Roger Scruton.[2] However, the debate about the status and value of animals reaches back into Antiquity. It is of course also in large part a debate about our humanness, about the nature of our difference from the rest of the animal kingdom. When we ask how we should treat animals and what are our responsibilities towards them we are involved in drawing crucial boundaries, boundaries which define what it means to be human.[3] Animals are objects of scientific study, but they also carry cultural meanings and values. The ways in which animals are represented are not fixed and the meanings which are ascribed to them change in order to reflect differing understandings of the relationship between nature and culture. The issue of human duty to animals is most sharply raised when it is a matter of whether animals should be killed and eaten. It is important to recognize that vegetarianism does not possess a unitary meaning. The case for abstaining from meat can be made on a variety of grounds: moral, medical, religious, philosophical. The refusal to consume flesh can be the expression of a reverence for the sacredness of life in all its forms. On the other hand, abstinence from eating meat can signify the intention to reject the corruption of matter by mortifying the flesh and repressing sensual desire. The rich and varied history of vegetarianism has been explored by Colin Spencer in his book *The Heretic's Feast* (1993), which provides a very helpful overview of the question.[4]

Until fairly recently interest in the animals issue has not been as great in France as it has been in Great Britain or in North America. The situation is, however, changing. Special issues of important journals such as *Critique*, *Le Temps de la réflexion* and *Autrement* have been devoted to the subject of animals.[5] In 1993 Janine Chanteur published *Du droit des bêtes à disposer d'elles-mêmes* and in 1997 there appeared a collective volume entitled *Les droits de l'animal aujourd'hui* which reprinted essays by, among others, Florence Burgat, Philippe Diolé and Théodore Monod.[6] In 1996 Eric Baratay published a history of the Church's attitude to animals from the seventeenth century to the present.[7] A considerable impetus to further work in this field was provided by the publication of Luc Ferry's polemical study, *Le Nouvel Ordre écologique* in 1992.[8] Subsequently Ferry – in collaboration with Claudine Germé – brought out an anthology of philosophical and legal texts on the animals question, *Des Animaux et des hommes* (1994).[9] In 1998 a weighty volume of essays edited by Boris Cyrulnik – *Si les lions pouvaient parler* – received widespread attention in the daily and weekly press.[10] This was followed by the publication of Elisabeth de Fontenay's *Le Silence des bêtes: la philosophie à l'épreuve de l'animalité* (1998), which situates the animals debate within an historical perspective which embraces philosophers from Plotinus to Derrida.[11] Over the years scholars from outside France have made significant contributions to our understanding of how French culture has viewed the value and status of animals. Two contributions stand out: the studies by George Boas and Hester Hastings which examined attitudes during the seventeenth and eighteenth centuries respectively.[12] More recently Kathleen Kete has described the significance of pet ownership within the increasingly urbanized Parisian society of the nineteenth century.[13] In a stimulating article of 1981, Maurice Agulhon emphasized the complex nature of the new sensibility towards animals which emerged in France towards the middle of the nineteenth century and which created the climate of opinion in which the Loi Grammont – which outlawed acts of cruelty towards domestic animals – was passed in 1850.[14] A number of the texts reproduced in *Des Animaux et des hommes* alert us to the importance which the animals issue had for writers and thinkers such as Michelet, Proudhon, Comte and Hugo. Much work none the less remains to be done on the French nineteenth century. My aim in the present chapter is to indicate the main shifts in attitude to the animals question between the late eighteenth and the early twentieth centuries. My intention is to show the extent to which the French discourse on animals – and the issue of whether or not to abstain from eating meat – was embedded in broader political and social concerns. When it came to implementing, reversing or resisting radical transformations of society, the question of vegetarianism was less marginal than one might at first suppose.

In France widespread feelings of sympathy towards animals emerged after 1750, significantly later than in England.[15] In France the animals issue was one element within the broader philosophical debate which surrounded the validity of Cartesianism. Descartes, as is well known, had taken the view that animals were machines, automata. His followers concluded that no cruelty or injustice took place when pain was inflicted on wild or domestic animals. For some the suffering endured by animals simply did not constitute a matter worthy of discussion. Before 1750 a writer such as the Abbé Pluche, who held a sentimental view of nature, was very much an isolated voice. A different situation prevailed across the Channel, where cruelty towards animals was perceived as an issue worthy of public interest. In England there also developed a growing awareness among medical practitioners – exemplified by figures such as George Cheyne, William Lambe and Joseph Ritson – of the physical benefits which accrued from the adoption of a vegetarian diet. In France the intellectual climate gradually changed and in the years which followed 1750 expressions of sympathy for the plight of animals became much more common.[16] Hastings cites examples from Rivarol, Voltaire, Bernardin de Saint-Pierre, Dupont de Nemours, Delille de Sales and other writers. In his *Traité des animaux* (1755) Condillac argued that animals were sentient creatures that possessed a spiritual – albeit mortal – soul. Charles Bonnet's understanding of the chain of being alerted readers to the affinities which existed between humans and the animal kingdom. In 1753 Maupertuis was the first eighteenth-century author to set out plainly the reasons why it was wrong for humans to kill or inflict pain on those animals which represented no threat to human life. Objections to vivisection were made by Voltaire, Bonnet, Delille and others. The morality of hunting was questioned by Rouher in *Les Mois* (1779). Seneca's claim that cruelty towards animals led to cruelty to men was revived by L.-S. Mercier, who contended that exposure to acts of butchery damaged human beings. Rousseau's valorization of pity as a spontaneous emotional response to the spectacle of suffering increased sympathy for the plight of animals. Interest in vegetarianism was on the increase albeit not to the same extent as in England. A French translation of Porphyry's *De Abstinentia ab esu animalium* appeared in 1747. Rousseau famously advanced the argument that humans were originally herbivorous, and in *Emile* (1762) he articulated the case for abstaining from meat. Bernardin de Saint-Pierre was for a time a vegetarian, as indeed was Rousseau himself.

For the liberal Enlightenment, rethinking the animals question was part of a humanitarian agenda which broadly reflected Rousseau's identification of the natural with the good. Human societies were held to be in crisis because they were artificial, factitious inventions which disrupted the order and harmony of nature. To suggest that animals possessed feelings, and perhaps a degree of intelligence and moral sense as well, not only undermined the notion of an unambiguous mind–body dualism, it also called into question the traditional Christian view that humans had dominion over the animals. There was clearly more at stake here than the straightforward issue of the treatment of animals. The animals question went to the heart of the liberal Enlightenment's challenge to the political and religious establishment. The recognition of similitude between humans and animals was potentially subversive of the religious definition of humanity which underpinned the existing social order. Indeed in the wake of 1789 some of those who espoused vegetarianism and the cause of animal welfare took on a political role as militant, anticlerical republicans. A commitment to end the suffering of animals was far from being synonymous with pacifism or non-violence. In 1791 John Oswald (1730–93), a former officer in the Indian army, published *The Cry of Nature: an Appeal to Mercy and to Justice on behalf of the Persecuted Animals*.[17] Oswald interests us because he was a convinced vegetarian who was involved in radical politics on both sides of the Channel. He knew well the leaders of the Revolution, men such as Brissot and Danton. He was a member of the Jacobin Club. In 1792 he published *La Tactique du peuple, ou nouveau principe pour les évolutions militaires*, a practical manual which aimed to teach the people how to fight effectively without their having to take the risk of seeking the advice of regular soldiers. In 1793 Oswald was killed, together with his two sons, fighting on the republican side in the civil war in the Vendée. In the English context Oswald's vegetarianism and revolutionary political beliefs mark him out as a precursor of Shelley. However, the extent to which Oswald's views coincided with those of his Parisian associates should not be underestimated. He was closely allied with Nicolas de Bonneville and the 'Cercle Social'.[18] The fusion between vegetarianism and animal welfare, on the one hand, and radical egalitarian politics, on the other, took place in an effervescent Parisian intellectual milieu marked by illuminism, freemasonry and a passion for Oriental religions.

The events of 1789 and of 1793 had a traumatic effect on the French nation. Opponents of the Revolution placed the blame for the violent overthrow of the *ancien régime* squarely on the shoulders of the liberal Enlightenment which, they claimed, had encouraged the corrosive spirit of individualism which had undermined the social bond and plunged society into chaos.[19] For reactionary Catholic writers such as Joseph de Maistre (1753–1821) and Louis de Bonald (1754–1840) materialism and individualism had produced anarchy, hedonism and revolutionary violence. The Enlightenment had legitimated the individual's quest for fulfilment and happiness, but this emancipation of desire had only served to release destructive anti-social passions. In order to save French society it was necessary to reassert the principle of authority and to reaffirm the primacy of society over the individual. Original sin alone explained the human condition. Humans were on earth to suffer and expiate, not to search for happiness. What appalled the counter-revolutionaries was the tendency of certain thinkers of the Enlightenment to present man as a more intelligent form of animal. In *Les Soirées de Saint-Pétersbourg* (1821) de Maistre insisted on the essential differences between humans and animals.[20] He was in no doubt whatsoever that animals were inferior beings and that humans had rightful dominion over them. Humans possessed intelligence and self-consciousness, whereas animals lived on the plane of the senses and were unable to form abstractions or general ideas. De Maistre gave the example of taking a dog to witness a public execution. The dog might well respond to the scene of violence and to the spilling of blood but, according to de Maistre, it would never grasp the moral dimension of the event because animals, unlike humans, were unable to relate their perceptions of external reality to an inner world of innate ideas. It was wrong to see animals other than as brutes. De Maistre did not sentimentalize animals and neither did he depict the natural world as the expression of order, harmony and purpose. Nature was red in tooth and claw and the processes of life on earth were all to do with competition and violent death as creatures, driven by their sensual appetites, sought to devour one another. De Maistre was not shocked by this. The animals merited their fate. It was also legitimate for humans to kill animals for food, clothing, even for purposes of amusement. Humans were divinely appointed agents of destruction whose 'tables [were] covered with corpses' (p. 221). However, for de Maistre the death of animals was not enough; the earth cried out for more blood and the bodies of men and women, guilty or innocent, were destined endlessly to be immolated, 'until the death of death' (p. 223). In this way while de Maistre insisted on the unbridgeable distance between self-aware humans and other forms of life, his uncompromisingly violent picture of cosmic extermination ultimately united humans and animals within a vision of expiation without real redemption.

De Bonald's opinions on the animals question were expressed in a late work, *Recherches philosophiques sur les premiers objets des connaissances morales* (1830).[21] His position was less extreme than that adopted by de Maistre, but his writing lacked the latter's literary quality. De Bonald rejected the arguments advanced by philosophers from Condillac to Cabanis to the effect that profound affinities existed between humans and animals. De Bonald stood by the argument that an animal was fundamentally a machine, albeit a machine organized by God. Humans possessed understanding, reason, imagination, will, judgement and an immortal soul. Animals relied on instinct and were fundamentally passive, fixed and unchanging. Humans were imperfect but perfectible in and through social relations. This was the vital point. Unlike the animals, individual human beings were not sufficient unto themselves. They could not rely solely upon the power of instinct. Humans only really existed as social beings: 'the brute, which finds within itself everything that it needs to know, is always alone, even when it lives with its fellow creatures, because it can learn nothing from its communications with them; Man, once he has acquired knowledge of God and man, is never isolated; he has known society, and he lives in its midst through his thoughts, even when he is physically isolated from it' (p. 179). The implication of de Bonald's argument was clear: by tending to collapse the distinction between humans and animals, the philosophers of the Enlightenment had promoted a current of egoistic individualism. The pursuit of virtue had been confused with the quest for happiness understood as the fulfilment of private projects and personal desires. Individuals had competed aggressively with one another in order to satisfy their sensual desires, and the consequence of this process had been the dislocation of French society during the Revolution. By emphasizing the bonds between man and nature the Enlightenment had released the destructive passions which it was society's duty to repress. De Bonald's response was twofold. First, he underscored man's status as a superior being who occupied a divinely ordained position at the centre of the universe. Second, he affirmed the primacy of society over the individual. Humans were not to be confused with animals.

Animals, however, were not simply metaphors for aspects of human nature. Animals existed in the real world. Did humankind have any significant obligations towards them? De Bonald took note of the fact that humans experienced feelings of affection for animals; he observed that they even addressed animals as if they understood speech. Since God had intended that humans should work alongside animals, de Bonald was willing to countenance such behaviour. However, as a general principle he severely disapproved of all attempts to invest animals with human qualities and emotions. He did not deny that animals suffered pain but he claimed that they did not experience unhappiness. An animal which experienced sensations of pleasure or pain was likened to a plant being watered or a piece of wood being placed on a fire (p. 180). It was not in society's interest for humans to be gratuitously cruel towards animals. On the other hand, to treat animals with the respect and affection normally extended to humans was either a sign of puerility or an act of profanation (p. 191). Animals reproduced themselves in order to be of service to man. For de Bonald they possessed no intrinsic value.

Despite the best efforts of the counter-revolutionaries Enlightenment attitudes towards animals were not however abandoned. Advocates of animal welfare continued to argue that to inflict pain on animals was fundamentally wrong. Social reformers still insisted that the behaviour of citizens would improve if cruelty towards animals were reduced. In 1814 Ménard published *L'Ami des bêtes ou le défenseur de ses presque semblables*.[22] He believed that it had been God's intention for man to live in peace and concord with the animals. However, man had become a murderous monster, 'l'égorgeur universel' (p. 28). Ménard placed most of the blame for this moral decline on rapacious kings and ambitious priests who in primitive times had employed animal sacrifice in order to entrench their authority. The priest-kings transformed naturally frugivorous man into a carnivore, they 'sanguinized' him, turning him into 'the universal devourer of living nature' (p. 17). This was profoundly wrong. Ménard explained that humans did not need to eat meat in order to be healthy. The Creator, the 'God-Mother' (p. 11), had never intended men to kill and eat animals, all of whom ultimately fulfilled a purpose in Creation. By way of response Ménard urged reform and championed a non-sacerdotal 'religion of the heart' (p. 79), inspired by the ideas of Pythagoras.

Ménard's approach remained very much that of the eighteenth-century humanitarian. A similar intellectual background conditioned the approach taken by France's most important vegetarian thinker of the period under discussion, Jean-Antoine Gleizes (1773–1843). In 1841–42 he published *Thalysie ou la nouvelle existence* which is a landmark text in the history of vegetarianism in France.[23] Gleizes argued passionately that eating animal flesh perverted the senses, disfigured the body and harmed the mind. He believed that whereas vegetarians drew physical and moral strength from their diet, carnivores were but a short step from practising cannibalism. Gleizes held that the slaughter of innocent animals was a crime and he forbade hunting on his estates in the south of France. In his view the genuine vegetarian aspired to purity, to harmony with nature. To eat vegetables was to enter into communion with the divine. Gleizes grounded his faith in the regenerative powers of a non-meat diet in a grand philosophical scheme which explained the universe in terms of the progressive spiritualizing of matter. Man's destiny beyond the grave involved transmigration to a vegetarian paradise on another planet. *Thalysie* appeared between 1840 and 1842 but the project had been conceived at the turn of the century when Gleizes had been part of the 'méditateurs' group which had met at Passy and whose members had been deeply interested in the ideas of the Pythagoreans.[24] Vegetarianism was the obsession of Gleizes's life. It figured in his published work as early as 1800, in *Les Nuits élyséennes*. In 1821 he brought out a shorter version of *Thalysie*. In a pamphlet of 1830 he argued that the true message of Christianity was vegetarianism. It is significant that Gleizes himself had only become a vegetarian at the age of 21 and that the change of diet had been occasioned by the bloody course taken by the French Revolution. Gleizes, a moderate republican, had been shocked to the roots of his being by the violence of the Terror. Why had the Revolution lost its way? Gleizes came up with a novel explanation and put the blame on the carnivorous diet of his compatriots. The consumption of meat was thus ultimately responsible for 1793! Gleizes concluded that in order for French republican virtue finally to triumph the nation had to change its diet. In his eyes vegetarianism had almost magical powers: it produced equality, fostered liberty and engendered social harmony!

Gleizes's ambition to regenerate humankind corresponded to the intellectual mood of the July Monarchy but, as I have suggested above, the argument of *Thalysie* was more in tune with the illuminist speculations of the late eighteenth century than with the science, philosophy and medicine of the mid-nineteenth century. The book had little impact within France although it subsequently made its mark when it was translated into English[25] and German. Reviewers of *Thalysie* found it difficult to take Gleizes's arguments entirely seriously. One such reviewer was Alphonse Esquiros (1812–76), who wrote an article on Gleizes for the *Revue des deux mondes* in 1846. Esquiros, a socialist thinker with Christian allegiances, sympathized with some aspects of Gleizes's argument, but took him to task for paying too much attention to the suffering of those animals destined for slaughter, while failing to take enough account of the findings of modern science regarding man's indubitably omnivorous nature. According to Esquiros the plants and animals became part of the very being of the humans who consumed them. Consequently the act of eating should itself be viewed as 'a vast and perpetual metempsychosis'.[26] Esquiros belonged to an emerging generation of philosophers and social reformers whose views were informed by more recent developments in biology, embryology and comparative anatomy, developments which were altering the understanding of man's relationship with animals and challenging traditional definitions of instinct and intelligence. Esquiros discussed what he termed the 'future of animals' in a chapter of his remarkable study, *Paris ou les sciences, les institutions et les mœurs au XIXe siècle* (1847).[27] He was clearly moved by the unnecessary suffering inflicted by humans on animals. He denounced owners of livestock who were more interested in profit than in the health of their animals. He compared exotic animals in zoos to prisoners condemned to a life of solitude and boredom, degraded by their incarceration. He regretted the heartless manner in which a rider used his whip on his horse. He fulminated against the force-feeding and mutilation of birds and animals. He concluded that what was required was a new, softer attitude; henceforward animals should be perceived as humankind's guests at nature's feast. The notion of embryonic recapitulation – the idea that in the womb the foetus proceeds through stages analogous to the general development of life of earth – appeared to throw new light on our closeness between humans and the animals. However, while modern science had relocated man within nature this did not mean that humanity's position at the pinnacle of creation was undermined. According to Esquiros 'unbridgeable limits' (p. 226) separated humans from animals. In his view animals had real value in so far as they participated in humankind's mission to transform the planet. Esquiros did not dream of recapturing a lost rural world of harmony. On the contrary he espoused the cause of technology and his text was a hymn to the achievements of modern science and industry. In his view nineteenth-century science and philosophy had effectively reconfirmed man in his position as God's representative on earth, as master of nature. Despite all the talk of humanizing and civilizing the animals Esquiros was really interested in legitimating humankind's rightful dominion over them. We should not confuse Esquiros's stance with that of Gleizes. Animals were to be treated without cruelty, but it was still quite legitimate to use them and to eat them.

In the years before 1848 many social reformers in France adopted an anthropocentric stance similar to that taken up by Esquiros. They expressed concern regarding the mistreatment of domestic animals but did not seriously question the notion that human beings were by nature carnivorous.[28] As is well known, the followers of Charles Fourier loved food and tended to enjoy hunting and were thus unlikely to be converted to the delights of vegetarianism.[29] Those who did abstain from meat found themselves the butt of jokes and dismissive comments.[30] At the same time, however, the existence of a new sympathy for animals was not to be denied. This new mood created the climate in which the Grammont Law was passed and the Société protectrice des animaux was founded. A major influence on the emergence of this new sensibility was the historian Jules Michelet (1798–1874), who first espoused the cause of the animal kingdom in the course of *Le Peuple* (1846) and enlarged on this theme in later publications such as *L'Oiseau* (1856), *L'Insecte* (1857), *La Mer* (1861) and *La Bible de l'Humanité* (1864). Michelet was the self-proclaimed spokesman of the poor, the oppressed, the victims of arbitrary authority. He considered that his duty as an historian compelled him to speak out on behalf of those who had been denied a voice, to defend the cause of the labouring masses, of women and children. By the 1840s Michelet had come to the conclusion that Christianity had failed. In its place he proposed a religion of France grounded in the values represented by the people: spontaneity, authenticity, sympathy and community. In the face of the social collapse wrought by modern industrial society Michelet argued in favour of a return to the values associated with the land – which he identified with the true spirit of the French Revolution. He objected to the Christian view of nature as fallen. He condemned the Church for holding that children were sinful. In Michelet's view the divine life expressed itself in and through nature, and the natural world was to be valued, not repudiated and denied. He felt sympathy for Hinduism which gave expression to what he termed the spirit of universal brotherhood. How unlike the Judeo-Christian tradition which, he claimed, branded animals as impure. Jesus, observed Michelet, died for mankind, not the animals, who were denied any form of salvation. Michelet set out to 'rehabilitate' animals, to restore to them the value which first the Church and subsequently Cartesian philosophy had denied them. Michelet wanted to change the status of animals in the modern world and like Esquiros he was an enthusiastic admirer of Isidore Geoffroy Saint-Hilaire's work on the domestication of animals and on the acclimatization of foreign species in France.[31] On the other hand, what makes Michelet's contribution to the animals debate markedly significant is the degree of compassion that he feels for those creatures that endure pain at the hands of humans: 'The animal! Dark mystery!... immense world of dreams and silent pains... All nature protests at the barbarism of man who misunderstands, degrades, who tortures his lower brother; nature accuses man in the presence of Him who created them both!'[32] In the works of natural history that he published during the 1850s and 1860s, Michelet mounted a spirited defence of the animal world. He objected to hunting, to all forms of unnecessary cruelty.[33] However, Michelet did not go so far as to become a vegetarian – although vegetarian propagandists subsequently cited him as an ally.[34] With regret he observed that humans could not avoid eating meat; their teeth and stomach were those of a carnivorous species. Men in particular needed to consume meat because it provided them with the energy that they required in order to build a better social world governed by law and justice. Women were different. In Michelet's view their destiny was to aid their husbands and bear them children; in their case a diet of vegetables was deemed to be quite appropriate.[35]

After 1850 a considerable number of republican sympathizers followed Michelet's lead and took up the cause of animal welfare.[36] In major works of the exile period we find Victor Hugo preoccupied with the relationship between man and the animal kingdom.[37] And in a poem such as *Le Crapaud* [The Toad] he specifically alerted his readers to the cruelty and sadism which characterized much human behaviour towards animals. Animals became the favoured subject matter of many artists and writers of the period. Indeed, several eminent political figures of the Third Republic were ardent zoophiles. Georges Clemenceau denounced bullfighting.[38] Raymond Poincaré derived deep personal consolation from his pet dogs and cats.[39] The link between animal rights, vegetarianism and progressive politics was most forcibly made by the anarchist geographer Elisée Reclus.[40] The impact of Darwinism led to new assessments of the relationship between animal instinct and human intelligence, for example by Henri Joly in *L'Instinct, ses rapports avec la vie et avec l'intelligence. Essai de psychologie comparée* (1873). In the second half of the nineteenth century the debate concerning mankind's relationship with animals became refocused around the issue of vivisection.[41] Vivisection had been practised earlier in the century without arousing great controversy. In books written for a broad public, scientists such as Paul Flourens had felt no qualms in describing experiments carried out on live animals.[42] However, by the 1870s vivisection had become a highly contentious issue, in large part due to the emphasis which Claude Bernard had placed on the need to carry out experiments on live animals. In France, as in England, opposition to vivisection became a central plank of the programmes of reform-minded social and political movements. Writing in 1896 Jules Bois remarked upon the presence of the animals issue in women's organizations of the *fin-de-siècle* period. He noted that since 1874 'leagues of mercy' organized by women had been active in denouncing vivisection, opposing hunting and criticizing conditions in slaughter houses.[43] The belief that unnecessary cruelty towards animals was a major cause of the violence that humans directed towards their fellows was the guiding principle of the early twentieth-century journal, the *Revue illustrée des animaux* (1904–15). This review gave expression to the views of the *Ligue pour la Défense des Animaux*. Among its supporters and contributors were leading politicians, intellectuals, artists, writers, scientists and university professors: these included Paul Adam, Ferdinand Buisson, Jules Clarétie, Paul Doumer, Camille Flammarion, Georges Clemenceau, Pierre Loti, Paul Reclus and Camille Saint-Saëns. The review drew its support from the great and the good and was published 'with the encouragement of the Ministry of Agriculture'. The covers of later issues were regularly adorned with photographs of celebrities with zoophile convictions, sometimes posing in the company of their favourite pets. However, the review had a serious purpose. Its motto was 'He who takes care of animals looks after his fellow human beings'. The review was written with young people in mind and it made a big effort to reach elementary school teachers. There were stories, competitions, poems and sentimental engravings of cats and dogs. Morally uplifting tales featuring domestic animals abounded. Articles on aspects of natural history and pet-keeping were included. Earlier texts by Lamartine, Michelet, Elisée Reclus and Hugo were reproduced for a new audience. There were also more substantial articles on the question of animal intelligence. Most issues contained pieces which denounced vivisection and the inhumane treatment of domestic animals. Objection was made to the use of animal products by the fashion industry, notably in the form of fur coats and hats which included the feathers of exotic birds. The review reminded its readers that the struggle to improve the condition of animals had an international dimension; numerous reports were published which related the activities of animal protection societies abroad. The 15 January 1907 issue contained an article by Clemenceau which was illustrated by a composed 'photograph' purporting to show the President of the Council of Ministers standing in a garden surrounded by five large birds. The legend accompanying the picture read as follows: 'M. Georges Clemenceau, President of the Council of Ministers, has a great love for animals. He has had his good friends – the birds which stand around him – brought from his garden at the Interior Ministry in the Rue Franklin. The birds allow him to forget the sadness which arises from human contact. We are particularly grateful for his collaboration.'[44] These comments are useful because they help us situate the review politically. The *Revue illustrée des animaux* reflected the spirit of moderate republicanism at the turn of the century. The editors – both significantly women – Ivone and Marguerite des Varennes campaigned relentlessly to improve the condition of animals. They were particularly concerned to get their message across to local mayors and to teachers in primary schools. To a large degree they felt that they were successful in achieving their goal. On the other hand, they expressed regret at the lack of involvement shown by the clergy in support of their activities. However, in the years which followed the Dreyfus Affair the Church was hardly likely to revise its teaching on the moral status on animals. And would members of the clergy ever feel at ease reading a journal which republished texts by Michelet and Elisée Reclus?

The stance adopted by the *Revue illustrée des animaux* was consensual, not radical or anticlerical. In a sense it was a journal of the enlightened new establishment. Extreme positions were eschewed. Limits were implicitly placed on the degree of compassion which it was legitimate to feel for the sufferings of lesser creatures. Thus when criticism was made of hunting, contributors focused more on the risk of accidents than on the morality of shooting animals and birds. Similarly, it was clearly felt that there was nothing inappropriate about publishing an article which extolled the advantages of a recently developed form of humane killer. A sign of the journal's reluctance to take up radical positions was the little space which was given over to vegetarianism and related issues. Firm condemnations of meat-eating were not absent from the pages of the review, but they were rare. In 1910 readers were given a stern warning: 'Beware lest the soul of the animal sacrificed for your feast should return to haunt your dreams, reproaching you for your crime of fratricide.'[45] Vegetarianism was therefore something to be encouraged. Furthermore readers were informed that the review's administrators were 'ALL vegetarians' and a promise was made to include vegetarian menus in subsequent issues.[46] Some years earlier, in 1905, Solange Pellat had contributed a strong piece on the force-feeding of geese, which had also included an attack on the practice of cooking live crustaceans and fish in boiling water. An editorial note appended to the article pointedly commented: '*No! It is not necessary to kill animals in order to feed oneself.* Butchery is not a necessity, it is simply an institution to which we have become accustomed.'[47] At this point the reader is referred to the example of Dr Georges Dujardin-Beaumetz, a military surgeon turned vegetarian. References to vegetarianism remain none the less rare in the pages of the *Revue illustrée des animaux*. The citing of Dr Dujardin-Beaumetz as an authority was an interesting choice since his advocacy of vegetarianism was based exclusively on medical and not on moral, religious, philosophical or social grounds. In February 1890 he had delivered a lecture on the matter to the members of the Société de Médecine Pratique. Dujardin-Beaumetz's own health had been restored once he had abandoned eating meat. He concluded that the adoption of a vegetable diet helped cure intestinal disorders and diseases of the kidneys. Instead of citing Dr Dujardin-Beaumetz as an authority, the *Revue illustrée des animaux* could have referred to the final figure I wish to examine in this essay: Dr Ernest Bonnejoy. He was the author of articles on vegetarianism which appeared in short-lived reviews such as *La Réforme alimentaire* and *Le Journal de la santé*; he brought out pamphlets describing the benefits of a meat-free diet and he finally provided a lengthy exposition of his views in a book entitled *Le Végétarisme et le régime végétarien rationnel* (1891).[48] Interestingly it was Dujardin-Beaumetz who penned a preface to Bonnejoy's study and in the course of this he drew attention to the differences which separated the two men despite their shared belief in the value of vegetarianism. Dujardin-Beaumetz portrayed himself as a straightforward medical man. He contrasted his approach with that of Bonnejoy, who inscribed the therapeutic benefits deriving from a meat-free diet within a much grander project of social reform. Bonnejoy was presented by his medical colleague as a man with a mission, driven by a proselytizing zeal.

Ernest Bonnejoy looked enviously across the Channel where the expanding vegetarian movement was being taken seriously. In Britain there were vegetarian newspapers and magazines, vegetarian restaurants and hotels. In Germany and Switzerland too the movement was growing. Why were the French lagging so far behind? What was to be done? Bonnejoy's answer was to promote a form of vegetarianism which he claimed was particularly appropriate for his contemporaries in France. This he termed 'Végétarisme rationnel français', which he helpfully abbreviated to VRF. French Rational Vegetarianism was unlike religious or philosophical forms of vegetarianism in so far as the emphasis was firmly placed on the tangible benefits which accompanied a meat-free diet. Bonnejoy explained that he personally had not become a vegetarian for philosophical or moral reasons but because – like Dujardin-Beaumetz –

abstinence from meat had cured him of an ulcerous condition which modern medicine had failed to treat successfully. He recognized the importance of the moral questions posed by the suffering of animals, but he insisted that henceforward the case in favour of vegetarianism could be made on scientific and medical grounds alone. In his view Louis Pasteur's recent work on microbes lent vegetarianism conclusive support. He explained that while vegetables and fruit contained within them a life-giving principle, raw or cooked meat contained toxins and microbes which were responsible for a range of ailments such as cancer, scrofula, apoplexy, gravel, rheumatism, gout, kidney disease and dyspepsia. Eating vegetables boosted the body's immune system and increased resistance to disease, whereas eating animal flesh exposed the body to the dangerous bacteria which started to proliferate from the very moment that an animal was killed. It was absurd to believe that life could proceed from death, from the debris of corrupt and decaying flesh. To feed meat to a sick patient was seriously to compromise his or her chances of recovery.

Rational French Vegetarianism was presented as an advanced doctrine built upon the firm foundations of modern medical science. Unfortunately, the French – unlike the English, the Americans, the Swiss or the Germans – remained emotionally attached to their nefarious carnivorous habits. Bonnejoy set out to show the French the error of their ways. He argued that children were instinctively vegetarian and that humans had the teeth of frugivores, not carnivores. He pointed out that vegetarians were no weaker than meat-eaters. He gave examples of healthy and resilient vegetarian populations from around the world. He observed that the healthy condition of the French peasantry demonstrated the benefits of involuntary vegetarianism. He went so far as to claim that a vegetarian diet produced not only a healthy but also a beautiful body. However, as Bonnejoy recognized, he had to struggle against a deeply ingrained national prejudice in favour of meat-eating. Recent precedent was less than encouraging: the Société Végétarienne de Paris, established in 1880 by Hureau de Villeneuve had rapidly been renamed the Société Végétarienne de France by Goyart but that had not prevented its prompt dissolution as early as 1882. The French appeared determined to remain what Bonnejoy termed 'des nécrophages', consumers of death. But why was the situation in France different from that which prevailed elsewhere in Europe? Bonnejoy provided an inventive historical explanation which centred on the meaning of the Revolution. He argued that during the *ancien régime* meat-eating had weakened the members of the aristocracy both physically and morally to such an extent that they were unable to resist the popular uprising of 1789. However, the victors of the Revolution were themselves unable to throw off the entrenched national prejudice in favour of meat. Bonnejoy placed most blame on the moral climate of the Directory. He claimed that in the wake of the events of 1793 a new ruling class had emerged which espoused materialism and indulged in sensual pleasure. Writers such as Grimod de la Reynière, Berchoux and Brillat Savarin encouraged excessive consumption of meat. The temper of the times favoured gluttony. It was hardly surprising that in the nineteenth century the French population remained deaf to the strictures of dietary reformers such as Gleizes.

Two common threads of vegetarian propaganda figured strongly in Bonnejoy's writing. First, he argued that the consumption of meat was harmful to physical health. Second, he claimed that the mental state and the moral condition of meat-eaters were also adversely affected by their diet. However, Bonnejoy was careful to adopt a posture less extreme than that which had been taken up by Gleizes. He did not anticipate that all members of society would rapidly see the light and definitively abandon meat-eating. He was even willing to joke that the consumption of a single 'saignant' steak was hardly going to transform a man into a murderer. None the less, the longer-term consequences of eating meat were not to be denied. Bonnejoy cited experiments by Liebig which proved that placid herbivores became aggressive when fed on animal products. The inference was clear: gradually meat-eaters would come to display the 'vices' of the animals that they consumed – stupidity, cowardice, selfishness, insalubrity, mental instability. Furthermore, we should bear in mind that when Bonnejoy published his study there was a considerable degree of moral panic in the air. In the closing years of the nineteenth century minds across Europe were preoccupied with ideas of decadence and degeneration. To a large degree Bonnejoy's French Rational Vegetarianism was a response to such concerns. Meat-eating was not only held to be unnatural and unhealthy, it was also viewed as the cause of more general processes of physical and mental decline. In Bonnejoy's opinion humans were drawn to the consumption of meat because it acted as a stimulant. However, he believed that carnivores rapidly became habituated to the effects of consuming flesh and that they soon moved on to take additional drugs, some alcoholic, some alkaloid. He claimed that all levels of French society were caught up in this process. The lower classes were affected by a rising tide of alcoholism. The members of the intellectual elite (scientists, doctors, journalists) were turning to morphine, opium and cocaine. The physical and moral health of the nation was in peril. Alcoholism caused by meat-eating was leading to the break-up of families and to wider social disintegration. French Rational Vegetarianism was called upon to reverse the process of decline. It was presented as thoroughly modern and scientifically respectable. It constituted a practical programme for reforming the health of the nation and was therefore in no way to be confused with religious utopianism or with dreams of metempsychosis. Nevertheless, this did not mean that Bonnejoy repudiated earlier forms of vegetarianism. He preferred to present his own brand of vegetarianism as the culmination of the doctrines of the past. Many pages of *Le végétarisme et le régime végétarien rationnel* were given over to a consideration of the role played by vegetarians in the history of philosophy and of religion. Attention was paid to the attitude towards diet displayed by Hinduism, Buddhism, Judaism, Pythagoreanism and Christianity (like Gleizes, Bonnejoy did not hesitate to declare that Jesus was a vegetarian). Bonnejoy quoted from numerous authorities including Plutarch, Porphyry, Ovid and the Church Fathers. He referred to writings by Voltaire, Rousseau, Gleizes, Lamartine and Michelet, as well as to more recent publications by vegetarian sympathizers such as Anna Kingsford in England and Edouard Raoux in Switzerland. His aim was clearly to build a consensual view of vegetarianism as something central to religion and philosophy throughout the ages. At the same time the reader has the impression that Bonnejoy was never hugely in sympathy with philosophical or religious vegetarianism. His attitude recalls the approach taken by the philosophers of the Enlightenment when they studied the history of religion: the elite could understand the truth in the light of reason but the masses were swayed by superstition and needed a form of positive religion. In Bonnejoy's judgement religious and philosophical arguments in favour of abstinence were usually devices employed in order to persuade the population at large to follow a healthier diet. Abstinence during Lent helped the body recover from the preceding winter. Similarly the Pythagorean notion of metempsychosis was described as a useful 'philosophical fiction' (p. 133).

Rational vegetarianism apparently had no need of such philosophical fictions. Bonnejoy's intention was to present vegetarianism as the sensible, rational way forward for the nation at large and not as the doctrine of a small group of eccentric individuals who believed in the transmigration of souls and were obsessed with a sentimental view of humankind's relation to the animal kingdom. Bonnejoy could hardly be portrayed as a dangerous extremist: he insisted that eggs and dairy products formed part of a sound vegetarian diet, paid little attention to the moral status of animals and avoided dealing directly with the thorny issue of vivisection. He believed that all members of society would benefit from a change of diet. Vegetarians were not weak – their sobriety was allied with physical and moral strength. Neither were they hostile to enjoying their food – Bonnejoy reprints the menus of meals consumed at various vegetarian festivities. The practice of vegetarianism was allied with other causes which aimed at improving the health of the nation: Bonnejoy urged his readers to avoid the dangers of adulterated foods by growing their own vegetables, baking their own bread and brewing their own beer and cider. A healthy vegetarian diet had the additional advantage of reducing the need to take medicines of dubious value. But how could the majority of the citizens of the Third Republic be persuaded to abstain from meat? Bonnejoy was a realist. He did not believe that vegetarianism could be imposed from above, and he disliked the authoritarian, 'Protestant' streak that he detected in Gleizes's theories. What then was to be done? It was certainly difficult to change eating habits which were not only validated by cultural prejudice but reinforced by commercial interests. Bonnejoy concluded that the state had a crucial role to play by ensuring that dietary education was made compulsory in all French schools. In the fight against alcoholism and addiction it was imperative that information on matters of diet and nutrition be disseminated as widely as possible throughout the population. However, at the conclusion of his book Bonnejoy admitted that in practice vegetarianism was 'a superior doctrine destined for the intellectual elite' (p. 329). The masses needed to change their diet but they were incapable of properly understanding vegetarianism. Bonnejoy's immediate hope was that vegetarian societies would develop in France on the English and German model. Vegetarianism might be the salvation for France but it was going to be a long haul. Perhaps the French nation needed a 'philosophical fiction' after all.

What general conclusions can we draw from the foregoing discussion? First, it is clear that in France the nineteenth century witnessed the consolidation of the zoophile tendencies of the Enlightenment. While Catholic thinkers insisted on the crucial significance of the divinely instituted barrier between humans and animals, many thinkers on the left inscribed the animal kingdom within a general vision of life on earth as a progressive, upward movement which culminated in freedom and consciousness. A version of Michelet's inclusive vision of the unfolding of life on earth – rather than atheistic materialism or hard-edged Darwinism – was favoured during the early part of the Third Republic. Each creature fulfilled a purpose, none was to be despised.[49]

The divine expressed itself in and through creation and it was incumbent upon humans to treat their fellow creatures with respect. Cruelty towards animals was judged to be wrong, for while animals did not possess consciousness they certainly experienced pain. Moreover, it was believed by many that those who acted in a callous way towards animals were likely to behave in a similar fashion towards human beings. Michelet's notion of universal brotherhood suggested that on one level humankind and the animals formed a single community. Henceforward co-operation should replace domination. However, the battle for animal welfare was not easily won in France. The Grammont law was acknowledged not to be very effective and further legislation had to be passed in 1898 and 1929. Brutality towards animals remained widespread in the countryside despite the best efforts of the enlightened members of the urban animal protection societies. Moreover, the issue of cruelty towards animals needs to be seen within the broader context of French society's evolving attitudes towards different forms of social and political violence – we have seen how the memory of the Revolution was present in the writings of vegetarians such as Gleizes and Bonnejoy. The question of the justification of acts of violence was a major preoccupation of thinkers of the right and of the left, from Joseph de Maistre to Pierre Leroux.[50] The Third Republic set out to persuade young people that cruelty towards animals was wrong, but it only achieved limited success. In the words of Theodore Zeldin:

> Games in which turkeys were made to dance on red-hot platforms; or in which spectators hurled stones at pigeons placed in boxes, their heads stuck out to receive the blows; brutality in slaughter-houses, where for example sheep's eyes were gouged out before they were killed; these practices were accepted by many, and seemed appalling at first only to rather eccentric, over-sensitive people. Cruelty remained one form of voluptuousness; and some sports ... were a way of giving a new significance, new purposes and new controls to older rituals of violence.[51]

In these circumstances it was unlikely that a vegetarian diet would ever become truly popular among the French. None the less, from Gleizes to Bonnejoy, vegetarian propagandists continued to argue the case for a meat-free diet. We have seen how in the course of the nineteenth century vegetarianism in France moved away from mystical and religious concerns and became focused instead on issues of health and physical well-being. What remained constant, however, was the presence of a moral imperative, the conviction that by turning vegetarian humans were recognizing both their responsibility towards the natural world and their duties to their fellows. Such was the position of the anarchist geographer Elisée Reclus who held that arguments relating to physical well-being were only of secondary importance when it came to evaluating the advantages and disadvantages of vegetarianism: 'the important point is the recognition of the bond of affection and goodwill that links man to the so-called lower animals, and the extension to these our brothers of the sentiment which has already put a stop to cannibalism among men.'[52] Reclus believed that animals should be viewed 'either as respected fellow-workers, or simply as companions in the joy of life and friendship'. In order for humans to be treated with justice and fairness an attitude of respect needed to be extended to other creatures, to the birds and animals that 'please us better as friends than as meat'. Few, however, were tempted to follow Reclus's example, and when it came to vegetarianism he remained largely a voice crying in the wilderness. In the nineteenth century, French supporters of the cause of animal welfare often compared their situation unfavourably with that which prevailed across the Channel. To this day the practice of vegetarianism remains much more marginal in France than in Britain. In France issues to do with hunting, animal rights and vegetarianism take on different shades of meaning because they are debated within a political culture which takes its source in the Revolution of 1789.[53]

Acknowledgement

As a student of aspects of Anglo-French cultural relations I would like to record my debt to Douglas Johnson whose work on Guizot and on the Dreyfus Affair – not to speak of his enlivening literary journalism – has consistently reminded the members of the broader British French Studies community that their scholarly activity is not an ideal construct, but something which is itself embedded in the evolving pattern of cultural transactions between Britain and France.

Notes

1   S. Clark, *The Moral Status of Animals* (Oxford, 1984) and *Animals and their Moral Standing* (London, 1997); M. Midgley, *Animals and Why they Matter* (Athens, Ga., 1983); P. Singer, *Animal Liberation* (London, 1976). See also K. Tester, *Animals and Society* (London, 1991).
2   R. Scruton, *Animal Rights and Wrongs* (London, 1998).

3 See H. Ritvo, *The Animal Estate* (Cambridge, Mass., 1987). See also J. Passmore, 'The Treatment of Animals', *Journal of the History of Ideas*, XXXVI, 2 (1975), pp. 195–218.

4 See C. Spencer, *The Heretic's Feast. A History of Vegetarianism* (London, 1993). See also J. Barkas, *The Vegetable Passion* (London, 1975); D. A. Dombrowski, *The Philosophy of Vegetarianism* (Amherst, Mass., 1984).

5 *Critique*, 375–6 (1978); *Le Débat*, 27 (1983); *Le Temps de la réflexion*, 9 (1988); *Autrement*, 56 (1984).

6 J. Chanteur, *Du droit des bêtes à disposer d'elles-mêmes* (Paris, 1993); G. Chapoutier and J.-C. Nouet (eds), *Les Droits de l'animal aujourd'hui* (Condé-sur-Noireau, 1997). See also F. Burgat, *La Protection de l'animal* (Paris, 1997).

7 E. Baratay, *L'Eglise et l'animal* (Paris, 1996).

8 L. Ferry, *Le Nouvel Ordre écologique* (Paris, 1992).

9 L. Ferry and C. Germé (eds), *Des Animaux et des hommes* (Paris, 1994).

10 B. Cyrulnik (ed.), *Si les lions pouvaient parler* (Paris, 1998).

11 E. de Fontenay, *Le Silence des bêtes: la philosophie à l'épreuve de l'animalité* (Paris, 1998).

12 G. Boas, *The Happy Beast in French Thought of the Seventeenth Century* (Baltimore, 1933); H. Hastings, *Man and Beast in French Thought of the Eighteenth Century* (Baltimore, 1936). See also L. Rosenfield, *From Beast-Machine to Man-Machine* (New York, 1968 [1941]).

13 K. Kete, *The Beast in the Boudoir. Pet-keeping in Nineteenth-Century Paris* (Berkeley, Ca., 1994).

14 M. Agulhon, 'Le Sang des bêtes. Le problème de la protection des animaux en France au XIXe siècle', *Romantisme*, 31 (1981), pp. 81–109. See also V. Pelosse, 'Imaginaire social et protection de l'animal', *L'Homme*, 21 (1981), pp. 5–33, and 22 (1982) pp. 33–51.

15 See D. G. Charlton, *New Images of the Natural in France* (Cambridge, 1984), pp. 200–3.

16 My information is drawn from Hastings, *Man and Beast in French Thought of the Eighteenth Century*.

17 Information on Oswald is drawn from David V. Erdman, *Commerce des Lumières. John Oswald and the British in Paris, 1790–1793* (Columbia, 1986).

18 See P. Le Harivel, *Nicolas de Bonneville pré-romantique et révolutionnaire 1760–1828* (Strasbourg, 1923).

19 See G. Gengembre, *La Contre-révolution ou l'histoire désespérante* (Paris, 1989).

20 Subsequent page references are to J. de Maistre, *Les Soirées de Saint-Pétersbourg* (Paris, 1960).

21 Page references are to L. de Bonald, *Recherches philosophiques sur les premiers objets des connaissances morales* (Gand, 1830), vol. 2.

22 Page references are to Ménard, *L'Ami des bêtes ou le défenseur de ses presque semblables* (Paris, 1814). The author's Christian name is not given.

23 *Thalysie ou la nouvelle existence* (Paris, 1840–42, 3 vols). For a detailed consideration of Gleizes, see C. Crossley, 'Food and Salvation: Jean-Antoine Gleizes (1773–1843) and Vegetarianism', *Romance Studies*, 13 (1988), pp. 7–21, and 'Flowers, Fragrance and Woman: the Vegetarian vision of Jean-Antoine Gleizes (1773–1843)', *Nottingham French Studies*, 28 (1989), pp. 7–21.

24   See A. Viatte, *Les Sources occultes du romantisme* (Paris, 1925), vol. 2, pp. 154–8. For a challenging discussion of the links between socialism and esoteric currents of thought during the Romantic period see P. Muray, *Le Dix-neuvième siècle à travers les âges* (Paris, 1984).

25   See the abridged English version translated by C. H. Collyns and published as *Thalysie or the New Existence* (London, 1897).

26   A. Esquiros, 'Les excentriques de la littérature et de la science', *Revue des deux mondes* (1846), p. 487.

27   Page references are to A. Esquiros, *Paris ou les sciences, les institutions et les mœurs au XIXe siècle* (Paris, 1847), vol. 1. See A. Zielonka, *Alphonse Esquiros (1812–1876). A Study of his Works* (Paris, 1985).

28   For a discussion of the articles on *Meat* and *Butcher* which Jean Reynaud contributed to the *Encyclopédie nouvelle* (1835–41), see C. Crossley, 'A propos de l'animalité socialiste', in J. Birnberg (ed.), *Les Socialismes français* (Paris, 1995), pp. 23–34.

29   The followers of Fourier were carnivores. See for example M. Briancourt, *L'organisation du travail et l'association* (Paris, 1845), p. 102. For a discussion of Fourierist attitudes to hunting and eating animals, see C. Crossley, 'Alphonse Toussenel: la bataille contre l'animal', in P. Viallaneix and J. Ehrard (eds), *La Bataille, l'armée, la gloire* (Clermont-Ferrand, 1985), pp. 401–8.

30   See 'Jupille' in Champfleury [J. Fleury], *Les Excentriques* (Geneva, 1967 [1852]), pp. 191–211.

31   For the contribution of Isidore Geoffroy Saint-Hilaire, see M. A. Osborne, *Nature, the Exotic, and the Science of French Colonialism* (Bloomington, Ind., 1994).

32   J. Michelet, *Le Peuple*, ed. P. Viallaneix (Paris, 1974), p. 175.

33   See for example J. Michelet, *L'Oiseau* (Paris, 1867; 9th edition), p. 9, and *La Mer*, ed. J. Borie (Paris, 1983), p. 264.

34   For a fuller discussion of these issues, see Crossley, 'A propos de l'animal chez Michelet' to appear in the *Actes du colloque du bicentenaire de la naissance de Jules Michelet*.

35   Cf. C. Adams, *The Sexual Politics of Meat* (Cambridge, 1990).

36   A notable exception to the left-wing zoophilia was Proudhon in *De la justice dans la révolution et dans l'Eglise* (1860).

37   See A. Viatte, *Victor Hugo et les illuminés de son temps* (Montreal, 1942), pp. 180–96. See also C. Crossley, 'A propos de l'animal chez Leconte de Lisle, Victor Hugo et les fouriéristes', in E. Le Breton et al. (eds), *Victor Hugo and the Parnassians* (Oxford, 1985), pp. 73–88.

38   G. Clemenceau, 'Corrida', in *Le Grand Pan* (Paris, 1909), pp. 350–4.

39   See J. F. V. Keiger, *Raymond Poincaré* (Cambridge, 1997), pp. 11–14.

40   M. Flemming, *The Anarchist Way to Socialism: Elisée Reclus and Nineteenth-Century European Anarchism* (London, 1979), pp. 250–2.

41   Cf. N. Rupke (ed.), *Vivisection in Historical Perspective* (London, 1987).

42   P. Flourens, *Examen de la Phrénologie* (Paris, 1845; 2nd edition), pp. 22–6. See also by Flourens, *De l'instinct et de l'intelligence des animaux* (Paris, 1861).

43   J. Bois, *L'Eve nouvelle* (Paris, 1896), pp. 308–10.

44   *Revue illustrée des animaux*, 37 (1907), p. 963. The 'photograph' gives the distinct impression of being a 'scissors and paste' montage.

45   *Revue illustrée des animaux*, 65 (1910), p. 1476.

46 The promised recipes did not materialize on a regular basis.

47 *Revue illustrée des animaux*, 22 (1905), p. 561

48 References are to E. Bonnejoy, *Le Végétarisme et le régime végétarien rationnel: dogmatisme, histoire, pratique* (Paris, 1891).

49 During the *fin-de-siècle* period the followers of eccentric religious movements emphasized the need to respect all forms of animal life. See, for example, the case of the modern Essenes described by Jules Bois in *Les petites religions de Paris* (Paris, 1894), p. 170.

50 For a discussion of the question of violence and of the relationship between humans and animals see P. Leroux, *La Grève de Samarez*, ed. J.-P. Lacassagne (Paris, 1979 [1863]).

51 T. Zeldin, *France 1848–1945. Anxiety and Hypocrisy* (Oxford, 1981), p. 157.

52 Quotations are from Elisée Reclus, 'On Vegetarianism', *The Humane Review*, 1 (4) (1901), p. 321. See J. Cornuault, 'Reclus-des-animaux', *Les Cahiers Elisée Reclus*, 7 (1997), pp. 1–4.

53 See C. Crossley and I. Small (eds), *Studies in Anglo-French Cultural Relations: Imagining France* (London, 1988).

7

Religion and Nationalism in Late Nineteenth-century France

Brian Jenkins

The last quarter of the nineteenth century has often been identified as a crucial period in the development of French nationhood. In the four decades before the Great War, we are told, peasants became Frenchmen[1] and began to think of themselves as part of a national (as opposed to a regional or local) community. This process (the growth of national consciousness) is presented in linear terms as one of progressive integration, brought about by social and economic modernization, the communications revolution, rising levels of literacy, the practice of citizenship and the increasing influence of the central state.

However, at another level identification with the nation was shaped by values, by perceptions of what this national community represented in cultural and ideological terms.[2] Here we are in the force-field of nationalism proper, and this process was conflictual and dialectical. In this same period at the end of the nineteenth century, the question of what France 'stood for', of 'national identity', became deeply divisive, arguably for the first time.

In France as in other West European countries, the ideas of national sovereignty and self-government launched by the French Revolution were strongly associated in the first half of the nineteenth century with the popular democratic movement and political radicalism. The 'nation' was the dispossessed and disenfranchised 'people' ranged against the political and social elites that dominated the 'state'. Bonapartism had admittedly given these ideas a rather different gloss, which in retrospect may be seen to have prefigured the more conservative and authoritarian turn taken by nationalism at the end of the century. However, at least until the Second Empire (1851–70) the mainstream Catholic and royalist Right viewed Bonapartism and republicanism with equal suspicion, and saw 'nation' as a dangerous concept inviting mass political mobilization and threatening the social hierarchy on which the French Restoration and the European 'Old Order' was based.

*

This essay addresses a crucial period in the development of nationalism as a political ideology, when its amenability to a distinctively right-wing and eventually conservative agenda was conclusively demonstrated. This process, in the last quarter of the nineteenth century, was certainly not unique to France, but it was demonstrated there with peculiar clarity, perhaps because the country's revolutionary experience gave a particularly sharp focus to ideological divisions. In particular the essay will consider how the conflict between Church and state, between Catholicism and anticlericalism, which were defining characteristics of the Left/Right divide from the early years of the Third Republic until the 1905 Separation Law, became intertwined with the definition of rival national value-systems, especially in the period of the Dreyfus Affair.

The political dialectics of nationalism

The shift of nationalism from 'Left to Right' in this period was a widespread phenomenon in western Europe, and took place against a background of socially disruptive economic change, mass migratory movements and increasing international tensions, as Eric Hobsbawm has pointed out.[3] However, in the case of France the transition reflected a process of ideological realignment which cannot be adequately explained in these generic terms, and whose rationale is to be found in the political sphere. Central to this was the foundation of the Third Republic, the nature of the social and political consensus the new regime sought to construct, and by derivation the diverse and often contradictory forces that were ranged against it. The Boulangist movement of the late 1880s provided the first opportunity for these forces to coalesce in opposition to the parliamentary Republic, but it ultimately exposed the contradictions that divided them. The ensuing period saw a redefinition of the boundaries separating Left and Right, and these were revealed with symbolic force in the Dreyfus Affair, though the underlying contradictions were not entirely resolved, as we shall see.

From the perspective of 'nationalism', the first point to be made is that the Third Republic claimed finally to have given durable institutional expression to the republican concept of national self-government based on extensive formal citizen rights (adult male suffrage, parliamentary sovereignty, civil liberties). Whereas previously the radical democratic movement had deployed the concept of 'nation' *against* the prevailing 'state' forms, now supposedly France had a state that was fully representative of the nation. For Gambetta 'the revolution was over', the ideals of 1789 had finally been translated into practice. And for Clemenceau 'the revolution was a bloc': the Jacobin radicalism of 1792 had not 'transcended' the original aims of the Revolution, it had simply been a necessary phase in the defence of those aims. The liberal democracy instituted by the Third Republic was the culmination of this long process.

However, for those who had equated the republican movement with radical social reform, direct participative democracy and the patriotic-revolutionary struggle against 'reactionary' Europe (the Jacobin tradition of the *sans-culottes*, the 'June Days' of 1848 and more recently the Paris Commune), the new regime markedly failed to meet their aspirations. It was socially conservative, with a rural-peasant rather than urban-plebeian base, dominated by a bourgeois political class which hid behind the 'phoney' democracy of parliamentary institutions, and which was willing to accommodate a reactionary Germany rather than seeking revenge for defeat in the Franco-Prussian War. As Zeev Sternhell has argued,[4] the alienation of this constituency from the new Republic, and the fact that (for some of them at least) the populist-nationalist component of its ideology increasingly became its dynamic core, was the key conduit whereby nationalism moved from Left to Right in this period.

There was, of course, a very different response to the socially conservative character of the Republic in the growth of the mainstream socialist and labour movement from the 1880s, whose emphasis on international working-class solidarity provided an ideological counterpoint to the populist-nationalist overtones outlined above. Their shared hostility to the bourgeois Republic, and their overlapping social constituencies, led to opportunistic alliances in the period of Boulangism, when some socialist formations (especially those identified with Blanquism) briefly flirted with the likes of Déroulède's Ligue des Patriotes. Thereafter, but for a few renegade socialists like Rochefort,[5] the mainstream movement distanced itself from the excesses of the new nationalism, and indeed its Marxist internationalism increasingly provided a target for a much wider coalition of forces that would now begin to form around distinctly right-wing perceptions of the 'national interest'.

Royalists financed Boulanger, and their backing of a populist adventurer allegedly discredited the whole cause of monarchist restoration in France. The *Ralliement* of 1893, when the Catholic Church reversed its

1877 'outlawing' of the Republic, is seen as an acceptance of this failure. But the real significance of both events is a deeper one, the coming-to-terms of the conservative Catholic Right in France with the age of mass politics, the recognition that it was no longer possible to rely on popular deference, ignorance and passivity to sustain the traditional social order, and that a more aggressive ideological initiative was necessary to limit the inroads of democracy. The new version of populist nationalism, equally critical of parliamentary democracy and of socialist internationalism, flowed into this vacuum. In its 'pure' form it was, of course, anti-bourgeois and socially radical. But its impatience with liberal democracy, its emphasis on 'national' rather than 'class' solidarities, were amenable to a rather different kind of project. One which would oppose the Republic less because it was 'parliamentary' than because it gave the masses access to the political process. And which would oppose socialist internationalism less because it was 'internationalist' than because it was 'socialist'. The road was open, in other words, to a conservative reshaping of the new nationalism, which would soon lose most of the traces of its left-wing origins. And in this process the religious dimension was to assume crucial importance, as we shall see later.

Variants of French nationalism in the late nineteenth century: the 'Right'

At this stage it might be useful to reiterate the dialectics of the process of ideological realignment outlined above. The consensus politics of the Third Republic abandoned the social radicalism and Jacobin nationalism (France as the democratic liberator of Europe) of the earlier republican movement. The new regime had after all been founded on the brutal suppression of the Paris Commune, which had combined the themes of social emancipation and national liberation. Some of these disillusioned radical elements were mobilized against the Republic more by its acceptance of 'national humiliation' than by its neglect of social reform, and this was the basis of a new unconditional or 'integral' nationalism which progressively moved away from its left-wing Jacobin origins to assume racist and aggressively xenophobic overtones. The Boulangist movement, with its emphasis on revenge against Germany, provided an early opportunity for this new nationalism to attract a wider popular base, usually identified as urban-plebeian rather than proletarian, but undoubtedly attracting workers as well as the more classic petit-bourgeois clientele.

For the purposes of this essay however, it is the convergence of this new nationalism with the ideology of the conservative Catholic Right which arouses the greatest interest. After all, the latter tradition had always been deeply suspicious of the revolutionary appeal of popular nationalism, and royalists of both persuasions (Legitimist and Orleanist) had been advocates of peace and an early armistice with Prussia in 1870 and of suppression of the 'patriotic' Paris Commune in 1871. Furthermore, notwithstanding the Gallican strand in French Catholicism, membership of the international community of Catholic Christendom was not easily compatible with the *'France d'abord'* ['France first'] of integral nationalism. Convergence in the Boulangist movement may easily be seen as an opportunistic alliance which did neither side any good, a coalition of opposites which threw doubt on the social radicalism of the new nationalists, and which discredited the royalists as desperate losers. But, as indicated earlier, a more profound ideological osmosis was taking place, which allowed the new nationalists to escape from political marginalization and which provided the conservative Catholic Right with a new basis on which to seek popular support.

The Republic's anticlerical policies (albeit less far-reaching than those demanded by radical republicans) involved a purge of Catholic influence in education, the civil service and, crucially, the Army. Ironically, of course, General Boulanger himself had been appointed to 'republicanize' the upper echelons of the military establishment, a further justification for the charge of cynicism levelled against the royalist campaign in his favour. However, a decade later in the Dreyfus Affair the contradictions appeared to have been resolved. If for the new nationalists, unconditional support for the Army General Staff had become synonymous with defence of the 'national interest', for the conservative Catholic Right the same institution had become synonymous with the elitist traditions that needed to be defended. As the prospect of a restoration of the monarchy receded, so 'nationalism' filled the vacuum left by 'royalism'.

The counterpoint to all this was the ideological 'space' vacated by the Left's abandonment of radical democratic nationalism. In crude terms it may be said that this allowed the Right to occupy this terrain, to rework it ideologically, and to create a new populist constituency around notions of ethnicity and organic national identity (as opposed to its traditional appeal to social deference). Of course, this is not to say that the Third Republic was indifferent to the idea of 'nation'. Far from it. Through the agencies of compulsory primary education, military service, mass suffrage, not to mention symbolic imagery (the tricolour, the *Marseillaise*, the 14 July holiday, the cult of the 'lost provinces' of Alsace-Lorraine), the regime is widely seen as the key historical architect of French nationhood. However, this 'state-sponsored' version was geared to building a very broad consensus (as Suzanne Citron has argued, it was not even particularly 'republican', and often invoked the theme of '*la France éternelle*'),[6] and more crucially the Republic was cautiously *attentiste* in its attitude to *revanche* against Germany and the recovery of the lost provinces. This exposed the Republic to 'outbidding' by a more virulent counter-nationalism, which as we have seen eventually took shape on its Right.

Variants of French nationalism in the late nineteenth century: the 'Left'

At the same time, on the Republic's Left the emerging socialist and labour movement increasingly distanced itself from its flirtation with populist nationalism during the Boulanger Affair to adopt a more rigorous class-based internationalist stance in the 1890s. One effect of this was to provide the new nationalists with the counter-image of an 'internal alien', revolutionary socialists or anarchists in league with their German comrades in the Second International and dedicated to undermining France's defences through their fierce anti-militarism. And given the Republic's allegedly half-hearted commitment to the national interest, this could be rolled together into a composite picture of the whole 'Left' (both radical republican and socialist) as suspect in patriotic terms, in thrall to Jews, Protestants, freemasons and other cosmopolitan 'anti-national' forces. Indeed, integral nationalism saw parliamentary democracy itself as inherently weak and divisive, and therefore subversive of the organic unity of the nation.

It is, of course, true that working-class solidarity implies collaboration across national frontiers, and to the extent that industrial workers were often made to feel 'excluded' from the national community, the slogans of socialist internationalism were not without appeal. The Third Republic's record on labour and welfare reform before 1914 lagged well behind that of Germany and Britain, and indeed the regime's social constituency seemed to embrace the peasantry and the middle classes at the expense of the urban proletariat. However, there were also significant obstacles to the development of a genuinely 'international' working-class consciousness. Ordinary workers, unlike the socialist and trade union leadership, had limited cultural and geographical horizons. The issues that preoccupied them required mediation at the level of the nation-state, their organizations (whether reformist or revolutionary in intent) were geared primarily to achieving change at this level, the traditions and memories that inspired them were largely 'national' in character. This is not to say that a sense of involvement in a distinctive 'national' community is incompatible with transnational class solidarities. Social identities are 'multiple', and it is only in special circumstances that they conflict with one another and that difficult choices have to be made. The First World War was to create precisely such tensions between loyalties to 'class' and to 'nation'. However, for the purposes of this essay it needs to be recognized that in the intervening period an attempt had been made to integrate the domestic socialist and workers' movement *ideologically* into the broad sweep of French post-revolutionary history. It was the Dreyfus Affair that provided the opportunity for this enterprise, and the upshot was that the Left/Right polarization that occurred at the turn of the century was not between 'nationalism and internationalism', nor indeed between 'bourgeoisie and proletariat', but between two rival versions of French national identity (indeed, between two models of 'nationalism', as we shall argue).

This configuration of political forces, with a clear line of demarcation around Left and Right, was not permanent. Indeed, it was exceptional and in more 'normal' times the image of the pragmatic 'centre' versus the 'extremes' fits reality far better. Only at the time of the Popular Front victory of 1936, or during the Liberation shake-out of Resistance versus Collaboration in 1944–45, is the Dreyfus scenario replicated with similar idealistic sharpness. But none the less it remained (and remains) a constant reference-point, even when the complexities of political circumstance did not allow it to take shape.

When Jean Jaurès argued, and finally with success, that the French socialist and workers' movement should intervene and take a position in the Dreyfus Affair, he symbolically situated the movement in the broad radical-democratic tradition which reconnected with 1792, 1848 and 1871. There was both idealism and relativism in this enterprise. First of all it invoked the notion that there was an ideal Republic, more socially radical and more completely democratic than anything that had yet been achieved, which had inspired the French revolutionary struggle ever since its Enlightenment beginnings and whose ultimate translation into practice was indeed synonymous with the objectives of socialism.

But it also invoked and endorsed the achievements of the liberal-democratic Republic, recognized that relative to their era these were 'progressive', and that the (albeit limited) recognition of civil liberties and freedom of association (notably the legalization of trade unions), adult male suffrage, free compulsory primary education, and restrictions on the political influence of the Catholic Church, provided the socialist and workers' movement with valuable instruments for the achievement of proletarian emancipation. Jaurès's strategy was not conclusively successful. Revolutionary Marxist and revolutionary syndicalist circles remained deeply suspicious of a process which they saw as reformist, which necessarily involved inter-class alliances and compromises with the 'bourgeois' Republic. So when French socialists finally united in a single party, Section française de l'Internationale ouvrière (SFIO), in 1905, they endorsed the Second International's position of non-participation by socialist politicians in governments led by non-socialists. And at the CGT (Confédération générale du travail) Congress of 1906, French trade unionists voted not to affiliate to the socialist SFIO, thereby preserving their autonomy as a distinctively working-class movement free from the machinations of parliamentary and electoral politics.

However, in as far as the Socialist Party and indeed the trade unions were forced to operate largely on national terrain and to seek to influence national government, a process of 'integration' of the working class was underway, though it would be a long time before substantive social reform would entirely dispel the feeling of alienation and 'exclusion' from the national community.[7] A sense of 'citizenship' was beginning to take shape, and as Eric Hobsbawm has argued,[8] this would be sufficient in 1914 for the workers of different nations to believe that the advantages they had gained were superior to those enjoyed elsewhere, and were worth defending in the event of war. In other words, through what Hobsbawm calls 'state patriotism', workers were being incorporated into the 'nation' despite the internationalist logic of class-based solidarities. The human tragedy that ensued in the First World War was something that Jaurès had struggled to avoid, and which he never witnessed due to his assassination on the eve of hostilities. He could not have foreseen that his attempt to reconcile socialism with the French radical-democratic tradition would eventually allow Left patriotism to be drawn into the maelstrom of unconditional 'nationalism'.

Religion and nationalism

The remainder of this essay examines the significant role played by religious issues, the conflict between Church and state, between Catholicism and anticlericalism, in the demarcation of left- and right-wing conceptions of nation and nationhood in the crucial period from the 1880s until 1914. There is nothing new at all in recognizing that the first 35 years of the Third Republic's life, and in particular the period from the Jules Ferry education reforms of the early 1880s until the law separating Church and state in 1905, were dominated by the regime's efforts to reduce the 'anti-democratic' political influence of the Catholic Church. And the links between this process and the formation of rival national value systems at the end of the last century has also been recognized to a certain extent.[9] This essay will suggest, however, that the two processes were more decisively linked than has perhaps hitherto been acknowledged, and indeed that the religious question more than any other defined the battle-lines.

In advancing this hypothesis, I am seeking to remedy what I now recognize as a significant lacuna in my book on the history of French nationalism,[10] which was justly criticized for its relative neglect of the religious dimension and its rather undifferentiated view of French Catholicism. The significance of religion in nation-formation has recently been underlined by Linda Colley's emphasis on the importance of Protestantism in the construction of British national identity in the eighteenth century.[11] In contemporary resurgent nationalisms, religious affiliation continues to fuel ethnic differentiation, not least in France itself.

In responding to such developments both in the historiography and in the contemporary evolution of nationalism, I may be accused of having gone too far in my attempts to redress the balance. However, in the context that concerns us here this recognition of the centrality of the religious issue in turn-of-the-century France is triggered by what I hope are significant observations. The Dreyfus Affair was ostensibly about the role of the Army, and its emergence on the Right as a counter-symbol of the national interest in opposition to the civilian authorities of the 'Republic' remained a classic reference point in the definition of rival national value systems through the period of the 1930s and the Vichy regime, to the 'Algérie française' conflicts of the late 1950s and early 1960s.

However, the aftermath of the Dreyfus Affair produced, within the space of five years, the *Délégation des gauches* uniting Socialists and Radicals in parliament largely around the theme of anticlericalism, the Emile Combes ministry and its notorious purge of Catholics in civil and military administration (the '*Affaire des fiches*'), the Separation of Church and state in 1905, and indeed the emergence of the Action Française as the 'great didactic centre of a Right otherwise devoid of doctrine',[12] a movement which, despite the atheism of its leader Charles Maurras, espoused Catholicism as a necessary ingredient of its 'integral nationalism'.

In other words, was it the religious issue (rather than questions of class, militarism or allegiance to the Third Republic) which crucially divided Dreyfusards and anti-Dreyfusards? Was it the Catholicism (and attendant anti-Semitism) of the Army General Staff which provided the focal point for the anti-Dreyfusard coalition, and anticlericalism which provided the cement for those who rallied to Dreyfus, to the notion of an 'ideal Republic' which transcended what the Third Republic had so far achieved? It is conventional enough to acknowledge the conflict between the conservative Catholic and liberal free-thinking sections of the bourgeoisie as a key ingredient of the Dreyfus Affair. It is more contentious to see religion as the main agency which defined the rival national value systems, and which drew in the mass support of populist nationalists on the one side and socialists on the other.

This hypothesis would seem to justify some further investigation of previous developments. As suggested earlier, the starting point would seem to be the gradual 'osmosis' on the Right which brought conservative opponents of the emerging Third Republic in the 1870s into contact with the more plebeian Jacobin clientele who rejected the regime for very different reasons, but whose nationalism eventually provided a point of convergence. The Boulanger episode is thus crucial for understanding how this nationalism fused with Catholicism, and largely through the vehicle of anti-Semitism in the 1890s eventually created a conservative synthesis, which served as a basis for opposition to both the liberal and socialist Left.[13]

Similarly, largely in response to the above process, with the Boulanger episode again a decisive turning-point, the Left found in anticlericalism the key unifying theme which allowed it to defend the republican ideal, not only against those who for various reasons opposed the very principles of representative democracy but also against those who accepted its form rather than its substance, the so-called '*républicains de résignation*'. In short, against the combined forces of the conservative, populist and moderate Right.

'Osmosis' on the Right

It is often said that the Third Republic came into being fortuitously. Founded on defeat in the Franco-Prussian War, the fall of the Second Empire and the ruthless suppression of the Paris Commune, the regime was initially dominated by the monarchist majority of Legitimists and Orleanists who had supported the cause of peace and an early armistice in 1870 (popular outside Paris). The 'Republic' was seen as provisional, and only survived because the rival royalist factions were unable to agree on who should occupy the throne and what sort of monarchy it should be. This provided the opportunity for the regime to consolidate itself: its social conservatism reassured those who had hitherto identified Republics with revolutionary upheaval, and its political radicalism was directed mainly against a safe target, the most reactionary and clericalist elements in the Catholic hierarchy. Moderate Orleanist royalists increasingly resigned themselves to the republican form of government and abandoned root-and-branch opposition in favour of political influence.

Fundamentalist opposition to the Republic initially took the form of Catholic revivalism in the 1870s, which presented defeat in the Franco-Prussian War as divine punishment for the nation's godless transgressions since the 1848 Revolution. Despite the political impasse of these years, statistical indicators reveal a new dynamism in recruitment to seminaries, a renewed clerical militancy and a new fervour among the faithful (for instance the cult of Marianism), albeit against a background of declining religious observance. To borrow a phrase used by Hippolyte Taine in 1891, 'faith is increasing in the restricted group and declining in the large group',[14] or, put another way, Catholicism regained in intensity what it had lost in extent. So while the possibility of royalist restoration faded (and this was eventually conceded by the *Ralliement* of 1893), Catholicism remained a significant conservative resource. Hitherto it had not been tapped as an overtly ideological instrument of mass mobilization; this had not been seen as necessary in a society where religious faith still remained well established and where it was still possible to an extent to rely on associated patterns of social deference and political passivity. The advent and then the consolidation of a regime based on representative democracy forced the conservative Right to face up to the realities of mass politics, and to 'politicize' religious belief in an unprecedented way.

The Republic, once it had seen off the royalist challenge of the 1870s, decided to take on a Catholic establishment that was regarded as the enemy of democracy. Its hold on education was seen as an obstacle to the spread of democratic values, and its influence in the civil service and Army made it a potentially subversive force within the state itself. Ferry's educational reforms of the early 1880s (free, compulsory and *secular* primary education for all), though relatively moderate given the demands of more radical anticlericalism, were opposed by conservative

Catholics as a campaign of dechristianization. This decisively politicized the fault-line which had divided Catholics and anticlericals ever
since the Revolution, and for a period of at least 25 years the religious
issue became a socially pervasive bone of contention, the acid test of
political loyalty, the key fissure between Left and Right.

The Boulanger Affair of the late 1880s had provided an opportunity
for an opportunistic alliance between royalists and the new nationalists,
drawn together by joint opposition to the liberal parliamentary Third
Republic.[15] That proved ephemeral and temporarily discredited both
partners, but what followed in the 1890s, and culminated in the Dreyfus
Affair, was a much more substantial ideological convergence. This
involved the fusion of not royalism but a conservative brand of Catholicism with an exclusivist form of nationalism, which after Boulangism
progressively shed its Jacobin origins and socialist associations to
become a distinctively right-wing phenomenon. And the main vehicle
for this synthesis was a shared anti-Semitism, whose religious foundations were expounded by the most widely-read Catholic newspaper of
the 1880s, *La Croix*, at the same time as Edouard Drumont's more
distinctly biological version was given its first airing through the publication of his *La France juive* in 1886 and the appearance of his newspaper
La Libre Parole in 1892.[16]

Anti-Semitism was not a major ingredient of Boulangism, though it
was part of the sub-culture of the movement. This was even true of
some of the socialist groups which supported Boulanger (the figure of
Rochefort is emblematic in this respect), where the Jew was crudely
equated with capitalism. The mainstream socialist movement certainly
distanced itself from this racism in the 1890s, indeed the whole
Boulanger episode became a bad memory as the movement abandoned
populist nationalism for a more rigorous class-based analysis with internationalist overtones. The electoral breakthrough of 1893, which
admitted 40 socialist deputies to parliament, also paved the way for
accommodation with the Republic, as we shall see in the next section.

This realignment created political space for the ideological convergence in the 1890s of conservative Catholicism with the new nationalism now purged of its leftist associations. And the religious question
was central to the mechanics of the process. The regime's anticlericalism
had revived a well-established conspiracy theory which saw the Republic as controlled by occult forces, now more visible than ever before in
the well-organized anticlerical influence of freemasonry. The masonic
lodges were seen by conservative Catholics as the agency of dechristianization, and by the new nationalists as a vehicle for liberal, bourgeois,

cosmopolitan interests – in short, an identikit picture of everything that was 'anti-national'. And anti-Semitism was a key ingredient of this scenario. Freemasonry was supposedly the natural habitat of the Jew (and indeed the Protestant), its rituals were presented as Judaic in origin, and its network provided a sphere of political and social influence which, according to the stereotypical image, attracted the parasitic Jew as 'honey attracts the bee'.[17]

This composite picture of the Republic as the prisoner of occult forces was a powerful negative image capable of transcending the ideological differences between conservative Catholicism and populist nationalism.[18] It was systematically developed by key intellectual figures like Maurice Barrès and Charles Maurras, and crudely popularized by Edouard Drumont. When the Dreyfus Affair broke, it provided a powerful and exemplary metaphor. An (assimilated French) Jew had betrayed his country, thus demonstrating that Jews could never become true patriots. As Barrès had argued, nationality was in the blood, it could not be acquired by residence: it was culturally predetermined, not something that could be freely chosen by individuals as the Revolution had pretended. And when the Left eventually rallied to Dreyfus, this simply confirmed that the Republic was indeed controlled by alien cosmopolitan forces intent on undermining the nation. The true representatives of the national interest were not the political class of the hated Republic, but those institutions that embodied *la France éternelle*, the Army and the Catholic Church.

This was perhaps a surprising destination for the populist nationalists, above all for Barrès who endorsed the French Revolution for its nationalist vitalism and had formerly not hesitated to describe himself as a socialist, but also for Maurras whose personal atheism was cynically at odds with the instrumental Catholicism of his movement, the Action Française. But the dynamics of the process of political realignment were more powerful than considerations of ideological integrity. The blend of nationalism and Catholicism provided a new rationale for opposition to the Republic, one which offered conservative elites a popular base and radical nationalists a degree of political leverage, and in the end the Dreyfus Affair was the catalyst for the convergence of the entire French Right. As Pierre Pierrard suggests in his formula '*Catholique = anti-Dreyfusard*',[19] the Affair mobilized not only conservative Catholics formed in the 'tridentine' traditions of the Church hierarchy, but also those identified with the more recent experiments in social Catholicism and Christian democracy (though not without exception, as proved by the pro-Dreyfus position of figures like Charles Péguy). And in the end it even cut a swathe through the parliamentary ranks of the moderate Right, those *républicains de résignation* who had gradually come to accept the regime's institutions and its anticlerical reforms.

These so-called *'Opportunistes'* (those in the 'Orleanist' tradition who had rallied to the Republic in the preceding two decades) largely opposed any reopening of the Dreyfus case, and behind Jules Méline were drawn into the camp of the anti-Dreyfusards. This was more a reaction to what was seen as the social and political 'threat' of the emerging left-wing Dreyfusard coalition than any endorsement of the verdict against Dreyfus, and the bourgeois fear of socialism was at least as important as fears of rampant anticlericalism. But by 1898–99 the issue of Dreyfus's guilt or innocence had anyway become secondary to a debate over the national interest. For the anti-Dreyfusards any reopening of the case would discredit the Army, the key symbol of the nation, and play into the hands of those who, behind all their rhetoric about justice and human rights, were cynically manipulating the Affair for their own political advantage.

As we have said before, the Right/Left polarization of the Dreyfus Affair would seldom be repeated in Republican politics, and this right-wing brand of nationalism still contained internal contradictions and tensions between its more conservative and more populist elements. None the less, it became embedded in the mind-set of the French Right for several generations, and was arguably only dislodged when the institutional power of Church and Army decisively waned in the 1960s. At one level, it has been seen in terms of the deterministic and organicist version of nationhood which Barrès and others 'borrowed' from German traditions at the turn of the century, a version which emphasized ethnicity (*droit du sang*) rather than the civic rights bestowed by residence (*droit du sol*), and this of course has significant echoes in contemporary France.

But it should not be forgotten that this version of nationalism was mobilized as much against 'French' Protestants, liberals and socialists as against Jews and foreigners. It was based on a specific *political* identification of what constituted true French values, and these were equated at the time with a *Catholic* France. Popular anti-Semitism and the myth of masonic conspiracy, powerful *'anti-bourgeois'* symbols, thus led irrevocably to an identification with Church and Army, profoundly conservative institutions which ironically were seen by many *bourgeois* as ramparts against the advancing tide of democracy and socialism. This 'political instrumentality' of nationalism equally has contemporary echoes: the Front national's anti-immigrant campaign also allows it to taint by association the entire liberal political establishment whose *laxisme* allegedly demonstrates their betrayal of the 'national interest'. And the Front's nostalgic appeal to conservative Catholic values demonstrates the continuing hold of the synthesis engineered 100 years ago.

'Osmosis' on the Left

The processes that paved the way for the parallel emergence of a rival national value system on the Left have already been referred to in passing above, and in conclusion we will summarize these with a little further elaboration. The elements of the eventual Dreyfusard coalition were essentially the radical republicans (soon to become France's first nationwide organized political party in 1901), the mainstream socialist movement (still divided into rival factions until 1905) and the small rump of 'pro-Dreyfus' *Opportunistes* behind Waldeck Rousseau (who became prime minister after the Dreyfusard election victory of 1899). Collectively they may be described as *républicains de foi*, as supporters of '*la République idéale*', and in the context of the time they may legitimately be equated with the 'Left', though evidently this definition is teleological.

As we have seen, this 'Left' had not yet taken shape ten years earlier in the Boulanger Affair. Some socialists flirted with Boulangism, seizing the opportunity to undermine the 'bourgeois' Republic and to obtain the promised social reforms, but also thereby confirming the continuing influence in socialist circles of *Communard* populist nationalism, and indeed of residual anti-Semitism. Some radical-republicans too were tempted by the promise of constitutional reform (abolition of the conservative Senate) and by Boulanger's reputation as an anticlerical (he had been promoted to General to 'republicanize' the Army General Staff).

Both socialists and radicals quickly turned their back on the Boulanger episode as an embarrassing interlude, but this still did not presage the realignment which would take place ten years later. On the one hand, the Radicals abandoned the campaign for constitutional revision, moderated their social reformism, and began to build the largely rural rather than urban electoral base that would soon make them France's biggest and most influential parliamentary party. On the other hand, the socialists increasingly adopted the class-based discourse of the Marxist Second International, which seemed to preclude alliances with the 'bourgeois' radicals and to renounce the French revolutionary tradition in favour of the language of proletarian internationalism.

Other processes were however preparing the eventual sea-change. The 1890s were dominated politically by conservative Republicans (the

Opportunistes) and though the Radicals had regrouped around the theme of anticlericalism and increased their parliamentary strength, they were still excluded from government. Of equal significance was the socialist breakthrough at the 1893 general election (40 deputies), which raised the profile of parliamentary and electoral politics in the movement, gave socialists experience of working within the Republic's institutions (not least at local level following their success in the 1896 municipal elections), and inevitably involved collaboration with political allies (primarily the Radicals). This was admittedly a vexed issue which divided reformists, Marxists and syndicalists. And at the popular level, historians still disagree over the extent to which the 1890s saw the beginnings of the 'integration' of the French working class into the citizens' Republic.[20] However, the process had undoubtedly opened up a new political option, which Jean Jaurès was to seize with undoubted effect when the Dreyfus Affair was decisively politicized by Emile Zola's famous letter 'J'accuse!' in January 1898.

The convergence between Radicals and Socialists in the Dreyfus Affair thus opened the prospect of a left-wing alliance which would create a new parliamentary majority following the elections of 1898. Though in the event the Socialists refused ministerial participation, the creation of the parliamentary *bloc des gauches* inaugurated a period that would become known as '*la République radicale*' (1898–1914) and which launched the Radical Party as a near-indispensable component of governing coalitions for the next 40 years or so. However, this convergence was more than an electoral convenience, it had significant ideological implications. The Dreyfusard theme of justice and human rights relocated the Socialists in the broad tradition of French radical republicanism, but more specifically, because the enemies of Dreyfus were identified with the Catholic and military establishment, his cause found expression in an intensified anticlericalism.[21]

As we have seen, anticlericalism had by this time become the principal 'radical' plank of a socially moderate Radical Party. But the willingness of Socialists to endorse it should not be seen as a shallow compromise. After all, they were steeped in the historical culture of the French Left which had always regarded the Catholic Church as a key agent of *la réaction*. The popular influence of Catholicism had, ever since the Revolution, been regarded as a primary obstacle to the spread of democratic values which were themselves the precondition of socialism. The Separation of Church and state in 1905 was greeted by leading Socialists with ringingly poetic phrases that carried the force of conviction – 'we have extinguished lights in the sky that will never be rekindled' (René Viviani) and 'we have interrupted the old lullaby that nursed human misery, and humanity has awoken racked with tears' (Jean Jaurès).

The belief that 'religion is the opium of the people' (Karl Marx), that it continued to provide the outer defences of a social order of which capitalism had become the core, had of course nothing specifically 'French' about it. But it had a special resonance in a *Catholic* country where the revolutionary heritage had placed the struggle between Church and state at centre-stage for a whole century, where the religious establishment in the late nineteenth century had identified so strongly with the alliance between throne and altar, and where Leo XIII's liberal papal encyclical of 1891 (*Rerum Novarum*) and his subsequent efforts to persuade French Catholics to accept the Republic had eventually had so little effect. Anticlericalism was thus perceived as central to the broader issue of human rights raised by the Dreyfus Affair, a key element in the revolutionary democratic tradition which Jaurès now invoked as the true inheritance of French socialism. And in this context it is easier to understand why Jules Guesde, the representative of orthodox Marxism in the socialist movement and Jaurès's principal rival, could himself hail Zola's letter as 'the greatest revolutionary act of the century'.

In conclusion

The Dreyfus Affair was for the Left, as it was for the Right, a defining moment which became an ideological reference point, even though the vagaries of politics seldom again reproduced such unequivocal polarization. It bequeathed to the Left a national value system which in exceptional circumstances could take the form of a genuine nationalism, most notably of course in the liberation struggle of the Left Resistance under the Nazi occupation and the Vichy regime. Anticlericalism would remain an important element in that Left sub-culture, and it might be argued that the role of the Catholic Church between 1940 and 1944 justified that vigilance. But we are, of course, dealing with a dialectical process, and others might argue that the Left's hostility to Catholicism itself provoked the reactions it feared, in the style of a self-fulfilling prophecy.

It is certainly true that after the 1905 Law of Separation the religious question might have been expected to disappear from the political agenda, and that at least one of the things that kept it alive was the fact that anticlericalism remained the only substantive issue on which Socialists and Radicals could agree. By the interwar period it arose only as a secondary matter (the extension of the separation laws to the regained provinces of Alsace-Lorraine, educational reform) and frequently appeared to be artificially stimulated (the continuing need to 'purge' ministries – such as the Interior – of any remaining clerical influence). This was no basis for the traditional left alliance in a climate where economic, social and foreign policy was increasingly dominant, issues on which Radicals and Socialists fatally disagreed and which condemned their cooperation to failure. The pivotal position of the Radicals as a party of government in diverse possible coalitions, the emergence of the Communist Party which pulled the Socialists competitively to the Left, were additional complicating factors which undermined the viability of this 'republican' Left.

None the less, however irrelevant it increasingly appeared in terms of policy, the religious question remained entrenched in historical memory, an acid test of political allegiance which equated Catholics with the Right and anticlericals with the Left long after the circumstances which had originally set them apart. The Occupation left an ambiguous legacy – while the Catholic establishment was embroiled with Vichy's Révolution nationale, the Catholic Resistance spawned a range of progressive movements in the postwar period from the Christian-democrat MRP to the *marxisant* worker-priest experiment, from the radical Catholic peasant youth movement (JAC) to the increasingly leftist Catholic trade union confederation (CFTC, later CFDT).[22] However, under the Fourth Republic religion continued to be a dividing line when dictated by political convenience, as over the contractual obligations of Catholic schools. The same issue rumbled on into the Fifth Republic, was sufficient to forestall a centre-Left coalition including the Christian-democrats in the early 1960s, and as late as the 1980s was revived by the fiercely anticlerical education union FEN, under whose influence the Socialist education minister Alain Savary attempted to force through an ill-fated Bill designed to complete the incorporation of Catholic schools into the state system. Even today, the party system reflects the old divide – there seems otherwise little reason why the socially still relatively progressive Christian-democrat CDS should be aligned with the Right, or why the Socialists should have continued for so long to treat the dwindling remnants of the Radical Party as preferred allies.

Well into the 1960s the political map of Left/Right electoral affiliation coincided strongly with the map of religious observance. Even today the extremes of devotion and atheism correlate more strongly than other indicators with voting patterns. But none the less the umbilical cord appears to have been finally broken. Dechristianization has taken its toll,[23] non-practising Catholics are equally likely to vote for Left or

Right, the Socialist Party is firmly entrenched in former Catholic enclaves such as Brittany, and the Front national enjoys greater success among the irreligious than among the devout.

At one level, then, the religious question is no longer a key ingredient of *la guerre franco-française*. Indeed, the last great episode in the clash of rival national value systems, namely the Occupation period, did not give primacy to the issue. Arguably the Resistance-Liberation myth provided a new basis for 'Left nationalism' (George Ross describes the 1972 Common Programme of the Left as Chapter Two of the Resistance Charter),[24] and anticlericalism was very much secondary to the themes of economic, social and democratic 'national' regeneration. After all, the 'tripartite' governments that enacted the great Liberation reforms included the Christian-democrat MRP alongside the Socialists and the Communists. As for the Right, until the emergence of the Front national the legacy of Pétain was largely disowned, and eventually De Gaulle was embraced as the incarnation of a more respectable mythology. Gaullist nationalism was synthetic in character, incorporating both Right and Left traditions, but it was modernizing in its thrust, abandoned colonialism, put an end to the political influence of the Army, and was largely silent on the religious question which indeed began to dwindle in significance precisely during the decade of De Gaulle's presidency.

It is ironic, therefore, that the religious question has returned to haunt the present at a time when so many commentators announce that *la guerre franco-française* itself is over, or in the words of François Furet (echoing Gambetta a century earlier) that 'la Révolution française est terminée' [the French Revolution has finished]. This is perhaps symptomatic of an age where national identity is perceived as threatened by the forces of globalization and mass migration, the need to rediscover the past is a natural corollary of this national introspection. The willingness finally to unravel the myths surrounding the Occupation period has exposed among other things the complicity of the Catholic establishment in the crimes of the Vichy regime. At the same time, it has been tempting for a Socialist Party that has abandoned traditional socialism to return to its 'republican' roots, to transfer its radicalism from the economic into the political domain, and thereby to revive anticlericalism albeit in the muted form of Savary's education bill or Chevènement's outmoded attempt to reincarnate Jules Ferry when he replaced Savary in 1984.

The religious question has cast a long shadow, and its legacy has recently resurfaced in a much more significant way. Anticlericalism shaped education more than any other institution, and the secular traditions of *l'école républicaine* have now been challenged from an unanticipated quarter, namely France's Muslim community, over the right of girls to wear the *foulard* (an 'ostentatious' religious symbol) in school. This seemingly innocuous issue has become a powerful metaphor in the debate about the increasingly multicultural character of French society, and its compatibility with French republican traditions of assimilation, equal citizenship, and national homogeneity. As the architects of the secular Republic, but at the same time the presumed champions of anti-racism, the Left has been torn by this issue. Perhaps it is finally time for the Left to re-examine its 'national value system', to abandon its folk memories, and to draw a line under the Dreyfus Affair.

Acknowledgement

In 1979 Douglas Johnson was the external examiner of my PhD thesis, and I remember with gratitude how he turned the ordeal of the viva into a relaxed and rewarding experience. Ever since, he has been generous in his support of my efforts both as a journal editor and as an author, and I have greatly appreciated his thoughtful advice and wise comment. It is a privilege to be invited to contribute to this volume in his honour, and I hope that my choice of topic, which deals with the period surrounding the Dreyfus Affair, is an appropriate one given his own expertise on the subject; see D. Johnson, *The Dreyfus Affair* (London, 1965).

Notes

1 E. Weber, *Peasants into Frenchmen: The Modernization of Rural France 1870–1914* (Stanford, 1976).
2 B. Jenkins, *Nationalism in France: Class and Nation since 1789* (London, 1990).
3 E. Hobsbawm, *Nations and Nationalism since 1780* (Cambridge, 1990), p. 109.
4 Z. Sternhell, 'Paul Déroulède and the origins of modern French nationalism', in J. C. Cairns (ed.), *Contemporary France: Illusion, Conflict and Regeneration* (New York, 1978), pp. 46–70.
5 E. Cahm, 'Socialism and the nationalist movement in France at the time of the Dreyfus Affair', in E. Cahm and V. Fisera (eds), *Socialism and Nationalism in Contemporary Europe (1848–1945)*, 3 vols (Nottingham, 1978–80), vol. 2 (1979), pp. 48–64.
6 S. Citron, *Le Mythe national: l'histoire de France en question* (Paris, 1987), p. 68.
7 R. Magraw, *The Bourgeois Century: France 1815–1914* (London, 1983), pp. 283–317.
8 Hobsbawm, *Nations and Nationalism*, p. 89.
9 S. Hoffmann, 'La nation: pour quoi faire?', in S. Hoffmann, *Essais sur la France: déclin ou renouveau?* (Paris, 1974), pp. 437–83.
10 Jenkins, *Nationalism in France*.
11 L. Colley, *Britons: Forging the Nation 1707–1837* (New Haven, 1992).

12 H. Rogger and E. Weber, *The European Right: A Historical Profile* (London, 1965), p. 97.

13 R. Tombs (ed.), *Nationhood and Nationalism in France: From Boulangism to the Great War 1889–1918* (London, 1991).

14 R. Gibson, *A Social History of French Catholicism 1789–1914* (London, 1989), p. 231.

15 B. Jenkins, 'In Search of a "Fascist historiography" in France between the wars', *Journal of Area Studies*, no. 10 (Autumn 1997), pp. 48–65.

16 P. Sorlin, *La Croix et les juifs* (Paris, 1967), pp. 9–10.

17 P. Pierrard, *Juifs et catholiques français* (Paris, 1970), p. 30.

18 C. Amalvi, 'Nationalist responses to the Revolution', in Tombs, *Nationhood and Nationalism in France*, pp. 39–49.

19 Pierrard, *Juifs et catholiques français*, p. 81.

20 Magraw, *The Bourgeois Century*, pp. 283–317.

21 Just as the Right had its 'masonic' conspiracy theory, so the Left identified a 'Jesuit' conspiracy to place those educated in its elite seminaries in key positions, notably the Army General Staff.

22 See C. Nettelbeck, 'The Eldest Daughter and the *Trente Glorieuses*: Catholicism and National Identity in Postwar France', *Modern & Contemporary France*, vol. 6, no. 4 (November 1998), pp. 445–62.

23 H. Mendras and A. Cole, *Social Change in Modern France: Towards a Cultural Anthropology of the Fifth Republic* (Cambridge, 1991), pp. 58–72.

24 G. Ross, '*Adieu vieilles idées*: the middle strata and the decline of Resistance-Liberation Left discourse in France', in J. Howorth and G. Ross (eds), *Contemporary France: A Review of Interdisciplinary Studies* (London, 1987), p. 65.

8
Distorting Mirrors: Problems of French–British Perception in the *Fin-de-siècle*

Martyn Cornick

Throughout his career Douglas Johnson has striven long in both Britain and France for a better understanding of each other's history and culture. Indeed, it is largely to Douglas's credit that the enterprises which he helped to found in 1980, the Association for the Study of Modern and Contemporary France and its journal *Modern and Contemporary France* have attracted support from the Cultural Section of the French Embassy in London. It is the intention of this chapter to make its own modest contribution to this spirit of mutual understanding.

*

At the very end of the twentieth century, perceptions between France and Britain[1] continue to be dominated by stereotypes and seem occasionally still to be fed by prejudice. A glance at the printed media bears witness to this. The crude slurs meted out by the *Sun* on 1 and 2 November 1990 to Jacques Delors and the French have remained in the minds of many people in Britain. In early 1998, when the tabloid newspapers trumpeted their outrage at the supposed paucity of World Cup tickets available for English fans, among others the *Daily Star* (2 March 1998) resorted to terms of simplistic prejudice to characterize France and the French. More surprisingly perhaps, the quality press is not immune from this tendency. In its edition for 23 February 1992, *The Sunday Times* provoked anger when its colour supplement asked the question 'What's Wrong with France?'. The magazine reproduced photographs of social unrest over the banner caption 'the French malaise', singling out the 'failing French economy' and the 'rise and rise of Jean-Marie Le Pen'. On this occasion *Paris-Match* was not willing to stand aside and let such negative images pass unchallenged, and in early March 1992 the weekly responded in the spirit of 'those who live in glass houses should not throw stones' by emphasizing some of the more unsavoury aspects of British life, in particular football hooliganism and the alleged takeover of Normandy by English second-home buyers. 'Once again the English fired first!', ran its headline. On a different level, *The Guardian* (27 June 1998) could not resist a side-swipe at France by joining in the Alain Sokal controversy on intellectual pretentiousness, asking 'Is French philosophy a load of old tosh?' And in France in 1997, the editors of a high-circulation history magazine felt it was still worth contrasting one hundred years of entente with nine centuries of hostility.[2] In November 1998 it was reported that a member of the Gaullist RPR party had proposed that in the interests of continued European integration the London Eurostar terminus at Waterloo should now be renamed . . . It may no longer be 'politically correct' to talk openly in terms of relative national superiority or to discern supposedly distinctive facets of national character; none the less such shorthand notions, traces of what has now been termed 'banal nationalism',[3] remain evident in manifestations of popular culture, journalism and advertising on both sides of the Channel. What is remarkable for our purposes here is that these ideas date directly from the end of the nineteenth century, around the years 1890 to 1905, when they reached a crescendo during a period of jingoistic nationalism whose legacy, partly at least, is evidently still with us.

The topic of Franco-British perceptions in the nineteenth-century *fin-de-siècle* is complex and wide-ranging, so it is necessary to be selective. We shall concentrate on some of the key factors informing perceptions between the two neighbours: how and in what different ways did France and Britain see each other at this time? To answer this question, we shall examine the weight of history, the nature of the stereotypes and the colonial rivalry which shaped the context of these perceptions; explore notions of racialist superiority and inferiority; and review the real and imagined threat of conflict. In short, the chapter focuses on the related discourses of 'Empire, Race and War'.[4]

One preliminary remark about the *fin-de-siècle* is apposite: it was the age of imperialism. Imperialist ideology exercised such a powerful influence that not only were international affairs and relations between the Great Powers conditioned by it, domestic politics, educational policy and mass public opinion were also fully subjected to it.[5] As critics on both sides of the Channel have shown, the ideas, propaganda and discourse associated with imperialism made themselves felt at all levels of society; imperialism and colonial ideology permeated everywhere.[6] One inspiring interpretive approach to this question will be found in Tzvetan Todorov's analysis in *On Human Diversity* where the contention is that ideas – and the discourse in which they are articulated – are equally as important in shaping history as actors and events.[7] Indeed, during the period under scrutiny history teaching and literature were both mobilized in what was effectively a propaganda effort to the extent that they became myth, that is myth in the sense intended by Roland Barthes, 'Myth that justified empire'.[8]

Relations between the two countries during the two decades leading to the Entente Cordiale of 1904 underwent one of their most uncongenial phases. It was partly colonial rivalry which underlay these tensions. Furthermore, and this will be our starting point, the weight of history pressed down hard on the two rivals.

The weight of history

The mental baggage, otherwise the prejudice and misunderstanding underlying the perceptions of the one people for the other, may be traced back to medieval times. Writing in 1893, the famous French historian C.-V. Langlois strove to elucidate the contemporary 'falsehoods and superficialities' he observed around him. Langlois even used the image of the distorting mirror: 'In our humorous and satirical literature,' he wrote, 'the English can see themselves as they might in one of those convex mirrors that exaggerate the ugly features, but in which they may still recognize themselves.' Similarly, the French could learn lessons about their own character if only they would be minded to subtract the exaggeration and mischief-making from the old jokes which circulated about them abroad. Langlois followed a number of leads in medieval texts. To take one of his most picturesque examples, the French once considered the English to be so alien a people that an enduring tradition endowed their neighbours with an animal appendage: this was the legend of *les Anglais coués*, or 'the English have tails'. Originating in Latin texts dating from the mid-twelfth century, an endless succession of jokes, puns and insults based on this tradition assailed the English well into the seventeenth century, when it died out. Evidently people like this should never be trusted.[9] Over the same period and into the eighteenth century, the myth of 'Perfidious Albion' accumulated a large corpus of stock phrases and historical parallels, including that of likening republican France to early Rome in its struggles with treacherous Carthage. Because revolutionary and Napoleonic

France showed great enthusiasm for classical references in art and propaganda, and because historical precedents lent credence to such propaganda, all this was vigorously rejuvenated during the First Republic and the Empire.[10] Indeed, the Carthage/Britain analogy has had a long history: the collaborationist radio presenter Jean Hérold-Paquis, who published a collection of talks under the title 'England, like Carthage, will be destroyed', was forcefully contradicted by Jacques Debû-Bridel after the Liberation in 1945.[11]

According to events and as fashions waxed and waned, as Douglas Johnson himself has implied, throughout the nineteenth century France tended to oscillate between Anglophilia and Anglophobia.[12] The tenacious myth of 'perfidious Albion' was kept alive in a broad range of media. According to the general terms of this myth, Britain was selfish, rapacious and duplicitous, and would never rest until it had colonized the world for commercial benefit. This idea became a dominant one in France between the end of the Commune in 1871 and the Entente of 1904, principally because of colonial rivalry: on balance, relations were hostile rather than friendly, and even led to the recognition of Anglophobia as an ideological force. When the latter surged up it could appear as powerful and threatening as anti-Semitism, as Max Nordau noted: 'French Anglophobia is like anti-Semitism. . . . It is a precipitate of history, of legendary lore, or religious, aesthetic and patriotic emotions.'[13] And at a time when these patriotic emotions were being whipped up into a frenzy it is not surprising that explosions of mutual suspicion, hatred even, fanned by a new and sensation-seeking journalism on both sides of the Channel, should become common-place. The likely outcome of this rivalry, which was so often expressed in terms of the glorification of a chivalric or heroic past, and which took place in a general atmosphere of belligerence, was war, so much so that the outbreak of hostilities in 1914 was seen by many as a self-fulfilling prophecy.[14] However, by this time – although it would have been difficult to imagine before 1904 – France and Britain had become allies.

In Britain, suspicion of the continental neighbour had grown in the aftermath of the Franco-Prussian War. Although the myth of 'Perfidi-ous Albion' did not have so focused an equivalent in Britain, none the less the political history of the previous hundred years or so provided countless examples to suggest that the French were unstable, unreli-able and all too prone to taking to the streets to foment revolution. The Siege of Paris and the Commune of 1870–71, the Boulanger crisis of 1889 and the Panama scandal of 1892 were all sufficiently fresh in the British collective conscious to reinforce the sentiment that after the 1789 Revolution, the French could never be considered reliable. Ironically, many of the great Victorian reviews, whether liberal or conservative in their leanings, evinced a pronounced admiration for French culture and literature, as did their Francophile contributors. Partly under the influence of John Stuart Mill, essayists would explain French political instability as rooted in the 'national character'. And this 'national character', because of hardships and deprivation caused by the French Revolution and subsequent insurrections, had supposedly become damaged, and even represented a flawed genetic inheritance.[15] The fundamental problem was politics. Faced with humiliation in February 1871, even so passionate a Francophile as George Meredith felt moved to admit that France 'has always been the perturbation of Europe'.[16] And among those who were not Francophiles, criticism of these perceived deficiencies was much more unrestrained. Thomas Carlyle for instance, author of one of the most widely read histories of the French Revolution, felt the French were 'vapouring, vainglorious, gesticulating, quarrelsome, restless and over-sensitive',[17] and condemned to repeat their political experiments. Moreover, British interpretations of the Dreyfus Affair were often expressed in similarly exasperated terms, again blaming the 'innate' tendency of the French towards revolution.[18]

Constructions of national identity may be defined not only in relation to a nation's *intrinsic* characteristics, they may also be defined relative to the *extrinsic* characteristics of rival nations. Eugen Weber has shown how the Third Republic's project of modernizing rural France inculcated an intense patriotism to bolster its citizens' sense of national identity, in particular through the educational system. Millions of French children were taught 'that their first duty was to defend their country as soldiers'.[19] In the 1890s, the process of learning this civic duty left a profound impression on one such child, the nine-year-old Jean Paulhan who, in his speech to the Académie française in 1963, recalled how he and his classmates had been called upon to restore France to a rank second to none:

> I remember with some emotion the day when our history teacher, M. Lion, a Jew from Alsace (like all Alsatian Jews he was fiercely patriotic) taught us – to our deep dismay – that France was not the world's leading nation, but was bettered by England to the tune of several thousand square kilometres because of its colonies. Upon which were we asked to swear that we would win back France's

lost supremacy. And along with all my little classmates, I swore to do so.[20]

Put simply, *identity* may be defined negatively in relation to an *other*, whatever or wherever it may be. For the historian Michelet, writing from the 1830s onwards, the opposition between France and Britain was a fundamental one: 'it was in opposition to England that France was constructed and became aware of its destiny.'[21] In the mid-1840s Léon Faucher opened a lengthy study on his neighbours with the words: 'England is certainly a world apart.' Despite centuries of 'racial' intermingling, 'every child of Great Britain has its nationality emblazoned across its forehead'.[22] Thus Britain was all that France was not, they were mutually exclusive. By 1900, for nationalists such thinking had become commonplace: as Paris *député* Georges Berry put it, 'In a word, England is anti-France'.[23] To adapt an observation made by Robert MacDonald, the ideas one had about one's own identity resulted in what one may call a distorted mirror image of the Other.[24]

Stereotypes provided a shorthand means to reinforce notions of identity. 'Stereotypes, culled from history, would reinforce the tendency to relate national unity to national character, national character to natural forces that determine that character and the policies of the nation.'[25] In order to provide an idea of the force and appeal of these stereotypes, we shall pause to consider a couple of powerful examples. The British were treated stereotypically in Larousse's *Great Universal Dictionary* published during the 1860s. This reference work contains a long entry on 'Angleterre'.[26] When it came to explaining the English character, the Larousse included a section entitled 'England judged by Jacques Bonhomme', which reads today like an Anglophobe's charter.[27] The stock French peasant character Jacques Bonhomme – who is, significantly, 'afflicted by an incurable Anglophobia . . . brought on by the study of history' – launches a furious attack on John Bull, piling up a mountain of evidence to refute the views of those in France who liked to promote the British model as a paragon of liberal democracy. The essay opens with a sketch of the natural and climatic features which shaped the national character. Because the British inhabited an island, and because of the inhospitable climate, by concentrating on 'the natural energy of [his] national character' John Bull had achieved success through overseas adventure and expansion.[28] These were the features which helped to lay the foundations of the greatest colonial empire in the world. The most hostile comments were indeed reserved for what was taken to be Britain's role on the international stage. All over the globe and throughout history, John Bull had divided his subject-peoples in order to dominate them:

> This astonishing obsession with the legitimacy of your dominance over all points of the known universe is a feature peculiar to your race. The world is your oyster, the oceans belong to you, people are your enemies when they refuse to be your humble subjects. Wherever one of your people sets foot, wherever he can hold a drop of salt-water in the palm of his hand, he feels at home and says: 'This is mine'.

Bonhomme identifies the octopus as the creature in the natural world most suited to symbolize this character:

> Its tentacles, terrible arms covering its whole body, have the terrifying ability to eat up everything they touch. The wretch who falls into this machinery is smothered and devoured in an instant. Yes, England is this octopus, and its thousand tentacles stretch around the world and suck it dry.

Such imagery was subsequently taken up and used by many anti-British illustrators and propagandists in France, especially at the time of the Boer War as well as during the Occupation. To quote one example among many, in 1942, in a fit of wishful thinking about British wartime reverses, the editor of a high circulation magazine asked: 'How is it today that this unworldly monster, this octopus, with its bleary and fixed stare, its bloated belly, finds its tentacles sliced through one after the other, its life force draining from its great body which has gorged for so long on people's corpses?'[29]

The Larousse singled out selfishness as the determining British characteristic:

> The basic vice of your race, your egotism, characteristic both of your nation and of its individuals, explains not only your tendency towards invasion and exclusion, but also the confiscation of wealth and isolation. You never associate yourself with another race, either for your interests or ideas. [Cf. the reference to Léon Faucher in note 28.] Here your politics accords with your national temperament, because you have no interest in assimilating or civilizing, just owning and exploiting.[30]

Defined as opposing identities, as they had done so frequently before the two nations appeared to be always destined to clash in war:

> Undoubtedly we shall witness the violent awakening of race hatred and long-term rivalries which would amaze this indifferent and sceptical age. In current circumstances and with an adversary as powerful as you, this time there will be a duel to the death which will split the world into two camps, like the struggle between Athens and Sparta, like that between Rome and Carthage, and where we will see not only two sets of interests, but two races, two philosophies, and even two separate civilizations.

In sum, this essay in the Larousse not only confirmed the 'Perfidious Albion' myth, it also conferred credibility on it since it appeared in a publication that was a recognized pedagogical tool, and which has been studied as a prime 'site of memory' (*lieu de mémoire*).[31] It would certainly be used in future by those who wished to find historical 'evidence' to bolster anti-British sentiment.

Stereotypical images of the British flourished too in popular adventure stories. From Paul Féval's *Les Mystères de Londres* (1844) to the adventure fiction of Jules Verne, in much of the exotic adventurism of a Pierre Loti or Claude Farrère, countless portraits of the British character were drawn which were not always flattering. There is certainly scope for further research in this relatively neglected area of what one may call 'formative' literature, although Philip Dine, in a study on literary images of the French colonial Empire, helpfully surveys some of the material.[32] Earlier works had a popular appeal because like Féval's, they were first published in the form of cliffhanging serials, sold well and went through numerous editions. They tended to portray Britain as a centuries-old rival if not as the hereditary enemy. Active in the 1840s well before Verne, Paul Féval set his serialized *Mystères de Londres* (rivalling Eugène Sue's *Mystères de Paris*) in a fog-bound and rain-soaked London. One character, the adventurer-pirate Fergus O'Breane, is shown to have witnessed all the horrors of British colonialism during his circumnavigation, including the 'odious opium trade'. He even drops anchor at Saint Helena to visit Napoleon, 'vanquished of Waterloo', to draw on the infinite wisdom of the greatest French hero. The all-seeing eye of Féval's narrator ranges wide, and in a few lines sums up the living legacy of British colonial expansion: '[O'Breane's] anger grew ceaselessly. Everywhere on his travels he saw England abusing its power and seeking gold in people's blood or sweat. Leaving the Indian seas, at ever farther intervals he found the same hatred, only more concentrated, ever ready to explode. At the Cape, the Dutch Boers; in America and in the whole of the two Canadas'.[33]

Stock images of the British abound in the fiction of Jules Verne, although it is important to stress that Verne himself was no Anglophobe. In 1904 he stressed what he called the positive potential of the 'English character': 'The English, through their *independence* and *self-possession*, make admirable heroes; especially when, as in the case of Phileas Fogg, the nature of the plot requires them to be confronted with formidable and entirely unforeseen difficulties.'[34] Verne's work, whose educational and recreational value was highlighted by his publisher, draws many 'typical' British characters, from his earliest story *Voyages et aventures du Capitaine Hatteras* (1867), through best-sellers such as *Cinq semaines en ballon* (1877), to historical novels like *Famille-sans-nom* (1889). And Verne's prolific contemporary Alfred Assolant provides another example in his *Aventures merveilleuses, mais authentiques, du Capitaine Corcoran*. First published in 1867, this reached its eleventh edition in 1905, and even in the late 1990s remained available in a facsimile reprint. *Corcoran* appeared in a series for 'children and adolescents' and is paradigmatic of pulp adventure fiction. Set in India in 1856, at the time of the so-called 'Great Mutiny', it tells the story of the young Breton sea captain and Louison, his tigress. The narrative shows the British to be inhumane and devious colonizers, complete with a secret agent called Doubleface. Corcoran is certainly laden down with the baggage of history since his Breton background suffices in itself to explain his atavistic outlook towards the British and his willingness to help the Indian rebels: 'I've no more feeling for an Englishman than I have for a kipper or a sardine. I'm a Breton sailor, and that's it. There's no love lost between me and the Anglo-Saxon race.'[35] For Corcoran and his like, Anglophobia was in the genes.

Of race and superiority

The race idea had already done much to perfect and embellish the old classic jingoism. Imperialism – a word that no longer needs explanation – has given it renewed force. It has, so to speak, spread it all over the globe.[36]

To expand on what the historian Michael Howard has noted of the British case, it was widely assumed in both Britain and France around the turn of the century that the 'white races' were 'inherently superior to the brown and black'.[37] Here it would be difficult to overstate the influences on such thinking both of 'romantic history'[38] and of Darwinism which, from the 1860s–1870s onwards, had a profound impact on both sides of the Channel. It would be possible to provide numerous examples, especially in the work of Hippolyte Taine who, in turn, was influential in Britain on a number of writers including J. E. C. Bodley.[39] Notions of racial superiority underlay colonial ideology, the 'white races' believing that they not only had the right but the duty to carry civilization to the 'inferior' races, to govern them and raise them up. All this is relatively well known, and was summed up in Rudyard Kipling's 'the White Man's Burthen', a phrase whose ideological underpinnings and diverse modes of expression were appreciated in nationalist circles on the French side of the Channel.[40]

It is not so well remembered today, however, that there was also a perceived hierarchy among the *white* 'races', in particular the 'Anglo-Saxon' and 'Latin' races. Unsurprisingly such thinking did nothing to soothe Franco-British tensions. In one instance, Lord Salisbury rather tactlessly revealed it at the height of the Fashoda crisis in late 1898, when the French and British nearly came to blows over a relatively minor colonial incident. In his widely publicized Mansion House speech on 9 November, the prime minister made a reference to the supposed 'decadence' of the Latin races. Implicitly included among these was France, whose recent 'humiliation' at Fashoda seemed merely to prove the point. That such racialist assumptions extended also to the fledgling British Left is borne out by another speech ('Labour and Empire') given by the MP John Burns, who argued that 'The Latin and other races were beginning to see that the world-wide supremacy of the Anglo-Saxon race was imminent, if it had not already arrived'.[41] Reports of such speeches and observations sparked a number of angry exchanges. In France these judgements prompted belligerent talk (in the usually austere *Revue des deux mondes*) of how, almost a century on from Napoleon's failed attempt at cross-Channel invasion, it was now technically and logistically possible to mount a seaborne attack on Britain's southern shores, whilst in the *National Review* (at this time firmly committed to the Dreyfusard cause) the paper in the *Revue des deux mondes* was dismissed as fantasy and wishful thinking.[42]

The debate over 'Anglo-Saxon superiority' was conducted in the wake of the publication in 1897 of Edmond Demolins's book, *A quoi tient la supériorité des Anglo-Saxons* [Explanations of Anglo-Saxon Superiority].[43] On both sides of the Channel passions were raised and arguments expounded at length about the merits and demerits of the racially and socially determined character of the 'Anglo-Saxons' and the 'Latins'.

Demolins, a social scientist, had been impressed by the (admittedly rather narrow) band of the British educational system that he had observed at first hand, and by what he described as the generally less parochial, more extrovert and enterprising character of the British. Moreover France, he argued, was ill-adapted to producing the civil servants that a modern state needed; ambition was limited and the influence of the state too overbearing. What was worse was that the French system held back demographic growth, stifled enterprise and tended to impoverish what we would now call 'middle-class' families, thereby damaging economic activity and growth. In short, the social structures and organization of public and private life led to a situation in which the French showed a tendency towards 'general ideas', whereas the British excelled much more readily at 'practical applications'. The French, therefore, or so the argument went, were less effective at colonizing and at developing their colonies. The reactions to this book are telling.

In Britain it was received in a glow of self-satisfaction, especially by the *Edinburgh Review*, whose Darwinistic reviewer basked in the ambient sense of superiority felt in some Victorian intellectual circles. 'It is not by patching up what is weak,' crowed the *Edinburgh*, 'but by strengthening what is strong, not by assuming the defensive but by pushing a successful line of attack, that nations as well as individuals attain to success.'[44] Demolins was to be admired for the spirited nature of his work and for daring to address what was, naturally, a delicate subject. At the end, the reviewer could not help agreeing that the French were inferior when it came to colonial enterprises; they had no 'people who would go out and work in these colonies . . . taking risks on their own heads.' This was a widely held view. The conclusion was that 'Until the French character changes there is no probability that French history will change, and the radical characteristics of the Frenchman are caution in private life, temerity in public employments.' The implicit message, finally, was that in order to succeed the French would have to emulate, if they could not adopt, British characteristics:

> It sounds a paradoxical thing to say, but it is nevertheless profoundly true, that France is a nation in process of being ruined by the thrift and prudence of its citizens. To live poorly because it is so much easier to save money than to make it, to have no children for fear they should die of starvation, that is the summing up of the Frenchman's pennywise philosophy, and, if there is truth in logic, it is the individual Frenchman who is keeping France back in the race, just as

it is the individual Anglo-Saxon who is winning the battle for his community.[45]

In France reactions were much more critical. If Demolins was to be congratulated for his courage and for some of his critical insights, it was thought that he exaggerated his case. And with just as much exaggeration, one reviewer resorted to sarcasm:

> M. Demolins has amused himself by making us believe that our proud neighbours invented work, virtue and happiness. He has found pleasure in representing England as a radiant sun where one would look in vain for a blemish, and our France as a black hole. He wanted to shake our senses by persuading us that on one side of the Channel, everything is for the better, while on the other everything is going from bad to worse.[46]

Demolins had surely been too selective: what about the deadly British Sunday? 'One has never been able to measure the height, the depth and the length of it.'[47] Striking a more serious note, Demolins added that 'this insular people, which guards its independence so jealously, does have its weaknesses and its servitudes. More than any other, it is enslaved by its habits, by its national prejudices.' Indeed, the general inhumanity of the British colonial system came in for another cutting generalization; unlike the implicit advantages of the French system, the British never tried to assimilate their colonized peoples, they simply lived apart the better to exploit them:

> It is much easier for them [i.e. the insular British] to travel the world, to sail the seas, than to try and shed their skin and get to know other peoples. In this they are very different from the Romans, to whom they like to compare themselves; they are impenetrable and impervious, and live side by side with other foreign races but borrow nothing and give them nothing, and the distance between them will always be the same.[48]

Such was the outcry provoked by Demolins's book in France that at least three book-length replies were published.[49] Of these the most interesting is the one by the pseudonymic 'Anold'. Prefaced by the Anglophobe *député* François de Mahy, Demolins was lambasted for having confirmed the impression given by their enemies that the French were a 'decadent race'.[50] After a chapter suggesting that the British had had an interest in destabilizing France through the Dreyfus Affair, and another entitled 'The English Peril' alleging that they were doing their utmost to undermine French interests and weaken its identity (from the Fashoda incident to the actions of Methodist missionaries in Algeria), the book concluded by reversing the terms of Demolins's title, and claimed that if anyone was superior it was the French. The British might be the strongest race 'today', in 1899; however, in a passage that in retrospect has a ring of prophetic truth about it, in future, after other nations will have tired of British insolence and allied together against the 'common enemy', and after its colonies will have gained independence, Britain will only have a memory of this 'proud and inglorious past'.[51] The final paragraph of the conclusion is bleak enough, seeing a future 'England' as hated, friendless and alone:

> With her mad pride, with her vain lack of awareness [it is worth noting how in the 'mirror effect' each nation sees the other as 'vain'] England will, herself, bring about her own fall. The whole world is tired of her invasions, bad faith and cruel rapaciousness. Soon, and sooner than she thinks, she will be alone facing all the other civilized nations, of which none will wish to enter into an alliance with a people whose perfidy has become legendary. Reduced to relying solely on herself, having only her ships to defend her, a formidable power in appearance, but in reality only fragile, because she is her own worst enemy she will see her immense empire crash like a house of cards, until that day when the ironic prophecy I made as a warning at the beginning of this book will come true: there will be Englishmen everywhere; but there will be no more England![52]

In the end, wrote 'Anold', in a passage tinted with almost Gaullist tones, France would strive to rise again and, through renewed efforts on the part of the whole population, restore its role as a leading world power. With the outbreak of the Boer War in the autumn of 1899, such critiques of British arrogance must have seemed justified.

As far as the race question itself is concerned, there are signs that just after the turn of the century on both sides of the Channel the notions of relative superiority and inferiority of the Anglo-Saxon and Latin 'races' were beginning to be regarded by a few enlightened thinkers as fallacious or spurious. For one, the psychologist Havelock Ellis, in an article moving away from notions of racial superiority, even if he still admitted that Europeans could be grouped racially, argued in 1901 that the British were just as 'Latin' as the French, and that the French were, if anything, the most representative European people because they exhibited the most mixed characteristics.[53] To the terms 'two races' he preferred the phrase 'two civilizations':

> The Latin race, we say, is decadent; France, we assume is Latin; therefore, France is decadent, in striking contrast to the superiority of the 'Anglo-Saxons'. The fallaciousness of these flourishing beliefs has often been pointed out; but . . . the anthropological evidence which has lately accumulated enables us to expose them with complete precision. . . . France thus represents Europe in miniature, in a sense that no other country, great or small, can claim to do, and if we had to choose one country as representing the quintessential racial elements of Europe we should be compelled to select France. . . . The stereotyped phrase regarding the 'Latin race' of the French is wholly incorrect and meaningless.[54]

Part of the problem stemmed from the nature of the press, Ellis maintained, which sustained what we have called the 'distorting mirror' effect. 'National jealousies impart to the most commonplace matters of fact an almost startling novelty.' And he explained the effect thus:

> The opinions of the flimsiest and feeblest of French newspapers – reflecting a vulgar, if not unnatural reaction against the Pharisaic attitude of England towards France – are reproduced in the most ponderous of our own journals, so to gain a significance and resonance which otherwise they would never reach. Thus the vicious circle is completed, and the English man in the street who takes his opinions . . . from the newspapers, is hopelessly chained to prejudice and error.[55]

As the few examples referred to at the beginning show, in essence this aspect of Ellis's critique remains valid in our own times.

In France, the work of Jean Finot also argued against 'pseudo-scientific' conceptions of racial difference between the French and the British.[56] On the English side of the Channel, Finot's efforts to discredit racial differences found an influential propagator in W. T. Stead, editor of the widely read *Review of Reviews*. In 1911, not long before perishing on the *Titanic*, Stead prefaced a translation of Finot's *Death Agony of the 'Science' of Race*, underlining that until 12 years previously, 'the pseudo-scientific theory of the mental and radical differences of Races was almost universally accepted', and that Finot's work had proved to be as 'potent as the blast of horns in Jericho'. Stead applauded this new view that there was 'no such thing as race', that men were 'indivisible', and that Finot had 'revolutionized the scientific concept of the race question'.[57] Once the Great War had broken out, Finot nevertheless resorted to the discourse of racialist difference in order to contrast the conduct of the Entente allies over that of the 'barbaric' German race.[58] And conversely, for all the talk of the Entente Cordiale, when the war was going badly during 1916 and the French public were asking where the British were at the time of the battle of Verdun, one observer was willing to criticize his own national psychology, particularly 'our failure to see the effect of our action or inaction upon others. Psychology is not our strong point and never was'. And he added:

> It explains, in some measure, our success in ruling subject races. We treat them as children . . . we have no need to bother about fine shades of character and temperament and nice points in racial distinction. Now it is clear we have fallen into the same error in respect of our Allies. We have treated them as if they were Anglo-Saxons or even British. We have applied the 'wait and see' policy to them, and the French naturally resent an intellectual occupation of that sort. They are in no mood to wait and see the destruction of their country.[59]

Of war; or the 'patriotard wave'[60]

The latent threat of war between France and Britain was a prominent feature of the *fin-de-siècle* period around 1900. There are two major factors underlying this: the Fashoda crisis and the outbreak of the Boer War.

In itself, the Fashoda crisis of October–November 1898 was a minor incident; it was the nature of press reactions on both sides of the Channel which led to much noisy sabre-rattling. Even the more cool-headed commentators in both countries pointed to the Royal Navy's mobilization in the autumn of 1898 as inauspicious.[61] Furthermore, because of the simultaneous 'Dreyfus agitation' and the parlous state of politics in France, opinion-formers in Britain feared that in order to deflect attention away from its seemingly endless crises, France might risk 'a sea war', if such a conflict could be engaged 'with safety'.[62] Naval rivalry formed another dimension of this hostile state of affairs. As far as assumptions governing relations between Britain and France during the nineteenth century were concerned, Britannia had traditionally ruled the waves since Trafalgar. If French naval policy had stagnated during the 1880s, around the turn of the century there were signs that technical innovations – particularly in submarine warfare – were beginning to shift the balance back toward France.[63] Moreover, because of *German* naval expansion, Britain's 'Two-Power Standard' – according to which the British fleet was to be maintained at a size equivalent to the combined navies of the next two Great Powers – was placed under strain. The huge review of the fleet at the Spithead, however, organized for Queen Victoria's Diamond Jubilee celebrations during the summer of 1897, helped to reinforce the notion of British naval supremacy; French nationalists simply had to agree that the 'Anglo-Saxons' were superior on the high seas.[64]

The animosities and anxieties triggered by the Fashoda crisis were experienced much more deeply at the time than is now appreciated, and there were genuine fears regarding the possibility of a seaborne attack. From the British perspective, this was considered as the 'problem of invasion', and invasion narratives multiplied at the turn of the century.[65] Ever since the publication in 1871 of George Chesney's sensational pamphlet *The Battle of Dorking*, after the upheaval of the Franco-Prussian War the British were worried by this threat. In attempts to appeal to governments' better judgement and to improve coastal and military defences, interested authors resorted to imagined accounts of continental armies swarming over (or indeed under) the Channel to threaten London. Pseudo-technical articles and books published in France on these matters did nothing to reduce the tension.[66] After Fashoda, and as the Transvaal crisis deepened, the idea was entertained that France might well lead an alliance of Britain's enemies (usually including the Russians, to whom the French were tied by treaty) to cross the Channel, to land armies and take over a supposedly impotent state apparatus in order to wreak their terrible revenge.[67]

The event which turned such belligerent fantasies into reality was the outbreak of the Boer War in the autumn of 1899, just after the Dreyfus retrial at Rennes. The world's press, and certainly most press opinion in Britain, had condemned the renewed guilty verdict on Dreyfus as a shameful stain on France; British protests, which included a mass rally in Hyde Park, began to be defused only when Dreyfus was pardoned. The Boer War, however, risked turning into a Dreyfus Affair *à l'anglaise*, with Joseph Chamberlain, the Minister of Colonies, cast in the role of General Mercier; indeed it was rumoured that Chamberlain intended to make war on France once the Boers had been defeated.[68] Almost totally isolated in Europe, Britain was swamped by a wave of 'Continental Anglophobia',[69] and was caricatured as a merciless imperial lion devouring the fledgling republic in the Transvaal. The editor of the *Revue des deux mondes*, Ferdinand Brunetière, whom the editor of the *Quarterly Review* had invited to explain why French opinion on the war was almost universally anti-British, pointed directly to British treatment of the Dreyfus Affair because for two years the press had continuously lambasted French 'injustice'; now, Britain was only being repaid in kind:

> For two years the English papers, *The Times* at their head, have stopped at nothing to heap invective or insult every one of us who did not believe in Captain Dreyfus's innocence. I can speak with some authority on this, since I have here the treatment meted out to us on this subject in the *National Review* by a certain Mr Conybeare.[70] ... So let the British allow to me to reply: they have no idea of how much the violent, insulting and excitable intervention of the British press in the Dreyfus Affair has whipped up justified indignation against them in France. The British complain about the language in our papers: let them remember how their press talked about us for two years! Since they refused to judge the Dreyfus Affair from our nationalist standpoint, then they should not be surprised if we, in our turn, refuse to judge the South African war from the British imperialist point of view! And, having wanted to see the Dreyfus Affair only as a 'question of justice', as they used to say, let them in turn allow us to see the war they are waging against the Boers as a 'question of equity', with no regard to British interests![71]

According to this view, what was most intolerable to French opinion was that for all the talk of 'Anglo-Saxon superiority', Britain was engaged in nothing short of wiping out an entire 'people' in South Africa, actions which probably would now be termed 'genocide' or 'ethnic cleansing'. What made matters worse was that the Boers were joined by combatants from other European nations grouped together in a Legion of Foreign Volunteers, led by the Action Française militant Colonel Villebois de Mareuil, who perished under British fire at Boshof in April 1900.[72] Villebois's martyrdom was lavishly commemorated at Notre-Dame, and led to a wave of sometimes violent anti-British sentiment, particularly when the Boer president Krüger visited France in the winter of 1900.[73] And once again, in the wake of the South African crisis there came another surge of nationalistic adventure writing in France in which familiar anti-British sentiment was recycled in a fresh context.[74]

A final factor to be borne in mind is that British popular nationalism – jingoism – never attained such a fever pitch as it did at the time of the so-called Relief of Mafeking. Here we are fortunate to have Paul Mantoux's first-hand account of Mafeking night (15 May 1900); his view of this drunken saturnalia is full of insight into how the identities of the two countries were conceived at this time. André Chevrillon was another perceptive observer of how the English press, with the notable exceptions of 'Labouchère's *Truth* and Stead's *Review of Reviews*', was devoted to the propagation of an arrogant jingoism and disdain for the Continent among the newspaper-reading public.[75] Mantoux, who would join the French team negotiating the Versailles Treaty in 1918–19, spent much time in London around the turn of the century. For him jingoism was only the latest manifestation of English arrogance – *orgueil* is his word – a fundamental trait of the national character which had always existed: 'it is a brutal and haughty arrogance, which suffices quite unto itself'. (Michelet and others had pointed to this unappealing characteristic.[76]) The implication in Mantoux's analysis is that so long as nations define each other in mutually exclusive terms, that is, based on crude or stereotyped conceptions of national character and, worse still, pseudo-scientific racialist ideas, then one should not be surprised at expressions of mutual distrust and hatred:

> Since time immemorial, when confronted by the foreigner the jingo has displayed not exactly hatred, but a sort of scornful aversion. He regards him as a poor wretch, and unworthy of comparison with himself. The Frenchman, who appears boastful, flippant, lecherous, is good only as a cook or dance-master. The Italian is dirty, lazy and mendacious and prone to stabbing people. The German, heavy and cumbersome, knows no happy medium between obsequiousness and utter vulgarity. I am not too sure whether our own popular ideas about our neighbours are very different. This outlook has been supported recently by the development of so-called scientific theories on race. We shall never be able to gauge just how much damage these theories have done. They have added a philosophical and modern veneer to the most primitive, absurd and wild prejudices. It is so easy to strike a blow at one's enemies by treating them as an inferior race. And it is so agreeable to consider oneself as superior and predestined.[77]

Another explanation for British arrogance at the time of the Boer War, according to Mantoux, was the widely held belief that Britain always won its wars. Against France alone, the list was impressive: 'Crecy, Poitiers, Agincourt, Ramillies, Malplaquet, Trafalgar, Waterloo!'[78] This was perhaps the most important message behind the 'Island Story', and was widely propagated among the British public by the popular and romanticized fiction of G. A. Henty and others in late Victorian Britain, and is encapsulated in the jingo rhyme: 'One fat Frenchman/Two Portugee,/One jolly Englishman/Could lick all three.'[79] And as for British attitudes towards Africa, Mantoux believed that at the time of Fashoda, the British were spoiling for a war; summing up the public and press mood at the time, he noted: 'a new Trafalgar, what a glorious dream!': 'The crushing of a great power like France would be so much more glorious than miserable frontier wars around the fringes of the Sahara or the Pamir! The London jingoes celebrated in advance, glowing from their innermost being!'[80]

In conclusion

It would be too simplistic to say that with the arrival of the twentieth century, relations between France and Britain reached a turning point and improved. In one sense, new preoccupations made themselves felt. The outbreak of the Russo-Japanese war in February 1904 gave rise to widespread fears of the so-called 'Yellow Peril', or that of 'Asia as a Conqueror'; the printed media were full of racialist speculation on what the future might hold. Yet with the passing of Queen Victoria early in 1901 there was a conscious realization that an era had come to an end. One particularly fitting image that comes to mind is that of the Roman god Janus, whose representation was used as a motif by the review *Nineteenth Century and After*. Looking back at the previous hundred years from the vantage point of the new century, some in Britain concurred gloomily with *The Annual Register* that the first and last years of the nineteenth century had 'many points of resemblance'. On both occasions Britain had been without close allies and engaged in long-standing wars which had been 'misjudged' by its rulers.[81] The chief antagonist in all this had been and had remained, France; now, however, there was a new rival – Germany. It was time for Janus to look forward. For the French and British, the cold war of the *fin-de-siècle* would gradually give way to better relations, and in advance of the Entente Cordiale there came repeated calls for reconciliation and 'alliance'.[82] Although initially the Entente amounted to little more than a colonial trade-off, it was almost immediately seized upon as a sign of 'mutual goodwill', as an opportunity for the future, and as a shift away from the 'belligerence in the relationship between the two countries over the previous two centuries'.[83] Indeed, the Entente laid the bases for the military co-operation which would enable a British Expeditionary

Force to disembark at Boulogne once war broke out in August 1914. Tragically though, the experience of Armageddon quickly demonstrated the destructive potential of the rival racialist discourses we have examined.

Acknowledgement

I am grateful to the staff of the London Library for their assistance.

Notes

Unless stated otherwise, all translations from the French are my own.

1   Two points should be made about discrepant nomenclature: first, the French often refer to 'Angleterre' when Britain or the United Kingdom is meant. Following the example of Philip Bell (in *France and Britain 1900–1940* [London, 1996], p. 5) I use the general terms France, Britain and Franco-British, and use 'England' and 'English' only when translating from the French or when referring specifically to that country and its inhabitants. Second, however, as Beatrice Heuser notes, historically the rivalry was a very specific one 'between France and the kingdom of England', and during the period examined here it may be argued that imperialism and jingoism were primarily 'English' phenomena; see C. Buffet and B. Heuser (eds), *Haunted by History. Myths in International Relations* (Providence and Oxford, 1998), p. 159.
2   'Les Anglais nous détestent-ils?', *Historia* (May 1997), pp. 35–71.
3   M. Billig, *Banal Nationalism* (London, 1995).
4   See M. Howard, 'Empire, Race and War in pre-1914 Britain', in H. Lloyd-Jones et al. (eds), *History and Imagination. Essays in Honour of H. R. Trevor-Roper* (London, 1981), pp. 340–55.
5   For a flavour of contemporaneous discussions of this dominant feature of the *fin-de-siècle Zeitgeist*, see H. W. Wilson, 'The Policy of Jingoism', *National Review*, vol. 32 (January 1899), pp. 631–41, J. A. Hobson, *The Psychology of Jingoism* (London, 1901) and V. Bérard, *L'Angleterre et l'impérialisme* (Paris, 1900).
6   For Britain, see R. H. MacDonald, *The Language of Empire: Myths and Metaphors of Popular Imperialism, 1880–1918* (Manchester, 1994) and for France, see A. Ruscio, *Le credo de l'homme blanc* (Brussels, 1995). For a comparative study, see T. G. August, *The Selling of the Empire. British and French Imperialist Propaganda, 1890–1940* (Westport, Conn., 1985), especially chapter 1.
7   See T. Todorov, *On Human Diversity. Nationalism, Racism and Exoticism in French Thought* (Cambridge, Mass., 1993).
8   MacDonald, *The Language of Empire*, p. 49.
9   C.-V. Langlois, 'Les Anglais du Moyen Age d'après les sources françaises', *Revue historique*, vol. 52 (1893), pp. 298–315.
10   H. D. Schmidt, 'The Idea and Slogan of Perfidious Albion', *Journal of the History of Ideas*, vol. 14 (1953), pp. 604–16; also, N. Hampson, *The Perfidy of Albion, French Perceptions of England during the French Revolution* (Basingstoke, 1998), esp. chapter 7.

11 J. Hérold-Paquis, *L'Angleterre comme Carthage sera détruite* (Paris, 1944);
    J. Debû-Bridel, *Carthage n'est pas détruite* (Paris, 1945).
12 'Entente and mésentente', in F. Bédarida, F. Crouzet and D. Johnson (eds),
    Britain and France. Ten Centuries (Folkestone, 1980), pp. 265–73.
13 M. Nordau, 'Continental Anglophobia', *National Review*, vol. 38 (February
    1902), pp. 838–53, quote from pp. 843–4.
14 See the sources in C. D. Eby, *The Road to Armageddon. The Martial Spirit in
    English Popular Literature, 1870–1914* (Durham, NC, 1987).
15 On the 'Idea of Character', and on Mill, see S. Collini, *Public Moralists.
    Political Thought and Intellectual Life in Britain 1850–1930* (Oxford, 1993),
    esp. chapter 3. British reactions to the upheaval of the Paris Commune in
    the spring of 1871 laid repeated stress on supposedly flawed character;
    see my chapter 'La vue britannique de Paris à l'époque de la Commune', in
    M.-C. Kok-Escalle (ed.), *Paris: de l'image à la mémoire: représentations artist-
    iques, littéraires, socio-politiques* (Amsterdam, 1996), pp. 135–56.
16 A. G. Meredith (ed.), *Letters of George Meredith*, 1, *1844–1881* (London, 1912),
    p. 223.
17 T. Carlyle, 'Letter to the editor', *The Times* (18 November 1870).
18 See M. Cornick, 'The Impact of the Dreyfus Affair in late-Victorian Britain',
    Franco-British Studies, no. 22 (Autumn 1996), pp. 57–82.
19 E. Weber, *Peasants into Frenchmen* (London, 1979), chapter 18 (p. 333).
20 J. Paulhan, *Œuvres complètes* (Paris, 1966–70), IV, p. 380.
21 See C. Crossley, 'Michelet et l'Angleterre: l'antipeuple?', *Littérature et nation*,
    18 (1997), pp. 137–52.
22 L. Faucher, *Etudes sur l'Angleterre* (Paris, 1845), I, pp. iii, iv.
23 G. Berry, *Français-Boers* (Paris, 1900).
24 'The other is a grotesque mirror of the self, a negation, an inversion, an
    antithesis'; MacDonald, *The Language of Empire*, p. 33.
25 E. Weber, 'Of Stereotypes and of the French', *Journal of Contemporary History*,
    vol. 25 (1990), pp. 169–203 (pp. 181–2).
26 Along with entries on 'Albion', 'Anglais', see *Grand Dictionnaire universel*,
    Tome premier (Paris, 1866), pp. 176, 361–2, 363–77.
27 *Grand Dictionnaire Universel*, Tome premier, p. 374 [col. 4ff].
28 Noticeably this essay relied heavily on some of the items included in its
    bibliography; for instance, it mentions Faucher's *Etudes sur l'Angleterre* which
    gives the following archetypal 'explanation' of the national character: 'This
    tendency towards exclusion and isolation may be explained, it is true, by
    British history. But fundamentally it derives from the national character.
    The English race [sic] never associates itself with any other, neither in its
    interests nor in its ideas; it is utterly incapable of and unsympathetic to
    efforts of assimilation . . . ' (p. v).
29 J. de Lesdain, *L'Illustration* (2 May 1942).
30 *Grand Dictionnaire universel*, Tome premier, p. 375 [col. 2].
31 P. Ory, 'Le "Grand Dictionnaire" de Pierre Larousse', in P. Nora (ed.), *Les
    Lieux de mémoire* (Paris, 1997, 'Quarto' edn), vol. 1, pp. 227–38.
32 P. Dine, 'The French Colonial Empire in Juvenile Fiction: from Jules
    Verne to Tintin', *Historical Reflections/Réflexions historiques*, vol. 23 (1997),
    pp. 177–203.
33 P. Féval, *Les Mystères de Londres* (Paris, 1998 [1844]), pp. 358–9.

34   Interview in *Temple Bar* noticed by the *Review of Reviews* (1904), I, p. 591; my italics.

35   A. Assolant, *Aventures merveilleuses, mais authentiques, du Capitaine Corcoran*, 2 vols. (Paris, 1905), I, p. 89.

36   P. Mantoux, *A travers l'Angleterre contemporaine* (Paris, 1909), p. 27.

37   Howard, 'Empire, Race and War', p. 340.

38   E. Weber, 'Nos ancêtres les Gaulois', *My France* (New York, 1991), pp. 21–39.

39   For Taine's 'racialist' ideas, see Todorov, *On Human Diversity*, pp. 114–18, 153–7. Taine's 'system' was discussed by W. F. Rae in *Notes on England* (London, 1873), pp. xxx–lix. For a British turn-of-the-century critique of Darwinism, see R. Balmforth, 'Darwinism and Empire', *Westminster Review*, vol. 158 (July 1902), pp. 1–13. As for Bodley, much remains to be done on his influence; see his *France*, 2 vol. (London, 1898) and the many reviews of this work on both sides of the Channel. E.-M. de Vogüé relates Bodley's 1891 visit to Taine's house in the rue Cassette in Paris in an anecdote contained in 'Un regard anglais sur la France', *Revue des deux mondes* (1 June 1901), pp. 677–93.

40   T. Bentzon, 'L'Armée anglaise peinte par Kipling', *Revue des deux mondes* (1 April 1900), pp. 512–54.

41   For both speeches, see *The Annual Register 1898* (London, 1899), pp. 178–9. Controversy surrounding Ambassador Monson's speech at the Chamber of Commerce in Paris also raised the temperature; for British perceptions of French unease, see also 'The Looker-On', *Blackwood's Edinburgh Magazine*, vol. 164 (1898), pp. 851–64.

42   Cf. XXX, 'Les descentes en Angleterre', *Revue des deux mondes* (15 March 1899), pp. 275–302, and H. W. Wilson, 'The Invasion of England', *National Review*, vol. 33 (June 1899), pp. 653–63.

43   E. Demolins, *A quoi tient la supériorité des Anglo-Saxons* (Paris, 1897; republished 1998).

44   'The Success of the Anglo-Saxons', *Edinburgh Review*, vol. 187 (January 1898), pp. 130–50, quote from p. 131.

45   *Ibid.*, p. 150.

46   G. Valbert, 'La supériorité des Anglo-Saxons et le livre de M. Demolins', *Revue des deux mondes* (1 October 1897), pp. 697–708, quote from p. 699.

47   *Ibid.*, p. 700. Taine's complaints about the British Sunday had evidently left a deep impression in France: 'Sunday in London in the rain: the shops are shut, the streets almost deserted; the aspect is that of an immense and well ordered cemetery.... After an hour's walk in the Strand especially, and in the rest of the City, one has the spleen, one meditates suicide.' See Taine, *Notes on England*, p. 9 and the rest of chapter II.

48   Valbert, 'La supériorité', p. 707. Thus the views expressed much earlier in the nineteenth century by such as Léon Faucher enjoyed a long currency; cf. note 28 above.

49   See Anold [pseudonym of J. Philipp], *A quoi tient la supériorité des Français sur les Anglo-Saxons* (Paris, 1899); C. Crespin, *Les Français sont-ils inférieurs aux Anglais? A propos d'un livre de M. E. Demolins* (Paris, 1898); L. Bazalgette, *A quoi tient l'infériorité française* (Paris, 1900).

50   Anold, *A quoi tient la supériorité des Français*, pp. i–xii.

51   *Ibid.*, pp. 185–7.

52   *Ibid.*, pp. 199–200.
53   H. Ellis, 'France and Great Britain: two civilizations', *Contemporary Review*, vol. 79 (1901), pp. 574–88.
54   *Ibid.*, pp. 575, 577–8.
55   *Ibid.*, p. 587.
56   Finot's work includes *Français et Anglais. L'Angleterre malade – médecins et remèdes* (Paris, 1902) and *Le Préjugé des races* (Paris, 1905).
57   W. T. Stead, preface to J. Finot, *Death Agony of the 'Science' of Race* (London, 1911), trs. C. Grande, pp. 4, 6, 7.
58   J. Finot, *Civilisés contre Allemands* (Paris, 1915).
59   C. Dawbarn, 'Some Causes of Misconception. The Impressions of an English-man in Paris', *Nineteenth Century and After*, 79 (1916), pp. 1280–90, quote from pp. 1280–1.
60   T. Barclay, *Thirty Years of Anglo-French Reminiscences, 1876–1906* (London, 1914), chapter XII.
61   For contemporary press treatment of the issues surrounding the Fashoda crisis, see R. Arié, 'L'Opinion publique en France et l'affaire de Fachoda', *Revue d'histoire des colonies* (1954), pp. 329–67; M. Hugodot, 'L'Opinion publique anglaise et l'Affaire de Fachoda', *ibid.* (1957), pp. 113–37; and T. W. Riker, 'A Survey of British Policy in the Fashoda Crisis', *Political Science Quarterly* (March 1929), pp. 54–78.
62   Cornick, 'The Impact of the Dreyfus Affair', pp. 63, 68.
63   Cf. C. I. Hamilton, *Anglo-French Naval Rivalry* (Oxford, 1993), pp. 317–18, with 'D'Armor', *Les sous-marins et la guerre contre l'Angleterre* (Paris, 1899) and H. Chassériaud, 'Politique navale', *La Nouvelle Revue* (15 January 1899), pp. 216–31.
64   F. Charmes, 'Chronique de la quinzaine', *Revue des deux mondes* (1 July 1897), pp. 229–35.
65   See Capt. W. E. Cairnes, 'The Problem of Invasion', *National Review*, vol. 36 (1900–1), pp. 341–60, and the list in I. F. Clarke, *The Tale of the Future* (London, 1978), pp. 13–29.
66   'Les descentes en Angleterre', *Revue des deux mondes*, art. cit., and 'Lieutenant X', *La guerre avec l'Angleterre* (Paris, 1900); for an idea of the anxieties these provoked, see *Review of Reviews*, vol. XXI (1900), *passim*.
67   See especially 'Vates', 'The Revenge for Fashoda', *Fortnightly Review*, vol. 73 (May 1903), pp. 773–808.
68   M. Cornick, 'La réception de l'Affaire en Grande-Bretagne', in M. Drouin (ed.), *L'Affaire Dreyfus de A à Z* (Paris, 1994), p. 580, and C. Andrew, 'France and the Making of the Entente Cordiale', *Historical Journal*, vol. X (1967), pp. 89–105.
69   Nordau, 'Continental Anglophobia', *passim*.
70   Frederick Conybeare, an Oxford don, was one of the most outspoken British 'Dreyfusards'; see Cornick, 'The Impact of the Dreyfus Affair', pp. 64–5.
71   F. Brunetière, 'France, Foreign Opinion', *Quarterly Review*, vol. 191 (April 1900), pp. 572–82, quote from pp. 574–5.
72   On the exploits of Villebois de Mareuil, see R. McNab, *The French Colonel* (Oxford, 1975), and for an idea of the deep impression he made in France, see J. Caplain, *Villebois de Mareuil, son idée, son geste* (Paris, 1902) and P. Deschamps, *La Mort d'un héros. Hommage à la mémoire de Villebois de Mareuil* (Paris, 1901).

73   On Krüger's visit, see Daragon, *Kruger en France. Marseille, Dijon, Paris* (Paris, 1901).

74   See R. Tholoniat, 'De Fachoda à l'Entente cordiale: Romans de la guerre anglo-afrikaner (1899–1907)', *L'Information historique*, 52 (1990), pp. 89–96.

75   Mantoux, *A travers l'Angleterre contemporaine*, chapters I and II, and A. Chevrillon, 'L'Opinion anglaise et la guerre du Transvaal', in *Etudes anglaises* (Paris, 1901), pp. 247–357.

76   Crossley, 'Michelet et l'Angleterre', p. 151.

77   Mantoux, *A travers l'Angleterre*, pp. 25–6.

78   *Ibid.*, p. 32.

79   For 'The Island Story', see MacDonald, *The Language of Empire*, chapter 2.

80   Mantoux, *A travers l'Angleterre*, pp. 37–8.

81   *The Annual Register 1900* (London, 1901), pp. 1–2.

82   Cf. Andrew, 'France and the Making of the Entente' with T. Barclay, 'A Lance for the French', *Fortnightly Review* (February 1900), pp. 173–7, P. de Coubertin, 'The Possibility of a War Between England and France', *ibid.* (May 1900), pp. 719–29, and H. Vaughan, 'A Plea for an Anglo-French Alliance', *Westminster Review* (December 1901), pp. 613–19.

83   R. Millet, 'La lutte pacifique entre la France et l'Angleterre', *Revue des deux mondes* (15 June 1904), pp. 765–801.

9

After Dreyfus and before the Entente: Patrick Geddes's Cultural Diplomacy at the Paris Exhibition of 1900

Siân Reynolds

> The diplomatists, the journalists have had their Peace Congress at the Hague ... But here is a more real Peace Congress ... a gathering of workmen and architects and artists have silently made peace more interesting than war.
>
> (Patrick Geddes, 'Thoughts at the Exposition')[1]

Not everyone took such a positive view of the Universal International Exposition of 1900 held in Paris. Maurice Barrès described this apotheosis of progress under the aegis of the Republic as 'lemonade and prostitution'.[2] However in the collective mind, the 1900 World's Fair came to symbolize that 'instant of suspended time [when] the material universe of need was sublimated into the magic of spectacle'.[3] This was the largest yet of the great exhibitions, and the most spectacular: 50 million visitors, over 300 acres of space, and a vast range of displays. In its honour, the French government built the Grand and Petit Palais; the Gare d'Orsay; and the Pont Alexandre III, a combination of advanced steel technology (to impress the Germans) and florid French statuary (to impress everyone else). The whole thing had been planned to circumvent the rival possibility of a 1900 World's Fair in Berlin. France intended to confirm its claim to be 'in the avant-garde of civilization' and to deprive Germany of such claims.[4] Behind the spectacle lay political tension. Political and cultural history do not always coincide: this article explores the intersection of biographical, cultural and political history in 1900.

Earlier exhibitions had been marked not only by Franco-German industrial rivalry but by obsessional Franco-British emulation. The 1889 Paris Exhibition was officially boycotted by Europe's crowned heads including Queen Victoria, since it commemorated the regicidal French Revolution, but industrialists did not stay away. In 1900, Vickers Maxim guns were back on display. The fair was, like its predecessors, a curious mix of international co-operation and national assertiveness. Thematic international displays of science or agriculture were installed on the Champ de Mars, while a series of striking 'national identity pavilions', known as the 'rue des Nations', appeared on the left bank of the Seine.[5] Our starting point is the series of attempts in 1900 by one of Europe's genuinely frontier-crossing intellectuals, Patrick Geddes of Edinburgh, to foster transnational co-operation at a signally unpromising moment in Franco-British relations.

The 1890s had provided several serious irritants to an already tense relationship. Republican France had in 1894 formed an alliance with Britain's rival in the Far East, Imperial Russia. Some British politicians thereafter toyed with the idea of a rapprochement with Germany. In 1898, the Fashoda incident in the Sudan, although resolved without bloodshed on the spot, had created extreme ill feeling on both sides of the Channel and even a war scare of sorts. During the Dreyfus Affair British opinion was massively pro-Dreyfus. When in September 1899, Alfred Dreyfus was once more found guilty 'with extenuating circumstances' at Rennes, Queen Victoria was not the only person to express her 'stupefaction'. There was talk of boycotting the coming Exhibition (one possible factor in the government's decision to offer Dreyfus a pardon). Hardly was that obstacle out of the way than the 'Transvaal War' broke out. French opinion, already irritated by the Jameson raid (1895), was vociferously pro-Boer. French guns were used against British troops in South Africa, and the French press was outspoken in condemning Britain. The Relief of Mafeking (May 1900) was greeted with no enthusiasm in Paris. A sensible observer would have concluded in that summer of 1900 that Franco-British relations were at an all-time low.[6]

As Jean-Baptiste Duroselle has pointed out, France in 1900 was a country from which *relatively* few people had emigrated or even travelled abroad, nor had it yet received many foreign immigrants.[7] 'Isolation', splendid or otherwise, was a term which might have been applied as well to the French as to the globally involved and emigration-prone British. But at elite level, there had been a number of cordial contacts, less diplomatic than academic and scientific. A striking feature of the 1900 Fair was the unprecedented number (127) of conferences which

were planned – ranging alphabetically from Abuse of Tobacco to Viti-
culture, by way of Ornithology and Tramways. Although some of them
were all-French affairs, many were thoroughly international and can be
seen as the context for the three quite distinct initiatives which Geddes
launched or pursued.[8]

Geddes as cultural entrepreneur

Patrick Geddes (1854–1932) was a man almost unclassifiable under any
label. His most recent biographer calls him a 'social evolutionist and
city planner'. Today he might be called an environmental philosopher.
His career was idiosyncratic and international. Born in Aberdeenshire,
he studied natural sciences in Edinburgh, London and Paris, before
making Edinburgh his home base, while holding a Chair of Botany at
Dundee. Early notoriety greeted his book *The Evolution of Sex*, co-authored
with J. Arthur Thompson (1889), but his writings are scattered in a
range of forms, and his greatest fame is probably as a town planner,
with followers in India, the Middle East and the United States. His wife,
Anna Morton, was closely involved in all his schemes. In 1900, he concentrated on his Paris initiatives and the family moved there for the
duration of the Exhibition.[9]

Geddes's links with France went back to his youthful *stage* at the
Roscoff marine biology station in 1878. Visiting the Paris Exhibition of
that year, he decided that Paris was 'everything London was not, a city
in which the university had a leading role . . . where culture and city life
had reached new levels of integration'.[10] Influenced by the social theories
(though not the conservative politics) of Frédéric Le Play, the pioneer of
the in-depth survey, he remained in touch with Le Play's follower
Edmond Demolins, met French scientists and academics, including
Louis Pasteur, and later formed close ties with the geographer (and
anarchist) Elisée Reclus. Simultaneously in the 1880s and 1890s, Geddes
had been working on three projects in Edinburgh, all of them relevant
to (but not identical with) his Paris 1900 projects.

First, he ran a series of summer schools known as the Edinburgh Summer Meetings. For about ten years, with a peak in the early 1890s when
over 100 students were attending, these offered courses to teachers and
others. They were run on unconventional lines for the time, being
based on outdoor fieldwork, mixing arts and science, and also mixing
the sexes. (A number of marriages resulted from these informal but
intense gatherings.) Several French lecturers lent their expertise to the
courses. Geddes's second project was the creation in the Outlook Tower,

near Edinburgh Castle, of what he called an 'index museum', designed to locate the city and region within its national and global contexts, a sort of *catalogue raisonné* of the environment. The third project, in which he was associated with a Scottish lawyer living in Paris, Thomas Barclay, was the Franco-Scottish Society, originating, yet again, in a Paris Exhibition, this time that of 1889. Starting with university exchanges, the society brought together a number of the great and the good, primarily from Paris and Edinburgh, in its founding meetings of 1895 and 1896.[11]

Bearing these in mind, it is easy to see the inspiration for Geddes's plans in 1900. The first in time, with similar aims to the Outlook Tower, was Elisée Reclus's Great Globe project, which Geddes enthusiastically espoused. The two men probably met in the early 1890s. Reclus (1830–1905), one of three famous brothers, was notorious in France. His career was as idiosyncratic and even more cosmopolitan than that of Geddes, with the additional spice of extreme political views. As a simple member of the *garde nationale* for the Paris Commune against Versailles in 1871, Reclus was imprisoned and sentenced to deportation, commuted to ten years' exile after appeals from British colleagues. His anarchist views developed in Switzerland, alongside another geographer, Peter Kropotkin, in the 1870s, while his *Géographie Universelle* gained him an international reputation. Following the 1890s wave of anarchist bomb attacks in Paris, with which they were not associated, the Reclus family again came under suspicion. Elisée was prevented from taking up an official post in Brussels, so he taught at the 'alternative' Université Nouvelle there. In 1893 and 1895, he lectured at Geddes's Summer Meetings in Edinburgh. His nephew Paul Reclus (1858–1941) took refuge in Britain from probable arrest in France and was taken under the wing of the Geddes family, under the assumed name Georges Guyou. The two families remained in constant touch for many years.[12]

Elisée had conceived the idea of building a three-dimensional relief globe for the Paris 1900 Exhibition. For the relief to be scientifically accurate (and not massively magnified as it is in most three-dimensional relief maps), the scale would have to be 1:100,000, making it an enormous structure – 26 metres across. It would be at the cutting edge of geographical knowledge, incorporating the surveys and explorations being energetically pursued all over the earth. (It is hard to appreciate now how little of the globe had been accurately mapped by 1900.) Reclus specified its design in a technical pamphlet. But it was more than a scientific project. As Geddes later put it, its purpose was 'primarily human, the unity of the world now the basis and symbol of the brotherhood of man upon it: sciences and arts, geography and labour uniting into a reign of peace and goodwill.' Reclus himself had written that 'geography is history in space, history is geography in time'.[13]

Despite Reclus's sulphurous reputation, the project did get some way towards realization. It was clear that it would be very costly: several million francs at least. Nevertheless it was looked on with favour in principle by the Paris city council of the day, which put its own spin on the project:

> It used to be said that the Frenchman could be recognized by his small moustache and his absolute ignorance of geography. M. Elisée Reclus and his collaborators will give talks which will attract and influence many. Our colonial possessions, so little known, ... will certainly be the subject of public lectures.[14]

Financially, the only hope was private funding from explorers or eccentric philanthropists. This was where Geddes came in, throwing himself with characteristic energy into fund-raising, and hoping in particular to interest American backers. Reclus's politics may account for the reluctance of millionaires to come forward. Reclus himself typically appealed to their idealistic sentiments: 'this will represent a useful work, indispensable if humanity is to achieve perfect knowledge of its environment – and we know, alas, how many futile and criminal projects have wasted human resources in the past.'[15] Until early 1899, both men were hopeful that finance could be found in time, but had to give up as time ran out.[16]

Geddes's second, and more successful, project was inspired by the Edinburgh Summer Meetings, and by a similar French venture at the 1889 Exhibition. This was for a grand-scale Paris 1900 Summer School, known in French as the Ecole Internationale de l'Exposition and in English as the Paris International Assembly (see Figures 9.1 and 9.2). Liaising with the French committee, Geddes was both the dynamo of the English-American section, and in practice the moving spirit of the whole. Both the British Association for the Advancement of Science and its French counterpart were involved in forward planning, and discussions had taken place in Edinburgh, before the twinned meetings of the two Associations at Dover and Boulogne in September 1899. Even then, the Dreyfus verdict threatened to derail the scheme: 'in the suspense of the Dreyfus Case and then in the movement against France and the Exposition as a whole which followed the judgement, all attempts towards promoting the proposed International Assembly were delayed,' as Geddes wrote, but the scientists wished to make a stand against

GROUPE FRANÇAIS
DE L'ECOLE INTERNATIONALE DE L'EXPOSITION
Programme du Lundi 10 Septembre au Samedi 15 Septembre

	LUNDI 10	MARDI 11	MERCREDI 12	JEUDI 13	VENDREDI 14	SAMEDI 15
Arts		M. P. PLAUSEWSKI. *Les applications artistiques et ornementales de la plante* Conférence suivie d'une visite aux spécimens exposés au Pavillon de la Ville de Paris. Rendez-vous : Petit Palais des Champs-Elysées, Salle de l'Ecole Internationale, à 10h. 1/4.	M. Paul SEGUY. *Nos poètes en musique Victor Hugo.* (avec auditions) Rendez-vous : Petit Palais des Champs-Elysées, Salle de l'Ecole Internationale, à 3h. 1/4.		M. Julien LECLERCQ, Rédacteur à la Gazette des Beaux-Arts *Les Ecoles étrangères de peinture au Grand Palais des Champs-Elysées.* Conférence-visite. Rendez-vous : Petit Palais des Champs-Elysées, Salle de l'Ecole Internationale, à 10h. 1/4.	M. DE SOLENIERE. *La Musique française d'aujourd'hui* (avec auditions) Rendez-vous : Petit Palais des Champs-Elysées, Salle de l'Ecole Internationale, à 4h.
Industrie ⎯ **Economie Sociale**	M. Lucien LE FOYER, Président du Groupe de Paris de l'Association de la Paix par le Droit. *Le mouvement pacifique à l'Exposition* Conférence-visite à l'exposition internationale des sociétés de la Paix Rendez-vous : Grande Salle du Palais des Congrès, à 4h.			Visite au PAVILLON DU CREUSOT. Rendez-vous : Pavillon du Creusot, à 10h.1/4.		
Histoire et Traditions Populaires					M. MARILLIER, Maître de Conférences à l'Ecole des Hautes-Etudes. La littérature orale en Basse-Bretagne. Rendez-vous : Petit Palais des Champs-Elysées, Salle de l'Ecole Internationale, à 4h.	Visite au PAVILLON DE LA FINLANDE. M. PESSLEFF *Industrie et Art de la Finlande.* M. Marc LEGRAND La littérature finlandaise Rendez-vous : Pavillon de la Finlande, à 10h.1/4.

LUNDI, MERCREDI & JEUDI de chaque semaine, **VISITE GENERALE DE L'EXPOSITION**, spécialement organisée pour les **délégations ouvrières**, sous la conduite de M. GEHIN. Rendez-vous au **Secrétariat Français**, Quai inférieur du Palais des Congrès, à **9h.1/2 du matin.**

Le prix de la Souscription donnant droit d'assister à toutes les conférences.

Figure 9.1 Groupe Français—De l'école internationale de l'exposition: Programme du Lundi, 10 Septembre au Samedi, 15 Septembre.

PARIS INTERNATIONAL ASSEMBLY
(Ecole Internationale de l'Exposition)

Programme for week from 25th to 30th June

MONDAY, JUNE 25th

10.10–11a.m. Lecture: «*Elements of Progress in the Exposition. I. Neotechnic Elements (Electricity, Art etc.)*» PROF. GEDDES.
Lecture Room, Petit Palais.

11.0 a.m.–12.30 p.m. *Visit to Esplanade des Invalides.* PROF. GEDDES.
Rendez-vous, South end, Pont Alexandre III.

3.30–4.30 p.m. Demonstration and Causerie: «*The International Institute of Bibliography*». MR. LA FONTAINE.
Palais des Congrès, first floor.

TUESDAY, JUNE 26th

10.10–11a.m. Lecture: «*The Problem of Poverty*». MR. ROBERT ERSKINE ELY.
Sous-sol, Palais des Congrès.

11.0 a.m.–12.30 p.m. *Visit to Social Economy Section.* MR. ELY.
Rendez-vous, Sous-sol, Palais des Congrès.

4.0 p.m. Conference: «*La Paix*». MR. CHARLES RICHET. Arranged in conjunction with the French Group.
Lecture Room, Petit Palais.

WEDNESDAY, JUNE 27th

10.10–11a.m. Lecture: «*Elements of Progress in the Exposition, II. Geotechnic Elements (Country and Town,)*» PROF. GEDDES.
Bureau de l'Ecole Internationale, Classe I, Champ de Mars.

11.0 a.m.–12.30 p.m. *Visit to Agricultural Section, Champ de Mars.* PROF. GEDDES.
Rendez-vous in the Section.

3.30–4.30 p.m. Lecture: «*The Problem of the City Child*» DR. W.H. TOLMAN.
Comparative Sculpture Lecture Room, East end, Trocadero.

5.0–6.0 p.m. *Visit to Section of Sciences and Letters. Champ de Mars.* MR. T.R. MARR.
Rendez-vous, Bureau de l'Ecole Internationale, Classe I.

THURSDAY, JUNE 28th

10.10–11a.m. Lecture: «*Current Social Ethics*». MISS JANE ADDAMS.
Sous-sol, Palais des Congrès.

11.0 a.m.–12.30 p.m. *Visit to Rue des Nations: Architecture, etc.* MR W. LAW BROS.
Rendez-vous at Servian Pavilion, Pont de l'Alma.

Afternoon Excursion to SÈVRES AND BELLEVUE. Leader: MR. T. R. MARR. Visit the National Factory of Porcelain at Sèvres and the Terrace at Bellvue. By boat from Pont de la Concorde at 2 p.m.

FRIDAY, JUNE 29th

10.10–11a.m. Lecture: «*Elements of Progress in the Exposition, III. Evolutionist Elements. (Education, Environment, etc.)* PROF. GEDDES.

11.0 a.m.–12.30 p.m. *Visit to Education Section, Champ de Mars.* PROF. GEDDES.
Rendez-vous, Bureau de l'Ecole Internationale, Classe I.

3.30–4.30 p.m. Lecture: «*Industrial Betterment*». DR. W. H. TOLMAN. Illustrated by lantern photographs.
Comparative Sculpture Lecture Room, Trocadero.

5.0–6.0 p.m. RECEPTION in *Sous-sol, Palais des Congrès. Tea, etc.*

8.30 p.m. *Visit to view the illuminations.*
Rendez-vous, Tracadero end, Pont d'Iena.

SATURDAY, JUNE 30th

Whole day excursion to BEAUVAIS. Leader: MR. W. LAW BROS. *Visit the Cathedral, Base-œuvre, The National Factory of Tapestry, etc.* Photographers should take their cameras.

It is possible that a party may remain overnight at Beauvais, thus giving two days to the town. Names of those who hope to join the excursion should be given to Mr. Law Bros or the Secretaries, not later than Friday at 5p.m.

Figure 9.2 Paris International Assembly (Ecole internationale de l'exposition). Programme for the week 25–30 June 1900.

'reactionary British opinion which condemned all of France because of the Dreyfus scandal', and backed the scheme.[17]

Geddes enlisted two Scottish patrons, the Liberal MP James Bryce and the geologist Sir Archibald Geikie, and received generous financial backing from the Scottish industrialist, Sir Robert Pullar (owner of Pullar's of Perth dyeworks). The president of the French section of the School was the former prime minister, Léon Bourgeois. Others involved included the director of the Pasteur Institute, Emile Duclaux; the educational reformer, Louis Liard; the rector of the Sorbonne, Octave Gréard; and the historian Ernest Lavisse (all of whom had attended the meetings of the Franco-Scottish Association and had visited Edinburgh).[18] A large American Committee was composed of names drummed up by Geddes's American tour in January 1900, mostly university professors, but also private individuals, many of them women. Geddes made a point of writing to his American co-secretary Robert Ely to the effect that 'the ladies', as he knew from his Edinburgh projects, were worth having on board:

> Wretched man! why have you left out the ladies? . . . Put them in again if not too late. They were our initial members in Chicago. And get many other capable ladies who are coming over. We will get far more real work and real help from them than from many Presidents![19]

The aims of the school were to provide information about the various scientific congresses; to offer expert lectures on arts and science related to the displays; and to offer to delegations of workers and to parties of students an educationally tailored guide to the exhibition: 'The Assembly has a popular as well as an academic side, offering privileges not only to men of science but also to the intelligent public'. German and Russian groups were also formed, and premises were found, in the Palais des Congrès basement, near the Place de l'Alma. In mid-April, Geddes reported that the Commissaire Général of the exhibition, M. Picard, had agreed 3,000 francs for fitting it out.[20]

The English group began lectures on 14 May, a month after the Exhibition opened, and carried on right through until the end of October. The French group began on 4 June. As can be seen from sample weekly programmes, the courses were wide-ranging and ambitious, covering fine arts, science and social science, with guided visits to displays, local churches or museums. There was a solid core of regular lecturers in both English and French, with specialists coming in for one-off sessions. Core lecturers were often personal associates of Geddes such as T. H. Marr from the Edinburgh Outlook Tower, John Duncan, the Edinburgh artist, and Jane Addams from the Chicago Settlement Hull House.[21] As for Geddes himself, his colleagues described him as 'perpetually in the breach', lecturing daily, taking parties round and liaising tirelessly with his co-secretaries. His correspondence shows that he was handling hundreds of individual requests and queries, on matters both important and trivial. He lectured virtually every day at ten o'clock on a range of general topics, including 'The True France', in which he argued that the fact that the Dreyfus Affair had become public was to the honour of the French; in Britain or America, he suggested, there might have been a hypocritical cover-up.[22] His energy kept him going through a gruellingly hot Paris summer, marked too by a series of personal tragedies, including the sudden death of the young colleague, Robert Smith, who was replacing him at Dundee for the summer term.

After all this energy, it was a little disappointing that the English language sessions had comparatively few takers, and even the French ones were slow to take off at first. The organizers wondered whether visitors to the Exhibition came just for fun rather than improvement. 'Things are still going along rather quietly,' T. R. Marr wrote to Sir Robert Pullar in August, but 'the number of students attending has shown a distinct increase.' Some of Pullar's money had been used to provide tickets for a group of 'workmen students of the Passmore Edwards Settlement in London ... and it is really quite encouraging to hear how much they have enjoyed visiting some of the sections ... under the charge of our lecturers.'[23]

By the time the exhibition closed however, enrolments had picked up and the quadrilingual school had given an impressive number of lectures and classes. A total of 100 lecturers, eight secretaries and ten other staff had worked over the four months, and some 300 formal lectures, 800 talks and 450 guided visits had been given to a total of visitor/students running into tens of thousands. 'Many a university does not offer as much in a semester,' said the international secretary (Geddes).[24] The official report on the Exhibition acknowledged 'the generous and high-minded work of this institution ... In carrying out its educational mission, the School neglected no opportunity to spread a spirit of fraternity among peoples.'[25] In retrospect it was the high point of Geddes's personal involvement in summer schools. Although he attempted something similar at the Glasgow Exhibition of 1901, it was less successful and thereafter his work took other directions. But such as it was, the summer programme during the Paris Exhibition was an unprecedented example both of Franco-British cooperation (which lay at the heart of the enterprise) and of multinational gatherings, in the basement hall which became one of the Exhibition's 'international centres'.

The third scheme which Geddes undertook in 1900 was more quixotic, though potentially further-reaching than either of the others. Had it succeeded, as one of his biographers remarks, it would have given him a job for life. When the Exhibition closed its doors in November 1900, the national pavilions (built by Britain, Hungary, Finland, Germany, Sweden, etc.) were due to be demolished. They had had a 'marvellous effect', 'turning the Seine into a Venetian canal', as Picard put it, and public opinion was not unfavourable when a scheme suddenly emerged to preserve them if not for posterity, at least for the foreseeable future. The idea, reproduced in Picard's official report, was to save them as an avenue of *international* museums: a museum of peace, a museum of the sea, a museum of experimental science, a museum of geography, and so on. The reader will have had no difficulty recognizing Geddes's hand at work. What was surprising, even when one remembers that he had one of the best address books in Paris, was the whole-hearted support he received from some very prominent people in France (including four ministers) and other countries' embassies and consulates. Geddes's plan had truly international dimensions: it amounted to no less than the partial dismantling of the national symbolism of each pavilion, using them to house thematic displays from all over the world, and adding up to a Geddesian 'Index Museum'. Based on the concept of the Outlook Tower, it was a countervailing idea, in an age when new museums were tending to become more national rather than less, more identified with national heritage and perhaps less with science and ideas. One of Geddes's disciples described it as a forerunner of the work of UNESCO. Picard called it 'a fine programme, a generous conception, based on the authority of eminent men'.[26]

Among the staunchest supporters of the scheme, who naturally included the French committee of the summer school, was Emile Duclaux, director of the Institut Pasteur. He wrote to Geddes on 2 November, begging him not to leave Paris, since the project depended heavily on him for two reasons: first, 'You have mastered the elements of the whole scheme better than anyone . . . nobody can replace you but yourself.' The second reason was a diplomatic one:

It is best that the project should be conceived and realized by someone who is not French, and who is supported as you are by a committee with a majority of foreigners. That way, the proposals will have the character of an act of international courtesy, which will make negotiations easier and less costly...For myself, although I am snowed under with work, I will do all that is in my power.[27]

Despite this high-level support, the scheme came to nothing. Geddes's biographers suggest that it almost succeeded, and that had it not been for a legal deadline on 1 January, which bound the municipality of Paris, it could have gone through. In fact there were a number of reasons, practical and political, which incline one to think it was unlikely to succeed. One reason was money: Sir Robert Pullar, whose pocket was deep but not bottomless, wrote in November 1900:

> I have heard more than once from Professor Geddes, who ... is interesting himself in this grand idea of utilizing the 'rue des Nations' for more permanent purposes. I trust this proves to be a more successful scheme than the 'international assembly', but the public appear to be difficult to move in any educational scheme where money is required.[28]

There were also physical problems, since the site was wanted by both the railway company and the river shipping authority. One important factor was the state of the buildings which, although impressive-looking, were temporary structures. The British pavilion, for example, was a replica of an Elizabethan manor house, designed by Lutyens and furnished with arts and crafts. A disdainful memo from the Foreign Office dated 10 November 1900 poured cold water on the scheme, ostensibly for practical reasons:

> It is evident that those who advocated the scheme as far as the British Pavilion is concerned did not ascertain the real state of affairs from any of the British commissioners on the spot ... It appears that the building is a mere make-believe without any real foundations and hardly weatherproof. The actual shell alone was the property of the Royal Commissioners and now no longer belongs to them as it was decided ... that the most economical way of dealing with it was to ask Messrs Aird who put it up, to take it away ... The ... fittings in the interior belong to the various tradesmen who placed them there ... whilst the valuable loan collection of English Art and Pictures is in the process of being returned to its owners in this country. In these circumstances, the realisation of the scheme is obviously impracticable, but were it otherwise, it is improbable that the offer of such a building would be taken as a compliment by the Paris Municipality.[29]

Other countries consulted were more enthusiastic, although possibly less well informed. An official architect's report concluded that most of the pavilions were either already solid enough for short-term preservation or could be made so at a cost. But neither the Paris city council nor the full French government were in favour. Picard was informed on 26 November that the French cabinet, while 'paying homage to the initiative', had voted against keeping the rue des Nations.[30] Without government or municipal financial support, preservation was clearly out of the question, and eventually even Geddes had to call it a day.

Politics, culture and peace: Geddes's address book and the liberal network

Politics had more to do with the scotching of the rue des Nations venture than might first appear, and this chapter will conclude with some reflections on how Geddes and his projects can be contextualized in political and cultural history. Historians must opt for one of a number of narrative genres – diplomatic history, biography, cultural history and so on. So the same period or events can be approached in ways that do not necessarily overlap. The only place Patrick Geddes's ventures are usually mentioned, for instance, is in biographical narratives which treat him as a pioneer, ahead of his time – and he was indeed a visionary. But these pay scant attention to the context in which he operated. Political histories on the other hand give priority to the outstanding events of the period: the Dreyfus Affair, the Boer War. Franco-British attitudes and relations meanwhile are treated either as a problem in diplomatic history, to be studied via official state papers, or as a question of cultural difference, to be studied via views of the other country published in articles or books.[31]

Geddes's international ventures deserve to be taken beyond the biographical; yet they are obviously not important enough to form part of a mainstream political narrative. What can be attempted, however, on the necessarily small scale of this chapter, is an approach inspired by the work of Christophe Charle and others, namely the cultural and prosopographical. By looking very briefly at the individuals around Geddes – politicians, intellectuals, journalists and academics – and replacing them in the political and cultural 'field' of the time, we may gain some insight to the workings of a transnational network, willing to create cross-Channel links in unfavourable circumstances.[32]

We have already noted the difficult context for Franco-British relations. The summer of 1900 brought further complications. In the French local elections in May 1900, the Parisians rejected their radical city council and voted in a nationalist (i.e. anti-Dreyfusard) majority, as fallout from the Affair continued. The Boxer rebellion in China caused panic for a while. In September, Lord Salisbury called a snap election (the so-called 'khaki' election), and defeated the divided Liberal opposition against the background of support for British troops in South Africa. For all these reasons perhaps, British visitors did not hasten to the Exhibition, as is attested by the low numbers attending the Summer School, and other evidence. Thomas Barclay, who organized a banquet for the French and British chambers of commerce in September 1900, claimed that hardly any British tourists came to the Exhibition before that. His account may exaggerate a little but is borne out by the official report of the Royal Commission.[33] Bryce, writing to Geddes in November, stated baldly that 'from first to last, there has never been any interest in England in the Exhibition'.[34] (Conversely, it was reported as late as 1903 that some English prizewinners at the Exhibition had never received their medals, because the official engraver was a pro-Boer Frenchman.)[35]

The network of Geddes contacts in 1900 – arbitrarily defined here as people who helped in any way with the summer school or who backed his rue des Nations project – naturally consisted of those who resisted such prejudices. The British representatives were chiefly radical Liberals, and the French almost all committed Dreyfusards. Geddes himself was not involved in party politics, but it can readily be guessed that his friends would be fairly 'advanced' in their views, even if we set aside the case of Elisée Reclus, whose politics Geddes neither endorsed nor condemned. Geddes's own reaction to the Boer War, for instance, was deeply to regret its impact on peaceful international links.[36]

The British associates who helped him were all more or less out of step with their own government and with public opinion, especially in relation to the Boer War. Their views appear most strongly in connection with the rue des Nations project, for which Geddes needed government approval, in order to save the British Pavilion. His key contact, James Bryce (1838–1922), was a respected academic, jurist and historian, and former President of the Board of Trade. But he was also an opposition MP, who had expressed hostility to the Boer War in the House of Commons during the summer, though he retained his Aberdeen seat in the khaki election. As he wrote to Geddes in July, 'there has . . . been a sad reflux not only in what are called "Liberal principles" but . . . in social and moral ideas also'. He rightly guessed that he would carry no weight either with Salisbury or with Lansdowne, the new Foreign Secretary.[37] Another of Geddes's supporters, the radical journalist W. T. Stead

(1849–1912), was outspokenly opposed to the Boer War, and his intervention on behalf of the plan to save the British Pavilion was positively
disastrous. The socialist aristocrat Lady Warwick had written to the
Prince of Wales, whose friend she was, at the request of Anna Geddes.
She reported back to Mrs Geddes, however, that 'the real stopper to the
whole question is that Mr Stead wrote to Lord Lansdowne and as you
know, he is in very bad books just now!... I am thoroughly of your
mind in the matter, but what can you expect from our present Government.' The Prince of Wales, on advice from Lansdowne and others, did
officially veto the plans.[38]

Thomas Barclay too (later to become a Liberal MP himself) was struggling against what he saw as a hostile environment. A propos the rue
des Nations project, he wrote to Geddes: 'I certainly think that Great
Britain should give up the habit of acting the spoil-play.'[39] Barclay is a key
figure in the Geddes network. Best known for his campaign in favour of
Franco-British rapprochement, he explicitly claimed in his memoirs
that the Franco-Scottish Society was his first step in a deliberate
sequence leading to the Entente Cordiale of 1903–4. Christopher Andrew,
no doubt rightly, judges that Barclay exaggerated his role in the moves
which led to the Entente: *realpolitik* over Morocco and Egypt plus military
complementarity were the key issues underlying the about-turn in foreign
affairs decisions in London and Paris.[40] But Barclay's significance in
helping to overcome elite prejudice was not negligible. He and Geddes
between them knew many of the leading republicans, political and
academic, as can be seen if we turn to some of the Frenchmen with whom
they had close links, almost all of whom were present at what Christophe
Prochasson calls the 'Republican establishment in session', the conference on the social sciences held during the Exhibition in September
1900.[41]

Not all of the Geddes contacts were Dreyfusards 'de la première heure':
Ernest Lavisse, for example, who had been an early member of the
Franco-Scottish Society, was one of the few academics who did not take
sides and appealed for conciliation during the affair. Léon Bourgeois, a
'closet Dreyfusard' in the spring of 1898, made up his mind comparatively late.[42] But as a central figure in French politics, Bourgeois was of
special significance. He was involved in the Franco-Scottish Society, was
an active president of the summer school, and backed the rue des
Nations venture. His philosophy of Solidarism was the ideology of the
French radicals now in office (albeit somewhat precariously at this
stage), and had been quoted by the French President Emile Loubet
when he opened the Exhibition in April.[43] Significantly too, Bourgeois

had been the French envoy to the 1899 Hague conference. Other names whom we find regularly recurring in Geddes's correspondence and writing in his support are Octave Gréard, rather old by now but still rector of the Sorbonne (he took Summer School students on a tour of the 'Nouvelle Sorbonne'); Louis Liard, the administrator behind the French university reform of the previous two decades; and one of the most conspicuous of Dreyfusards, Emile Duclaux (1840–1904), director of the Institut Pasteur. Duclaux is oddly neglected in most accounts of the Dreyfus Affair, yet he was the key scientist who had argued that the case should be judged scientifically with all the evidence in the public domain. Immediately after Zola's 'J'Accuse!', he helped found the Ligue des Droits de l'Homme, and threw his immense prestige into the fight.[44] Douglas Johnson tells us that Duclaux could not sleep until he had expressed himself in public on the Affair.[45] In truly scientific manner, his students when they met nationalists in the street, used to shout 'A bas les microbes!'[46] Duclaux as noted earlier was a committed backer of the rue des Nations project, and his name (along with other Drey-fusards) was enough to spur Edouard Drumont to write against it in *La Libre Parole*.[47]

This indeed is a kind of confirmation *a contrario* that the Geddes network had at its core a group of like-minded men – liberals, pro-Dreyfusards, sometimes freemasons (Bourgeois), academics who already had personal or professional links with their cross-Channel equivalents. The occasional names that stick out like sore thumbs in the lists supporting the rue des Nations project, or indeed the Franco-Scottish Society, are precisely those who are not 'the usual suspects'.[48] In order to look as if they were appealing across the board, both Geddes and Barclay readily admitted that they sought to attract prominent men of different political tendencies, and they succeeded in catching a few isolated names. Thus we find François Coppée and Prince Roland Bonaparte enlisted for the rue des Nations, and Melchior de Vogüé in the Franco-Scottish Society. Coppée had previously been persuaded to lend his name to the Ligue de la Patrie Française, although he was never a great enthusiast; Vogüé was explicitly associated with a literary cam-paign against the Dreyfusard icon, Zola. But significantly, neither of them appears anywhere else in the many records of meetings, corres-pondence and so on associated with Geddes.[49]

Somewhere in between the committed group of regulars and the isolated outsiders, comes the biggest catch of all for the Geddes circle, the French foreign minister, Théophile Delcassé. *Le Figaro* reported on 22 November 1900 that he was among four cabinet ministers who had signed their names in favour of the rue des Nations project (Millerand, Leygues and Baudin were the others). He had evidently not signed in isolation then, but neither can he have done so lightly. Delcassé was a moderate Republican, appointed to the quai d'Orsay just before Fashoda. And he played a long and devious game, leading finally to the Entente. At this stage, November 1900, he may have abandoned any secret thoughts of intervention in South Africa, but was still far from any open rapprochement with Britain. Together with his opposite number Lord Lansdowne however, he was to be a key player in moves towards the Entente. To find him among the signatories of Geddes's audacious project is perhaps indicative of a slight but significant shift somewhere in the log-jam of official Franco-British relations. It indicates that there were some links, fragile but real, between official diplomacy and the kind of cultural diplomacy we have seen at work through the Geddes circle, which was essentially composed of those in opposition to the British Tory government and to the nationalist movement in France.

In an article written to mark the end of the Exhibition, Geddes wrote, with special reference to his international museum project – and in words rather like those of his friend Reclus:

> An active policy of constructive peace, unattainable by mere vague aspiration, by passive negation of war, or even by legal adjustment, is yet gradually organizable, map by map. Let those who desire peace think of this. For as soldiers play their game of war upon the map, so it must be with the game of peace.[50]

All three of his 1900 projects were designed to appeal to those who sought to offer some resistance to the bellicose 'air du temps'.

Notes

1   May 1900, Geddes papers, University of Strathclyde Archives, T. GED 6/1/6; the Hague conference was held in summer 1899. This chapter is based on this collection (hereafter referred to as T. GED with its number) and on the Geddes correspondence in the National Library of Scotland (hereafter NLS plus MS reference). My thanks to the staff of both collections for help and permission to reproduce material. I also gratefully acknowledge the grant of Research Leave from the Humanities Research Board of the British Academy, which enabled me to prepare this chapter. All translations from the French are my own.

2   Quoted in P. Ory, *Les Expositions universelles de Paris* (Paris, 1982), p. 29.

3 Régis Debray in *Le Monde* (7 June 1990), quoted in M. Galopin, *Les Expositions internationales au XXe siècle* (Paris, 1992), p. 12.

4 On the history of exhibitions, an expanding research topic, besides Ory and Galopin, see B. Schroeder and A. Rasmussen, *Les Fastes du Progrès, les expositions internationales 1851–1900* (Paris, 1992), especially pp. 132–9; L. Aimone et al., *Les Expositions internationales 1851–1900* (Paris, 1993). Cf. official 1900 report by A. Picard, *Rapport général (Le bilan d'un siècle 1801–1900)* (Paris, 1906–7), 6 vols (copy in Musée social, Paris).

5 Aimone, *Les Expositions internationales*, p. 79, and *ibid.*, p. 6. The first 'national pavilions' were displayed in 1878.

6 On British reactions to the Dreyfus Affair, see M. Cornick, 'La réception de l'Affaire en Grande-Bretagne', in M. Drouin (ed.), *L'Affaire Dreyfus de A à Z* (Paris, 1994), pp. 575–80. The *Daily Mail* called the Rennes verdict a 'moral Sedan'. On reaction in France to the Boer War, see references in D. Brooks, *The Age of Upheaval, Edwardian Politics 1899–1914* (Manchester, 1995), chapter 1; J.-B. Duroselle, *La France de la Belle Epoque* (Paris, 1992), chapter 6; for a first-hand account, see T. Barclay, *Thirty Years, Anglo-French Reminiscences 1876–1906* (London, 1914).

7 Duroselle, *La France de la Belle Epoque*, pp. 240ff.

8 For a full list see Schroeder and Rasmussen, *Les Fastes*, p. 138.

9 There are several biographies of Geddes, three by Philip Boardman: *Esquisse de l'œuvre éducatrice de Patrick Geddes* (Montpellier, 1936); *Patrick Geddes, Maker of the Future* (Chapel Hill, NC, 1944); and *The Worlds of Patrick Geddes* (London, 1978). Helen Meller's analytic study, *Patrick Geddes, Social Evolutionist and City Planner* (London, 1990) is excellent but has little on Paris 1900 (pp. 113–117).

10 Meller, *Patrick Geddes*, p. 32.

11 See Meller, *Patrick Geddes*, chapter 4. For the origins of the Franco-Scottish Society, see Barclay, *Thirty Years*, pp. 129ff, 195, 359ff., and A. Steele, 'The Franco-Scottish Society and the Edinburgh University Department of French', in *The Franco-Scottish Society Bulletin*, Centenary Number, no. 24 (May 1995).

12 On Elisée Reclus, see M. Fleming, *The Anarchist Way to Socialism, Elisée Reclus* (London, 1979) on his politics; and the special number of *Hérodote*, no. 22 (1981) on his geography. Cf. articles on both Elisée and Paul Reclus in J. Maitron et al., *Dictionnaire biographique du mouvement ouvrier français* (Paris, various years).

13 Quoted by Y. Lacoste, *Hérodote* (1981), introduction. Cf. P. Geddes, 'A Great Geographer, Elisée Reclus (1830–1905)', *Scottish Geographical Magazine* (September–October 1905), pp. 1–14. On the globe project, see Elisée Reclus, *Projet de construction d'un Globe terrestre à l'échelle du Cent-Millième* (Paris, 1895), copy in T. GED 13/2/3 plus related documents; see also NLS MSS 10564, etc. The cost, originally an impossible 20 million francs, came down to about 4 million.

14 Cf. report by M. Thuillier, 25 December 1897; the council voted unanimously to allocate a site at the Trocadero, NLS MSS 10625. (Until May 1900 the Paris council was dominated by radicals, and included Reclus's old acquaintance, ex-anarchist turned moderate socialist, Paul Brousse, which may explain their sympathetic reaction.)

15 Reclus, *Projet*, pp. 8–9.

16 In February 1898, Reclus, still hoping to raise funds, spoke to the British Association. Geddes was trying to raise interest in the US, but reportedly gave up trying when one millionaire asked him, Monty Python-wise, 'But what's in it for me?' In the event there *was* a globe at the 1900 World's Fair, Galeron's less complex celestial globe which became a much-visited attraction; see Picard, *Rapport*, vol. 6, p. 223. For extensive correspondence between Reclus, Galeron, Picard and others, see the Reclus papers in Bibliothèque nationale, Manuscrits, NAF 22916. Reclus despaired at 'capitalist attempts to combine the two projects unscientifically' as a tourist attraction and gave up the idea. See Reclus's letter to Nadar, 18 April 1899, admitting defeat: 'our sons and grandsons will do greater works than we can', *Correspondance* (Paris, 1925), vol. 3, p. 210.

17 Memo by Patrick Geddes, n.d., T. GED 6/2/2; Boardman (1944), p. 211.

18 Full lists of patrons and lecturers in published prospectuses and reports of the Paris International Assembly, T. GED section 6; see catalogue entries.

19 NLS MSS 10509, copy of letter from Geddes to Ely, 27 April 1900.

20 Report on Paris International Assembly by Patrick Geddes, T. GED 6/1/1.

21 T. R. Marr (1870s–1940s) bore the brunt of organization too. His devoted work for Geddes included running the Outlook Tower until 1901. Cf. Meller, *Patrick Geddes*, p. 17 and other references. John Duncan, RSA (1866–1945), was beholden to Geddes for his Chicago post; Jane Addams (1860–1935) was a pioneer of the US Settlement movement, Hull House, Chicago; later a leading peace and women's rights campaigner and winner of the Nobel Peace Prize.

22 T. GED 6/2/16 and Boardman (1944), p. 219.

23 Copy of letter from Marr to Pullar 16 August 1900, T. GED 6/1/5.

24 Boardman (1944), p. 230.

25 Picard, *Rapport*, vol. 6, p. 29, and see copious material in T. GED 6.

26 Picard, *Rapport*, vol. 6, p. 288. See also Boardman (1978), pp. 184ff. Details of the rue des Nations scheme are scattered through a huge correspondence with newspapers, contractors, politicians and officials; T. GED 6/3/1–8.

27 Duclaux to Geddes, 2 November 1900, NLS MSS 10531.

28 Pullar copied out by Marr, November 1900, NLS MSS 10566.

29 10 November 1900, T. GED 6/3.

30 Picard, *Rapport*, vol. 6, p. 289.

31 And there is also a 'new cultural history', which often focuses on exhibitions, and is chiefly concerned with the representation of national identity or with semiological analysis, cf. notes 2–4, 6 and 9 above. For a classic diplomatic history see C. Andrew, *Théophile Delcassé and the Making of the Entente Cordiale* (Cambridge, 1968); see also C. Campos, *The View of France* (Oxford, 1964) and M. Cornick, 'Faut-il réduire l'Angleterre en esclavage? A case study of Anglophobia', *Franco-British Studies*, 14 (1992), pp. 3–19.

32 Cf. C. Charle, *Paris fin de siècle* (Paris, 1998) and many other works; C. Prochasson, *Les années électriques* (Paris, 1991). Geddes's 1900 correspondence has been used for names, only a few of them mentioned here. Private addresses listed by him in 1900 include for instance those of Jules Siegfried, Charles Seignobos, Paul Desjardins, Léon Marillier, who fit the pattern suggested below.

33 Barclay, *Thirty Years*, p. 184, quotes the Royal Commission report.
34 T. GED 6/4/21, letter of 12 November 1900.
35 Andrew, *Théophile Delcassé*, p. 201.
36 'This lamentable war . . . will put back all culture and may spoil our kettle of fish'; Geddes to Marr, 2 October 1899, NLS MSS 10566.
37 Bryce to Geddes, T. GED 6/4/21, July 1900 and 12 November 1900. On Bryce's opposition to the war see Brooks, *The Age of Upheaval*, p. 31.
38 Lady Warwick to Anna Geddes, 22 November 1900, T. GED 6/3/3.
39 Barclay to Geddes, 21 November 1900, T. GED 6/3/3.
40 Andrew, *Théophile Delcassé*, pp. 202–3, 209–10.
41 On the conference see Prochasson, *Les années électriques*, pp. 230–1. Millerand, Seignobos, Charles Gide and Durkheim all attended, and Geddes was on the 'standing commission'; T. GED 6/1/10.
42 On Lavisse and Bourgeois during the Dreyfus Affair, see J.-D. Bredin, *The Affair*, trans. J. Mehlman (New York and London, 1987) p. 235.
43 On Bourgeois see T. Zeldin, *France 1848–1945*, vol. I (Oxford, 1972), chapter on Solidarism, especially p. 294.
44 Duclaux's case raises the question of private cross-Channel connections: in January 1901, he married the English poet Mary Robinson, widow of the French Judaic scholar, James Darmesteter. Barclay's wife was French, as was Archibald Geikie's. Geddes's younger son married Reclus's granddaughter. This is, perhaps, *la petite histoire*, but not negligible as an indication of personal Franco-British links. Note too that Geddes's friend Demolins wrote the controversial *A quoi tient la supériorité des Anglo-Saxons?* (1897); see Cornick, 'Faut-il réduire . . . ', p. 15, and his chapter above; Liard was an authority on British philosophy, and the Comte de Franqueville, of the Franco-Scottish Society, was an expert on English institutions.
45 D. Johnson, *France and the Dreyfus Affair* (London, 1966), p. 220.
46 Prochasson, *Les années électriques*, p. 235.
47 In his anti-Semitic journal *La Libre Parole* Drumont attacked Geddes for cheek, and singled out Duclaux, 'whose role in the Dreyfus Affair made him disreputably famous', press cutting in T. GED 6/3/5. That the new nationalist city council was also opposed to this 'cosmopolitan' scheme can be surmised from Gaston Moch's letter to Geddes on 8 April 1901, T. GED 6/4/21 ('if only the old majority had been there'); see D. Watson, 'The Nationalist Movement in Paris 1900–1906', in D. Shapiro (ed.), *The Right in France*, St Antony's Papers, 13 (1962), pp. 49–84.
48 See *Le Figaro* (22 November 1900) for a leading article listing the supporters of the scheme (copy in T. GED 6/3/1). They included ministers Millerand, Delcassé, Leygues and Baudin, plus Herbette, Gaston Moch and the loyal Bourgeois (Léon and Emile), Liard, Lavisse, and so on.
49 On Coppée, see Bredin, *L'Affaire*, 1983, p. 260; on Vogüé, Charle, *Paris fin de siècle*, pp. 179ff.
50 'Closing the Exhibition, Paris 1900', *Contemporary Review* (November 1900), quoted Boardman (1978), p. 182. On Delcassé, see Andrew, *Théophile Delcassé*, passim.

10
A Left-wing Intellectual of the 1890s: Georges Clemenceau

David R. Watson

It may seem surprising to see Georges Clemenceau described as an intellectual since he is now remembered above all else as the statesman who led France to victory in the final year of the First World War, and then negotiated for his country at the Versailles peace settlement. Beyond that there may be some recognition of his first period as prime minister from 1906 to 1909, when he became notorious for his tough action against strikes, and against the anarcho-syndicalists of the Confédération générale du travail. His various sobriquets, the Tiger, 'Le Premier Flic de France' ('France's number one cop'), 'Père-la-Victoire', all suggest a down-to-earth quality. Certainly he had the common touch which provided the charisma that kept French soldiers fighting through the last bitter and desperate months of the war. He was known for his colloquial, even earthy language, used in repartee in Parliament and elsewhere. Brutal directness, used for humorous or simply for aggressive effect, was one of the characteristics of his style, spoken or written; for example, his comment on the death of the President of the Republic, Félix Faure, 'that does not make one man the less in France', or his undiplomatic rejoinder to the Austro-Hungarian foreign minister in 1918, 'Count Czernin has lied'. Even in his formal speeches, although not colloquial, his style was for the most part straightforward in contrast to the elaborate rhetorical flourishes of many contemporary orators, Jaurès for example.

All this is true. But there was another side to this complex character. It has to be admitted that he is not included in a recently published reference work on French intellectuals on the grounds that he was primarily a politician and not an intellectual.[1] Nevertheless he was seen at the time as an intellectual as well as a political figure, and I will argue that this characterization still has validity. First of all, he was a prolific writer. From 1893 to 1906, and again between 1913 and November 1917, he wrote an article for newspaper publication almost every day. In the last years of his life, from 1920 to 1929 he wrote an enormous amount. That is to say that from the age of 52 to his death, he was writing virtually all the time when he did not hold governmental office, producing an immense amount of journalistic writing, only a small part of which has been collected and published in book form. It may be argued that journalism in itself would not qualify him as an intellectual, although the author of the chapter in the standard collective history of the press in France on this period has stated that 'Journalism was then, more than at any other time, closely linked to literature. . . . From several points of view literature and journalism were intermingled.'[2]

Clemenceau's journalism is a good illustration of this point. A good deal of it was intended to be taken as literature, and after being published in the daily or weekly press was then brought out in book form. He wrote a novel *Les Plus Forts* which was published first in the weekly *L'Illustration* and then published as a book in 1898. Much of his other journalistic writing at this time took the form of short stories, of sketches and vignettes which are on the borders of fiction and non-fiction, and which was then collected and published as books: *La Mêlée sociale* (1895), *Le Grand Pan* (1896), *Au Pied du Sinaï* (1898), *Figures de Vendée* (1903) and *Aux Embuscades de la vie* (1903).

He also wrote a play, *Le Voile du bonheur*, that was published and performed in 1901, and in the 1920s he published studies of the Greek orator Demosthenes, of his friend the Impressionist painter Monet, and an immense two-volume work, *Au soir de la pensée*, a sort of universal intellectual history. In his youth, he had published the thesis that he wrote for his degree in medicine, which was in fact a philosophical work: at that time he translated John Stuart Mill's book on Auguste Comte and Positivism into French, and embarked on other studies in relation to Mill which were not published then, although fragments survived and were published after his death.[3]

One does not have to think that these works are masterpieces of creative writing or profound path-breaking philosophical studies to say that Clemenceau was in aspiration and in achievement an intellectual. Many Third Republic politicians moved in intellectual, artistic and literary circles. Banquets, salons, receptions, regular meetings of small groups of associates often over a meal, were an important feature of Parisian society, as described in Proust's novel and in many other sources. Clemenceau took part in this social and cultural scene, especially in the period 1885–1900, although he withdrew from it at the time of the Dreyfus Affair. Afterwards, his life centred on more intimate and private friendships. As an example of another political figure who moved in such circles, Pierre Miquel's biography of Poincaré can be quoted:

> He often visited Alphonse Daudet in the company of Clemenceau. On occasion he dined with Zola's circle, or with the young Pierre Loti. Paul Hervieu became a close acquaintance. He met the musicians Reynaldo Hahn, and Widor, the painters Clairin and Bonnat...[4]

Jean-Baptiste Duroselle's biography of Clemenceau has provided detailed accounts of his social life including his artistic and literary contacts.[5] But such frequentations would not be enough to substantiate the claim that he was an intellectual and it could not be argued that Poincaré was one. In Clemenceau's case, the claim is based on his writings, and as has been stated, the corpus is immense. As well as the purely journalistic writing, in the sense of articles which were commenting on current politics, there were three periods of his life in which his writing had more intellectual aspirations. These are the periods 1860–70, 1893–1900 and after 1920. In total he was engaged in literary activity for nearly all the time after 1893, except when in ministerial office.

Of these three periods, it is the period 1893–1900 which is of interest here, and it is at that time that Clemenceau set out most definitely to turn himself from a politician into a writer. His defeat in the 1893 election seemed to have ended his political career. He was deeply compromised in the corruption of the Panama Affair through his earlier close association with Cornelius Herz, the sleaziest figure in that scandal; the charges were unfounded but they still stuck. Even more baseless, ludicrous even, was the accusation that he was in the pay of the British Secret Service, the Norton Affair. Nevertheless, both sets of charges were believed by much of the public, and seemed enough to preclude any revival of his political fortune. So he set out to make a new career as a writer. He saw two paths in front of him as a writer. One was the purely literary path. He told Edmond de Goncourt in 1894 that he wanted to write a novel and a play if he could find the time. Léon Daudet tells us that he also went to see his father Alphonse to ask his advice on the technical side of novel writing.[6] But as well as creative writing, he wanted also to expound a broad political/philosophical programme. As he said, looking back on this time in a speech of 1906, 'What a mistake it would be to think that political action is confined to Parliament... Writers made the Revolution. He who has something to say is an invaluable force. The idea will find its way, in whatever form it is expressed.'[7] Thus his writing for the daily press in the years 1893–97 was hardly at all comment on the details of political life on a day-to-day basis. It consisted of discussion of the 'social question' as it was then called, and of reflections on human life and thought in the widest sense. Printed in the first place as newspaper articles, either in his own *La Justice* or in the Toulouse Radical paper *La Dépêche*, they were collected respectively in two volumes, *La Mêlée sociale* (1895) and *Le Grand Pan* (1896). I will refer here to the recent republication of *Le Grand Pan* in a series subsidized by the French government. This pretentious title is a reference to the Greek word which means *all*, and was also, in Greek mythology, the name of a god who represented Nature. By using this in his title, Clemenceau was affirming his anti-Christian pantheism and his introduction developed such ideas with a great display of classical erudition. I do this because in the introduction to this edition I am taken to task for having completely dismissed Clemenceau's intellectual efforts. My verdict is contrasted with that of J.-B. Duroselle, whose biography of Clemenceau praises his writing highly. J.-N. Jeanneney quotes from my biography the phrases, 'appalling example of bad writing in an elaborate tortured style', without stating that I applied this not to the whole of his writings, or even to all of *Le Grand Pan* (although I do admit to saying that 'it is difficult to swallow today'), but simply to its long introduction.[8] In fact, Duroselle says much the same about this introduction.[9]

I can quote in my support Maurice Barrès's verdict: 'Clemenceau lacks the modesty of the true intellectual.'[10] Clemenceau was equally dismissive of Barrès's intellectual pretensions, telling his secretary Martet: 'Barrès had nothing between the ears ... absolutely nothing.... And that means nothing. Just a few clichés. ... The sad thing is that one day he began to see himself as a thinker ... '[11] However, I did pass a more favourable verdict on some of Clemenceau's articles of this time, and did point out that he was taken very seriously by his contemporaries, except for Barrès. He was seen as a mentor by a large section of the educated youth of the time: this point is made by L. Jerrold in an article of 1906, explaining how Clemenceau reinvented himself after the disaster of 1893:

Few public men ever fell as heavily: probably no one has ever recovered with such a spring from such a fall. ... Clemenceau the wily, unscrupulous, worldly politician was heard of no more: two years later, Clemenceau the philosopher, Pantheist and praiser of life astonished a totally different world which had hardly spoken his name before ... *Le Grand Pan* charmed 'les jeunes' of literary Paris who had ignored Clemenceau up to then. . . . After winning 'les jeunes' he forced himself gradually upon the public imagination again, book by book, with leading articles day by day, steadily working from philosophy towards public affairs afresh, through the Dreyfus case, Church and State, and the Franco-German crisis.[12]

One of those young men has recalled his enthusiasm. Francisque Varenne wrote:

The formidable figure of Clemenceau began to excite my passionate interest. When I was less than twenty years old my cousin Alexandre Varenne, five years older than I was, read out to me, around 1896, the impassioned pages of *La Mêlée sociale* and of *Le Grand Pan* in which we found our joint aspirations magnificently expressed.[13]

Léon Blum was another young man of literary tastes and socialistic inclinations who was much influenced by Clemenceau at this time. He told Louis Lévy that before Lucien Herr converted him to the full version of collectivist socialism, he had had the same mentality as Clemenceau: 'Just imagine the Clemenceau of that time, a Clemenceau hesitating between socialism and anarchism, which was also Jean Grave's position then.'[14] This quotation is interesting, not only for its proof of Clemenceau's influence on Blum, but also for indicating how far to the left Clemenceau had gone at that time. In fact Herr himself, who from his position as librarian at the Ecole normale supérieure converted several generations of the intellectual elite to socialism, had been first of all under Clemenceau's influence until he discovered German Marxism and adopted the collectivist creed.[15] Charles Andler refers to Herr's admiration for and friendship with Clemenceau at a slightly earlier period:

Clemenceau's refined and solid culture in literature and art strengthened their bonds. However the affinity was above all political. Herr valued Clemenceau's great knowledge of the Anglo-Saxon world, and his anglophile liberalism. He learnt much from him.

But by 1893 Andler writes that Herr had vain hopes of converting Clemenceau to collectivist socialism: 'He probably dreamed of converting to socialism the boldest of the republican leaders, and firstly Clemenceau whom he admired . . . '[16]

A sign of Clemenceau's acceptance in the literary world was the prominent part that he played in a banquet of 2 March 1895, in homage to Edmond de Goncourt.[17] It was a grandiose and well-publicized occasion, with 300 or else 600 (sources differ) invited guests, all 'hommes de lettres' or 'amis de lettres'. The speakers were Poincaré in his capacity as Minister of Public Instruction and Fine Arts, de Goncourt himself and six others, including Clemenceau. The remaining five were all leading literary figures, whose names have endured: de Heredia, Céard, Daudet, Zola and de Régnier. Clemenceau's speech throws light on the way he looked at literature and elucidates his own efforts at creative writing. I will return to this later. But I will first mention a second banquet, of 4 April 1895, in which he also played a role and which created a great stir at the time. It was organized by the Union de la jeunesse républicaine in homage to the scientist and politician, Marcellin Berthelot, who had been attacked by the literary critic, Ferdinand Brunetière. Brunetière had just become editor of the *Revue des deux mondes*, and was thus a leading figure in the cultural establishment. In an article after a visit to the Vatican, he had proclaimed the bankruptcy of science, and the return of religious belief. Berthelot was taken up as a symbol of the opposite view, that science was not bankrupt, and that a scientific view of the world negated religious belief. This was, of course, Clemenceau's position.[18]

Clearly, then, Clemenceau set out to be, and was accepted as, 'a man of letters'. Can he be called an intellectual? Interestingly enough, it is precisely at this juncture that the term intellectual made its appearance in the French language as a noun denoting a certain type of person, and having a different and more specific shade of meaning than the older and broader terms of writer or man of letters. As Clemenceau himself played a part in the introduction of this neologism, it is worth looking at in some detail.[19]

It used to be thought that Clemenceau himself coined the term: in fact, he did not do so, and specifically denied that he had. Examination of his writing shows him avoiding it, using other terms such as 'hommes de pensée' and 'hommes de pur labeur intellectuel'. When he did use the term, on 18 May 1898, he put it in inverted commas. Nevertheless, its widespread adoption owed much to an incident of the pro-Dreyfus campaign for which he was primarily responsible. This was the publication in the newspaper *L'Aurore* of Zola's open letter to the President of the Republic on 13 January 1898, and the publication the next day of a protest letter supporting Zola signed by figures from the world of literature and learning. Although Clemenceau was neither the owner nor the editor of *L'Aurore*, he was its most eminent columnist, and he organized the publication there of Zola's letter, giving it the famous title 'J'Accuse...!' He also organized the publication of the supporting signatures which became known as the 'Manifesto of the Intellectuals'. Barrès referred to the 'protestation des intellectuels' in an article of 1 February 1898, and the term soon became universally accepted, and lost the pejorative connotation that it had in the eyes of Barrès and his friends. Examination of the background reveals that, although Renan had used the word in 1845–6, it was in notes which remained unpublished until 1906.[20] It was only from around 1885 that it begins to appear in its modern sense. This is perhaps surprising as it was used frequently in English from the second quarter of the nineteenth century. The *New Oxford English Dictionary* cites Byron's *Journal*, written in 1813, published in 1836, and several other early sources. It comes to mind that Clemenceau may have met it in English, in which language he was fluent. His library contains a high proportion of books in English, and he did a competent translation of John Stuart Mill's *Auguste Comte and Positivism*. However, there is in fact no sign of him using the word before its general acceptance. After its appearance in Paul Bourget's prose in the 1880s, it was given widespread currency by Henry Bérenger who employed it 35 times in four books published between 1890 and 1898. Bérenger, like Clemenceau, was an intellectual and a politician. His fame as a writer in the 1890s is now totally forgotten, and he is remembered only as the politician who outlined French policy towards the international oil industry and negotiated with the United States about French war debts in the post-First World War period. But at this time, he was a novelist and writer, who made a speciality of the concept of 'les prolétaires intellectuels', the title of one of his books. What he meant by that term were men who had had higher education, but who were without inherited family money, and thus found it difficult to support themselves financially. This concept was, of course, central to much contemporary fiction, notably Bourget's later work, and Barrès's *Les Déracinés*. Clemenceau himself, although coming from a wealthy family, had dissipated his resources, and could be seen, at this stage of his life, as being virtually himself a member of the intellectual proletariat.

Thus, at this stage, the term '*intellectual*' was used in a snobbish, derogatory way to categorize the defenders of Dreyfus, seen as poor and outside the bounds of respectable society. This was intended to be indicated by Barrès and the group who went on to found the anti-Dreyfus Ligue de la Patrie Française, itself full of names claiming intellectual or literary authority. It was typical of Clemenceau that he accepted the term, and turned an insult into a badge of honour, thus ensuring its acceptance into the language. A particular nuance deriving from its appearance in the language in this context is the idea that an intellectual is someone who has claims to authority from his education and/or intellectual achievement, and who uses this authority to take a stance on a political, social or moral issue. This, Sartre's definition, derives from the twist given to the meaning of the word by Zola and Clemenceau in 1898. It followed also from that date to the 1970s, that the intellectual would be left-wing.[21]

It can then be argued strongly that Clemenceau was not only the man of letters that he set out to be in 1893, but that he was an intellectual in the new sense of the term that appeared at this time, and which became accepted in France precisely because of the pro-Dreyfus campaign in which he played a leading role. But was he a fin-de-siècle intellectual? The answer to that question must depend on how restrictive is the interpretation of fin-de-siècle. In the *New Oxford Companion to Literature in French*, the reader is referred from fin-de-siècle to 'decadence' and the names given there are Lombard, Mirbeau, Lorrain, Rachilde, Huysmans, Loüys and Péladan. Certainly Clemenceau had nothing in common with their type of writing, although Mirbeau was a personal friend and admirer. The Decadents, with their interest in the weirder aspects of the Catholic religion, in mysticism and the occult, were at the opposite pole from Clemenceau's mentality. He outlined his own idea of literature in the speech at the Goncourt banquet referred to above. Talking about the Goncourt brothers' own writing, he quoted with approval their words from the preface to their novel *Germinie Lacerteux*, 'literary study = social investigation'. He went on to develop this idea, stating that it was exemplified in the best contemporary French literature, saying 'Our century has demonstrated the unity of human knowledge. Science must not any more be separated from its expression in art. Newton is a poet, Balzac is a savant.'[22]

His own novel, *Les Plus Forts*, had the subtitle 'roman contemporain' (contemporary novel) to indicate that it was written in this spirit. It was an attempt to produce a work in the Naturalist tradition of the Goncourts, and of Zola, painting a picture of contemporary society, and illustrating in fictional form the social problems he had discussed in the non-fictional articles of *La Mêlée sociale* and *Le Grand Pan*. *Les Plus Forts* now appears almost ludicrous, with its contrived plot, and characters who are no more than pasteboard figures. The plot turns on the struggle between an elderly impoverished aristocrat, Puymaufray, and the businessman Harlé for influence over the life of a young woman, who is legally the child of Harlé, but in reality the natural daughter of Puymau-fray. The novel ends with her supporting Harlé when he gets the army to fire on strikers at his factory, and marrying his chosen suitor, an unscrupulous and opportunistic politician. Clemenceau clearly portrayed himself in Puymaufray and the novel is interesting from that biographical point of view, but it is otherwise unreadable today. However, contemporary judgements were far more favourable. Léon Blum gave it a favourable review, as did other French critics at the time, whilst Ernest Dimnet, writing in 1907, found that it was 'a work of noble inspiration', although clumsy and sentimental, even maudlin.[23]

Clemenceau was far removed from the Decadent group, but his genre remained strongly present in the literary and intellectual world of the 1890s. Looking at a general account such as Albert Thibaudet's, it is clear that decadence was only one strand of the intellectual atmosphere. Thibaudet writes: 'Documentary Naturalism remained solid and flourishing ... A full account of the documentary novel would have almost the dimensions of an encyclopaedia.'[24]

Of course, he does characterize the intellectual world of the last decade of the nineteenth century as being one in which a new generation reacted against the great masters of the previous generation, Renan and Taine, with their advocacy of science, naturalism and rejection of religious belief. It was this current of thought that had been picked up by Brunetière in his article on the bankruptcy of science, and whose more eminent exponents were Bergson, Bourget and Barrès. This intellectual shift had its parallel in the political world, the *Ralliement*, 'l'esprit nouveau', the attempt to carry out a bilateral disarmament between the Republic and the Catholic Church. French Catholics were told by the Pope to abandon their royalism, and rally to the Republic: in response the dominant group of conservative politicians professed an 'esprit nouveau' which abandoned the militant anticlericalism of previous years. These were important components of the *Zeitgeist* of the 1890s, but they were far from being unchallenged, and Clemenceau was one of those who challenged them. The outcome was the failure of the *Ralliement*, the eclipse of 'l'esprit nouveau' and, with the Dreyfus crisis, a return to the most extreme form of the conflict between the Catholic religion and the Republic. Clemenceau played a leading role in all this, and it brought him back to the centre of the political stage. It also appeared to justify his overriding philosophical world-view. Another way of putting it would be to say that Clemenceau himself to some degree imposed his world view and was thus responsible for making the Dreyfus Affair the supreme and exemplary battle that it was. As Joseph

Reinach put it in his contemporary history of the Dreyfus Affair, 'Like an artist, he saw the Affair as a splendid drama. The child of the Encyclopaedists, he envisaged a great battle against the forces of the Middle Ages.'[25] I would argue that Clemenceau was an intellectual figure who counted during the years after 1893, and that his intellectual stance was one which counted for much, perhaps more in the eyes of contemporaries than did the literary Decadents who now seem to epitomize the idea of the fin-de-siècle. Certainly outside the realms of literary history, the last decade of the nineteenth century appears to be one of economic recovery and rapid technological progress. The world was recovering from the economic depression of the 1880s, and this period has been characterized by some economic historians as being a second Industrial Revolution. Many of the technological developments that were to make the twentieth century very different from the nineteenth-century world of coal and steam were already appearing, the telephone, electricity, the internal combustion engine, for example.[26] All of these were elements of the *Zeitgeist* as well as the doubts about science and the first stirrings of 'modernism' in the literary and cultural fields.

In the final assessment, of course, it is with good reason that Clemenceau's reputation in history is not that of an intellectual. For whatever his achievements in that area, they were eclipsed by others. It is not simply that his purely political achievements were so much more significant than his intellectual ones. It was that Clemenceau himself was more striking as a personality, as a human being, than the ideas he expounded. This had already been noted by Ernest Dimnet in 1907, before Clemenceau had the substantial achievements in the political sphere that he was to have later. Dimnet wrote:

It is curious that his literary achievements should be so seldom spoken of. M. Clemenceau has written nearly a dozen volumes on a variety of subjects, and some of them were undoubtedly successful if passing through six or eight or ten editions means success. Yet M. Clemenceau as a writer is not well known ... In the case of Clemenceau it is clear that the man, far from serving the writer, simply outshines him.[27]

Péguy's sympathetic portrait makes much the same point. He wrote of Clemenceau in 1904:

His political situation has almost always surpassed his political rank, and his political action has surpassed his official situation ... Even today he can inspire friendship and admiration among many quite young people, even among Socialists, who prefer his astringent Radicalism to the empty rhetoric of scholastic Socialism. . . . For he is not merely a representative of the last generation, he goes far back in the traditions of the French mind. He is a philosopher only in the eighteenth century sense: but in that sense he is exactly what was then called a *philosophe*. He is just enough aware of scientific and philosophic thought, without having studied it in depth, just well enough informed and just ignorant enough to give exposés of it. In the eyes of his friends and admirers, he is not quite a spoiled child, but something more amusing, a spoiled father or an old uncle, who has had his bad moments, but who in his good moments, charms everyone. The ordinary activities of his Parliamentary and political life would condemn Clemenceau: what saves him and wins him the sympathy of others, are his moments of forgetfulness, the extravagances, when the natural and true part of him comes to the top.[28]

Notes

1 J. Julliard and M. Winock (eds), *Dictionnaire des intellectuels français, les personnes, les lieux, les moments* (Paris, 1996).
2 C. Bellanger, J. Godechot, P. Guiral and F. Terrou, *Histoire générale de la presse Française*, III, p. 277 (Paris, 1972).
3 J. Martet, *Clemenceau peint par lui-même* (Paris, 1929), pp. 257–69. They consisted of a study, *La Femme*, intended as a refutation of Mill's plea for the emancipation of women in his *The Subjection of Women* (1869). See D. R. Watson, 'Clemenceau and Mill', *The Mill Newsletter*, VI, i (Fall, 1970), pp. 13–19.
4 P . Miquel, *Poincaré* (Paris, 1961), p. 147.
5 J.-B. Duroselle, *Clemenceau* (Paris, 1988), pp. 309–429. See also P. Guiral, *Clemenceau en son temps* (Paris, 1994), pp. 119–44.
6 E. and J. de Goncourt, *Journal* (Paris, 1935–6), vol. IX, p. 199.
7 D. R. Watson, *Georges Clemenceau, a Political Biography* (London, 1974), p. 139.
8 G. Clemenceau, *Le Grand Pan* (Paris, 1995), ed. J.-N. Jeanneney, p. 7.
9 Duroselle, *Clemenceau*, p. 328.
10 Watson, *Georges Clemenceau*, p. 142.
11 J . Martet, *Le Silence de M. Clemenceau*, (Paris, 1929).
12 L. Jerrold, 'M. Clemenceau', *Contemporary Review*, XC (July–December 1906), pp. 680–1.
13 F. Varenne, *Georges Mandel, mon patron* (Paris, 1947).
14 L. Lévy, 'Comment ils sont devenus socialistes', quoted by D. Lindenberg and P.-A. Meyer, *Lucien Herr, le socialisme et son destin* (Paris, 1977), pp. 137–8.
15 G. Ziebura, *Léon Blum et le Parti socialiste 1872–1934* (Paris, 1967), pp. 30–5.
16 C. Andler, *Vie de Lucien Herr, 1864–1926* (Paris, 1932), pp. 27 and 95.
17 His speech on this occasion was printed in *Le Grand Pan*, but wrongly dated to 1894; Clemenceau, *Le Grand Pan*, ed. cit., pp. 452–9. See the account in

Duroselle, *Clemenceau*, pp. 317–23, and E. and J. Goncourt, *Journal*, III, p. 1102; J. Renard, *Journal* (Paris, 1955), p. 182.

18 A. Compagnon, *Connaissez-vous Brunetière? Enquête sur un anti-dreyfusard et ses amis* (Paris, 1997), pp. 17–18. H. W. Paul, 'The debate on the bankruptcy of science in 1895', *French Historical Studies*, 5 (1968), pp. 299–327.

19 W. M. Johnston, 'Origin of the term intellectual', *Journal of European Studies*, IV (1974). J.-P. Honoré, 'Autour d'intellectuel', in G. Leroy (ed.), *Les Ecrivains et l'affaire Dreyfus: Actes du colloque de l'Université d'Orléans, 1981* (Paris, 1983), pp. 149–57; C. Charle, *Naissance des 'intellectuels' 1880–1900* (Paris, 1990); J.-F. Sirinelli, *Intellectuels et passions françaises* (Paris, 1990).

20 Watson, *Georges Clemenceau*, p. 149. J. Julliard, 'Clemenceau et les intellectuels', in A. Wormser (ed.), *Clemenceau et la Justice: Actes du colloque de décembre 1979* (Paris, 1983), pp. 101–12. Julliard confuses Renan's notes of 1845–6, which were published in 1906 as *Cahiers de Jeunesse* with *Souvenirs d'enfance et de jeunesse*, published in 1883. This famous text does not use the word 'intellectuel' as a noun; it is, however, to be found in the notes of 1845–6; E. Renan, *Œuvres complètes* (Paris, 1969), t. IX, p. 125.

21 Sirinelli, *Intellectuels et passions françaises*; P. Ory and J.-F. Sirinelli, *Les Intellectuels en France de l'affaire Dreyfus à nos jours* (Paris, 1986).

22 Clemenceau, *Le Grand Pan*, ed. cit., p. 456.

23 Watson, *Georges Clemenceau*, pp. 143–4. S. I. Applebaum, *Clemenceau: Thinker and Writer* (New York, 1948), p. 94. Léon Blum, article in *La Revue blanche* (15 August 1898), reprinted in *L'Œuvre de Léon Blum* (Paris, 1954), I, pp. 52–5. E. Dimnet, 'M. Clemenceau as writer and philosopher', *Nineteenth Century and After*, LXI (April 1907), p. 615. O. Mirbeau, *Les Ecrivains* (Paris, 1926), pp. 24–30, is a eulogistic review of *La Mêlée sociale*, written in 1895.

24 A. Thibaudet, *Histoire de la littérature française de 1789 à nos jours* (Paris, 1936), p. 437.

25 J. Reinach, *Histoire de l'Affaire Dreyfus* (Paris, 1901–11), II, p. 639.

26 Such aspects are discussed in M. Teich and R. Porter (eds), *Fin de Siècle and its Legacy* (London, 1990).

27 E. Dimnet, 'M. Clemenceau as writer and philosopher', p. 611.

28 C. Péguy, *Cahiers de la Quinzaine* (15 March 1904), reprinted in *Choix de Péguy* (Paris, 1952), pp. 30–4; quoted in Watson, *Georges Clemenceau*, pp. 147–8.

11
Seduction and Sedition: Otto Abetz and the French, 1918–40

Nicholas Atkin

Given the French historical profession's current zest for biography, coupled with its fascination with the Occupation, it is small wonder that many Vichyites have recently been placed under the spotlight. In the past decade, biographies have appeared on such luminaries as Pétain, Laval and Darlan, as well as many lesser lights, including Papon, Bousquet and Touvier.[1] Just as there will soon be a study on every department during the *'années noires'*, local history being another popular area of Vichy studies, so too will there be biographies galore on the many curious, colourful and contemptible characters who achieved prominence during the Occupation.[2] There remains, however, one man who, to date, has not received the attention he deserves: Otto Abetz, Hitler's so-called wartime 'ambassador' to France. Although he figures in diplomatic accounts on Franco-German relations, and although his 'kingdom' of collaborators has formed the focal point of important studies by Raymond Tournoux, Pascal Ory and Philippe Burrin,[3] in most works on Vichy he remains an elusive, shadowy and enigmatic figure, rarely stepping out of the footnotes into the limelight.[4]

Abetz himself was never slow to court publicity. During the 1930s, first as a youth leader and then as a member of the Dienststelle Ribbentrop, he was frequently in France where he cultivated the image of a genuine Francophile, deeply committed to the cause of international reconciliation, beguiling his hosts with his wit, generosity and charm, although such social graces did not prevent his 'expulsion' from France in 1939. As wartime ambassador to Paris, he wined and dined French politicians, writers and artists, arguing that he at least, if not Berlin, was committed to a friendship between their two countries. Captured in 1945 and tried at Paris in 1949 for war crimes, he deployed this double-game argument, so beloved among former Vichyites, claiming that he had deliberately thwarted the wishes of his Nazi masters by striving towards a genuine Franco-German *rapprochement*.[5] It was a theme he would return to in his memoirs *Das öffene Problem. Ein Rückblick auf zwei Jahrzehnte deutscher Frankreichpolitik*.[6] Shortly before his death in a mysterious car accident in 1958, almost certainly orchestrated by partisans of the French Resistance, and working as a journalist following his early release from prison in 1954, he heralded the election of the Strasbourg Parliament as a new beginning in the quest for European harmony. At Abetz's funeral, held at Düsseldorf's Stoffel Cemetery, Dr Ernst Achenbach, a key figure at the German embassy in Paris during the Occupation and later a deputy of the Strasbourg Assembly, reiterated this hope, describing his former superior as an astute diplomat whose love of France had been sincere.[7]

Historians, today, are wisely cautious of such claims,[8] yet the absence of any sustained and penetrating analysis of Abetz has meant that the disingenuous claims of the former ambassador and his supporters frequently go unchallenged. He still appears the hesitant Nazi and genuine Francophile, ready to disobey Berlin and entertain French overtures of collaboration. To be fair, a study of his early life reveals something of an idealist, whose dislike of bloodshed and commitment to Franco-German understanding was, to an extent, genuine. In the 1920s and early 1930s, he tried to put these ideas into action, organizing cultural exchanges between the two countries. Yet, even at this stage of his life, Abetz was bristling with arrogance and opportunism and, in the 1930s, he fell under the spell of Nazism. It is possible, as is sometimes claimed, that he hoped to civilize National Socialism from within; it is certain that fascism came to pervert whatever was left of his earlier idealism.

I

Otto Abetz was born on 26 March 1903 at Schwetzingen in Baden, an area close to Alsace-Lorraine; later in life he claimed that his upbringing near to the disputed provinces was one of the reasons why he felt an affinity for French culture. His father was of lowly means, a revenue official, the local *markgräfisch-badische Rentamtsmanns*.[9] As a young boy, Abetz attended the Reformsrealgymnasium at Karlsruhe, yet his education was interrupted by the outbreak of a war which profoundly altered his world view. On his way home from school, he witnessed Red Cross ambulances transporting the injured from the nearby front. In 1915, the Margrave's Palace at Karlsruhe was shelled.[10] In June the following year, while watching the Hagenbeck Circus in Karlsruhe, Abetz himself

was wounded in the leg, the result of an expedition led by Henri de
Kérillis, the future editor of the *Echo de Paris*, and a scourge of fifth column-
ists. According to British intelligence, in 1935 Abetz visited de Kérillis
and 'told him smilingly about the raid, and on the strength of it invited
him to come to Germany to give lectures'.[11] Kérillis wisely refused, and
in 1939 had no hesitation in denouncing Abetz as a 'spy'.

Although his career as a diplomat taught Abetz that war can never be
spurned as an instrument of policy, there is little doubt that the 1914–18
conflict left him with a dislike of bloodshed. According to his memoirs,
the war also shaped his world view in three other respects: in his dislike
of communism; in an early faith in social democracy; and in his desire
for Franco-German friendship. That Abetz despised the far Left cannot
be doubted. Although he professed an early sympathy for socialism, he
despaired of the civil disorder that prefigured and accompanied the
German Revolution. Yet whether his anti-communism conveyed him
into the social-democrat camp remains dubious. While his Catholicism
led him to dabble in personalism, he had little liking for the Weimar
brand of democracy and readily blamed the Republic's leaders for having betrayed Germany at Versailles.[12] Even his religious beliefs quickly
waned: André Weil-Curiel, an associate in the early 1930s, observed
that he had largely abandoned his faith and, during the Occupation,
Abetz steadily worked against the Catholic members of the Marshal's
entourage whom he regarded as a threat to collaboration.[13] Rather, as
Burrin reminds us, he was more a believer in 'direct action' and eagerly
threw himself into the youth movements springing up all over Weimar
Germany.[14] In 1927, he was elected president of the Central Committee
of the Youth Movement at Karlsruhe, and three years later travelled to
Paris where he encountered Jean Luchaire, the youthful editor of *Notre
Temps*, who espoused both the cause of pacifism and Franco-German
understanding.[15] Along with Luchaire, Abetz set up a series of Franco-
German youth camps, the first of which met at Sohlberg in the Black
Forest in summer 1930. The following year, as John Wallace records,
they gathered at Rethel in the Ardennes where Abetz encountered his
future wife, Suzanne de Brouckyère, Luchaire's secretary and later
an official in the Radical Party central office at a time when Daladier
was Party secretary. Although, in the words of British intelligence, 'a
beautiful and wealthy woman', these attributes would not prevent
Abetz from striking up a relationship with Luchaire's daughter,
Corinne, a film starlet, nor (allegedly) with Madame Bonnet who
reputedly handed over 'information from her husband's office', a charge
never substantiated.[16]

If Abetz himself is to be believed, his desire for Franco-German under-standing not only arose out of his youth activities, but also from his artistic training. Having travelled extensively in Europe, visiting Switzerland, Italy and Greece, in 1923 he 'entered the Baden Kunsthoch-schule, and three years later passed the state examination in drawing.'[17] That same year he obtained a position as an art teacher at a secondary school in Karlsruhe. Reflecting in his prison cell on his earlier study of art and literature, he perceived a shared 'cultural heritage' between France and Germany, between the 'Gothic styles of the Mont Saint-Michel and Marienburg Cathedral' and between the Latin and Teutonic influences on Balzac. As he himself remarked, 'Cannot Germany, rich in dynamism, and France, masterful in classicism, still complement each other more fully?'[18] To further this interest in the respective culture of these countries, he undertook lessons in French and succeeded in mastering both the grammar and pronunciation of that language, although his faltering delivery earned the mockery of Parisian journal-ists reporting his trial in 1949.[19]

While Abetz undoubtedly built up an empathy for French culture, how much emphasis we should place on such philosophical ramblings rehearsed in a set of memoirs designed to rehabilitate his reputation remains open to doubt. It should further be remembered that Nazis of all shapes and sizes manipulated art in defence of their cause. Rather, in the 1920s, can we not already see the faint contours of the diplomat who, during the Occupation, would hold court in Paris: the left-leaning pacifist and aspirant *homme de lettres*, lukewarm in his support for liberal democracy, yet keen on direct action and on the lookout to make his mark, a man who would be equally at home discussing affairs of state with Laval, and cultural matters with Brasillach, Châteaubriant and Céline.

II

Arrogant and ambitious, in the early 1930s Abetz was to be found at youth and veterans' rallies where he endlessly rehearsed his platitudes about the need for greater European reconciliation. However, the arrival of Nazism into power, and the accompanying international tension, created friction within the Franco-German camps; and it was probably this strain that accounts for their demise in 1934. Abetz himself was troubled by the arrival of Hitler, and in 1934 was ousted from his post as president of the Confederation of Youth. However, in May that year he relinquished his post as an art teacher in Karlsruhe to travel to Berlin where he became director of the Reichsjungenfuhrung, forging a close working relationship with Baldur von Shirach, in charge of Nazi Youth Organizations.[20] Exploiting the opportunities created by the Nazi takeover, Abetz relinquished youth affairs to devote his energies in two other directions – towards the Dienststelle Ribbentrop and the Comité France-Allemagne. Anxious to clip the wings of Neurath, who had been rewarded with the post of Minister for Foreign Affairs, and keen to undermine the position of Goebbels, who as Propaganda Minister was responsible for distributing pro-German propaganda in France, in 1934 Ribbentrop set up his own bureau, a type of unofficial foreign ministry, whose remit was to build up links with abroad, especially with France and Britain.[21] Anxious to recruit experts on French affairs, this new office inevitably called on Abetz whose job, in the words of Ribbentrop himself, was to promote 'good will' between France and Germany, building up links with French youth and ex-servicemen's organizations, and inviting French 'writers, businessmen, financiers and soldiers' to Berlin.[22] It was a task Abetz quickly warmed to. As Alfred Kupferman recounts, one of his first undertakings was a joint commemoration of 2 August 1914. Held at Sohlberg in 1934, this drew together both writers and youth leaders such as Drieu La Rochelle and Weil-Curiel. The following year, Abetz organized a series of veterans' rallies, involving the leaders of both the Union Nationale des Combattants and the Union Française des Anciens Combattants: Georges Lebecq, Jean Goy and Georges Scapini. In 1936 these men, under the auspices of Abetz, commemorated the anniversary of Verdun and celebrated the Olympic Games at Berlin.

The activities of the Dienststelle Ribbentrop drew Abetz into the circle of the Comité France-Allemagne (CFA), an organization established in November 1935 under the stewardship of Fernand de Brinon and Scapini.[23] Its ranks comprised: veterans, for instance Goy and Henri Pichot; journalists such as Paul Ferdonnet, later the voice of Radio Stuttgart; countless academics and businessmen; and a smattering of rightist politicians, among them Gaston Henry-Haye and René de Chambrun. Subsidized by the Quai d'Orsay until 1939, records Kupferman, the body invited several prominent Germans to Paris, including the film-maker Leni Riefenstahl, and published its own journal, *Les Cahiers franco-allemands*, which dealt with a wide range of artistic and literary matters. To complement these activities, within Germany Abetz revived the ailing Franco-German society, originally created in 1923 by the historian Dr Grautoff. In October 1935, the Deutsch-Französische Gesellschaft (DFG) was relaunched at a ceremony in the castle of Monbijou near to

Berlin, and attended by the French ambassador to the Reich, André François-Poncet. As Abetz himself recalled in 1945, it was the DFG and not the CFA, that was the more active, partly because it received far more enthusiastic support from government organizations and industrial bodies.[24] Presided over by two academics, Achim von Arnim and Friedrich Grimm, the DFG won the backing of the presidents of Lufthansa and Mercedes-Benz, and established branches in several large towns. It also had its own journal, *Deutsch Französische Monats-Hefte*, which duplicated articles published in its French sister paper.

As Abetz became more deeply involved in Franco-German links and became more influential in the Dienststelle Ribbentrop, questions must be asked about his feelings towards Nazism. At his trial in 1949, he made much of the suspicion with which he was viewed by the party leadership and the way in which he was blamed for setbacks in Franco-German relations, most famously the escape of General Giraud in 1942 which led to his temporary recall to Berlin. To an extent, this mistrust was genuine, especially among Ribbentrop's rivals. In 1937, his French contacts, in particular his wife, led him to be investigated by the SS. Writing in January 1942, Goebbels remarked that having a French spouse placed Abetz under 'a severe psychological strain', a point echoed by Hitler himself who feared that Abetz might betray secrets of state in the marital bed.[25] Yet, as John Fox has revealed, the Führer, who generally despised the diplomatic corps, had considerable time for Abetz, granting him several personal audiences and allowing him to sit in on meetings with heads of state, a privilege not lightly given.[26] Anxious to play down such confidences, Abetz in 1949 also made much of the fact that he did not join the Nazi Party until 1937. However, as Burrin reminds us, the reason why he enrolled so late in the day was because, in the period 1933–7, a ban was placed on all new applicants in order to prevent the NSDAP from becoming inundated.[27] Nor should it be overlooked that in 1935 Abetz entered the SS, rising to the rank of *Brigadeführer*.[28]

Perhaps the more apposite question to be asked is not whether Abetz was a Nazi, but what type of Nazi was he? In this respect, some of the most revealing evidence is provided by the novelist Jules Romains, in his *Seven Mysteries of Europe*, although even here it is necessary to read between the lines. Himself a champion of Franco-German *rapprochement*, Romains first met Abetz in 1934 and liked what he saw, describing his new friend as a 'cheerful man', 'a healthy fellow, with reddish hair, an open freckled face with frank, clear-cut features, a pleasant voice, often interrupted by laughter. He might well have come from French Flanders or Alsace.'[29] In conversation with Romains, Abetz professed an affinity with the western nations, the Belgians and northern French, yet expressed nothing but 'aversion' for the Prussians, hardly surprising perhaps given that Abetz himself hailed from the Catholic south of Germany. Mesmerized by a man who seemed to share his own aspirations, Romains concluded that Abetz was not an ardent Nazi, and was only attracted to the movement as a fad. The novelist thus accepted one of Abetz's invitations to Berlin where they again spoke about Nazism. In Abetz's eyes, the NSDAP comprised two contradictory elements: one which attracted fanatics who did not give a 'damn for German and western civilization', the other which comprised 'reasonable' people who viewed Nazism as a way for Germany to reform itself. Röhm and his followers belonged to the former element, and Abetz expressed regret that 'Hitler didn't kill them all', strange words from a man who abhorred violence.[30] Asked about Goebbels and Rosenberg, Abetz proved evasive. His admiration was for Ribbentrop, part of the 'reasonable tendency', and for Hitler himself. For Abetz, the Führer was 'enthroned above a cloud, immersed in reverie, a little absent-minded with regard to men and things below, perhaps even rather uncertain about the value of these men (his supporters) and about the decisions it would be best to make concerning them.'[31]

Although evidence remains impressionistic, it appears that in the early 1930s Abetz may well have had doubts about his new colleagues. As Burrin relates, he was not on the extreme right of the party and was subdued, for example, in his anti-Semitism, although as wartime ambassador he exerted considerable pressure on Vichy to assist the SS in the round-up of Jews. Yet, even in these conversations with Romains, it is apparent that Abetz was attracted to the energy and youthfulness of Nazism, and was greatly in awe of Hitler. For its part, Nazism had use of young men such as Abetz, relentless in their enthusiasm, and hungry for power and influence. It was thus no surprise that Abetz should have been drawn into the murky world of fifth-columnist activities as Berlin geared itself up for war.

III

As Nazi Germany became more and more daring in its foreign policy, remilitarizing the Rhineland in 1936 and precipitating the *Anschluss* in 1938, it recognized the need to monitor the behaviour of the liberal democracies, allaying fears whenever necessary, and fermenting tension if needed. It was a task Abetz excelled in, building up numerous contacts.

At the International Exhibition at Paris in 1937, he was entrusted with 80 million francs to distribute among German delegates.[32] At the Nuremberg conference that same year, he was charged with similar funds, and the reception of guests of the German government. When, in February 1938, Ribbentrop was appointed Foreign Minister, Abetz travelled to Paris as part of the official German delegation. When in Paris, which was often in the period 1938–9, he was frequently a guest at the chic society parties hosted at Versailles by Elsie de Wolfe. John Wallace recalls how, in the run-up to the Munich conference, Abetz was especially busy entertaining his French hosts reassuring them about the true German intentions towards the Sudetenland.[33]

Interrogated on these links by the French authorities preparing his trial in 1949, Abetz recounted what reads like a 'Who's Who' of Parisian political and intellectual life before the war.[34] Among youth leaders, he listed Marel Gautherot of the Jeunesses socialistes; Cécil Mardous of the Ligue des Auberges de jeunesse; and Marc Sangnier and Claude Popelin of the Scouts de France. Among veterans' leaders, he included Scapini, Goy and Lebecq as well as René Cassin, later to become de Gaulle's legal expert in the Free French. Among journalists and writers, he cited Guy Crouzet, Pertinax, Saint-Exupéry, Alfred-Fabre-Luce and Wladimir d'Ormesson. Among politicians, he named Herriot, Bonnet, de Monzie, Champetier de Ribes, Henry-Haye, Reynaud and Taittinger. From 'les milieux mondains', he mentioned the Comtesse de Casteliane and Monseigneur Mayol de Luppé, the future chaplain of the Légion des volontaires français contre le bolchevisme (Legion of Volunteers against Bolshevism). Curiously, the one man missing from the list was the one man with whom he would later do business during the Occupation: Pierre Laval. This is possibly because, at this juncture, Laval was on the fringes of national politics, spending much of his time pressing for the creation of a Pétain government, another man whom Abetz does not appear to have encountered before 1939. Whereas Abetz and the marshal would have had little in common, it is possible to see why Laval would get on so famously with the future ambassador: both men were from lowly backgrounds; both had been teachers, Laval albeit briefly; both had turned their backs on socialism; both disliked warfare; both had been involved in pre-war efforts to promote Franco-German cooperation; both were Anglophobes; both had an easy way with words and social charm; both frequently acted off their own backs; and both had inflated ideas about their own abilities and importance.

The other avenue through which Abetz attempted to influence matters was the press. Anthony Adamthwaite recounts how, on 12 July

1939, shortly after the expulsion of Abetz, two journalists – Aubin of *Le Temps* and Poirier, formerly of *Le Figaro* – were arrested, charged with having received German money for spying purposes.[35] Soon after, Kérillis alleged in *L'Epoque* that Berlin had poured 8 million francs into promoting pro-German propaganda, notably in the pages of *Je Suis Partout* and *Action française*, allegations echoed by the communist journalists Gabriel Péri and Lucien Sampaix in *L'Humanité*.[36] Such claims provoked consternation in the Chamber, as well as law suits from de Brinon and Welczeck. The failure of Kérillis to substantiate his claims was a disappointment, yet there can be little doubt that Abetz was engaged in what British intelligence dubbed 'corruption par l'éditeur', spending the equivalent of 2,000 a month in sterling influencing the press.[37] According to a senior official at the pre-war German embassy, Eugen Feihl, the funds available were superior to those of the ambassador who was often in the dark about the activities of Ribbentrop's emissary.[38] All were amazed at the way in which pro-French articles in German newspapers were quickly replicated in the Paris press. Such articles multiplied as war approached. Immediately after Munich, Abetz was fraternizing with Chautemps, and organizing a series of CFA magazines devoted to Franco-German friendship. When German troops marched into Czechoslovakia, Abetz toured provincial France, speaking to 'civic bodies, colleges and clubs', denouncing those opposed to Hitler's actions as 'bellicists'.[39] Suddenly, his activities ceased. On 1 July 1939, *Le Temps* announced that Abetz, who had recently stated Germany and Poland would soon be at war, had been requested to leave France.[40]

The full truth about Abetz's 'expulsion' will probably never be known as it is likely that the relevant papers were burned outside the Quai d'Orsay as German armour approached Paris in June 1940. It is significant that the *acte d'accusation* contains little about his pre-war activities,[41] suggesting that the relevant dossiers had indeed been destroyed, or that the judicial authorities were protecting those figures who had dallied too closely with Abetz. This was at least the view of *L'Humanité* reporting his trial in 1949.[42] That he was deliberately hoodwinking his French hosts about the real nature of German intentions and disseminating damaging propaganda through the press cannot be doubted. Nor can the fact that he was relaying sensitive information back to Berlin. Yet whether he was truly a spy remains open to conjecture. It must be recalled that, in summer 1939, France was suffering from 'spy fever', and an accusing finger was pointed at anyone who had German contacts. Adamthwaite has shown how many of these accusations, especially those directed at Madame Bonnet and the journalists Poirier and Aubin were without foundation.[43] Abetz's own demeanour also raises doubts about whether he was caught up in the world of espionage. His public and private behaviour was bombastic, brazen and arrogant, hardly the characteristics of a master spy; and it is likely that it was this conceit and willingness to abuse his position that tipped the scales against him. In talks with British embassy officials, Léger, secretary-general of the Ministry of Foreign Affairs, confided that 'no hard evidence' of spying had been uncovered. Had the expulsion been connected with espionage, he explained, 'Abetz would have been put in prison'.[44]

Rather it appears that by summer 1939, Ribbentrop's emissary had become an embarrassment to the French government, behaving in an increasingly outlandish manner. The problem was how to proceed against such a close supporter of the Reich's Foreign Minister at such a delicate juncture in international affairs. Matters in the east provided the excuse. Since early 1939, Hitler had been pressurizing the Free State of Danzig in an attempt to precipitate war with Poland. In late June, the story circulated that Hitler, closely followed by Goering, was to visit Danzig. As Donald Cameron Watt relates, the city would then 'fill up' with 'tourists' and, 'after Hitler and Goering had left, the free city would unilaterally declare its adhesion to the Reich. If Poland reacted, she would be the aggressor.'[45] Getting wind of this tale, and almost certainly acting on his own initiative, on 30 June Abetz entertained French and British journalists, and openly boasted that a pro-Nazi coup was to take place in Danzig 'over the weekend' and that the liberal democracies would do 'nothing'.[46] To his amazement, talk of a *coup* provoked the French into action. On 30 June 1939, Daladier informed the German *chargé d'affaires* that unless Abetz could be persuaded to leave France voluntarily, he would be expelled.[47] Abetz departed within 48 hours.

The 'expulsion' provoked various responses. The British Foreign Office considered the action 'strong', given that Abetz was 'so close a familiar of Ribbentrop'.[48] In Belgium, where Abetz had also been active, building up relations with Léon Degrelle, it sparked off a bitter row between left- and right-wing newspapers, the Rexist *Pays réel* taking an almost identical line to that peddled by the German press.[49] In France, as commented, it provoked spy fever. Yet, naturally, it was Germany that was most affected by French actions; Ribbentrop interpreted the exclusion as a personal insult. In a lofty telegram of 9 July 1939, in which he evoked his friendship with Abetz and their 'tenacious pursuit of understanding between France and Germany', he called for a 'speedy, energetic and successful démarche' to resolve the matter, and instructed the ambassador to Paris to seek an interview with Daladier.[50] This was not a task Welczeck, a diplomat of the old school, particularly warmed to. Long irritated by the unofficial diplomacy conducted by Abetz, he was upset that his own authority had been undermined in the process. Indeed, when Daladier broke news of the expulsion to German embassy officials, the *chargé d'affaires* was 'horrified', confessing that Abetz 'was more important than the ambassador'.[51] Yet eager to prove his credentials and anxious to dispel rumours in Berlin that he himself had been involved in the intrigues against Abetz,[52] Welczeck had little option but to follow orders. On 12 July, he reported that he had seen Daladier the previous evening and reminded his French host that Abetz was the 'friend and well-informed intimate collaborator of the Reich Foreign Minister', and was someone whom he could personally vouch for.[53] The expulsion, he continued, 'would have the worst possible effect' on Ribbentrop and was bound to arouse 'a strong reaction throughout Germany'. For his part, Daladier replied that he did not know Abetz, conceded that 'alarmist' reports were being spread, and promised to look into the matter. Welczeck, however, did not hold out much hope of success as he judged the current political opposition to Daladier would make it difficult for him to reverse the expulsion order. This might have given private comfort to the ambassador, but not to Ribbentrop. In a peremptory telegram of 13 July, Berlin made clear it thought Daladier was evading the issue: 'Abetz is to return to France in the next few days'.[54]

Thus it was that, on 20 July, Welczeck revisited the issue with Daladier, 'restating our arguments in the most energetic and insistent manner.'[55] 'Apart from withdrawing the expulsion order,' he continued, 'the French government also owed Abetz a declaration of his *bona fides*' in order to remove suspicions that he was a spy. This prompted Daladier to remind Welczeck that technically no expulsion order had as such been issued; Abetz had instead been 'courteously requested' to leave France. By this stage, Daladier had done his homework into the affair, remarking that Abetz's comments on Danzig had been 'alarmist' and that his 'activities had revealed propagandist designs'. Very reluctantly did he agree to re-examine the case. Putting the best possible gloss on this conversation, yet probably aware that there was no real possibility of the French changing their minds, Welczeck suggested that Daladier's response was evidence that Abetz would soon be able to return to Paris, and promised to make the necessary arrangements to house him in the embassy building. Buoyed by such confidence, on 2 August 1939 Berlin cabled the embassy to let it know 'in strict confidence' that Abetz was 'already in France' and would arrive in Paris shortly.[56] His visit, it was claimed, was a 'purely private one', concerned with a proposed lawsuit against Kérillis.[57] Two days later, however, it was disclosed that Abetz had been refused entry into France, being turned away at the border at Basle.[58] Once more, Berlin huffed and puffed, the Paris embassy again being instructed to make strong representations to either Daladier or Bonnet.[59] In the event, this diplomatic war of attrition did pay some dividends. On 11 August Welczeck reported that, although the French would not issue an entry visa, they had agreed to a public statement which would make clear Abetz was 'neither suspected of espionage, nor that he had broken French laws'.[60] In this way, his *bona fides* would be vouched for. When this 'half-apology' appeared in the French press on 14 August,[61] a memorandum in the German Foreign Ministry acknowledged that Abetz had been afforded some 'moral satisfaction', but it was obvious Berlin did not view the matter as closed.[62] 'Counter-measures', it was concluded, would have to be taken, although these 'have not been decided upon in detail'. It mattered little as soon France and Germany were at war.

Undaunted by the outbreak of fighting and indefatigable as ever in distributing pro-German propaganda, in late 1939 Abetz was in Amsterdam. Here, according to Morley Roberts of the *Daily Express*, he was playing his usual trick of 'Come into my parlour':

A suite of rooms at the best hotel, a lot of lovely ladies in constant attention; perfect dinner parties, friendly flowing conversation. 'We Nazis have been misjudged. This war is stupid. Germany can protect Holland and its wealth. We must be good neighbours – you help us, we help you. Waiter, more champagne.'[63]

Not surprisingly, his activities were little appreciated by the Dutch and he soon found himself back in Germany where he spent the remainder of the phoney war devising defeatist propaganda for French troops.[64] It was the defeat of the allied armies that paved the way for his return to Paris. In a move clearly designed to humiliate the French and reap revenge for the expulsion a year earlier, on 4 August 1940 Hitler appointed Abetz 'plenipotentiary of the Foreign Office with the Military Commander in France', *Bevollmachtiger des Auswartigen Amtes beim Wehrmachtsbefehlsbotshafter in Frankreich*.[65] As *L'Action française* of 7 August so helpfully pointed out to its readers, he could not strictly be called an 'ambassador' as no peace treaty had been signed with France.[66] It was of little importance. Neither the German bureaucracy in general nor Abetz in particular were constrained by titles. The new 'ambassador' soon became the principal political link between Berlin and Vichy, and he quickly transformed the German embassy on the rue de Lille in Paris into the 'banking house' of collaborationist activities.[67]

IV

In surveying the pre-1940 career of Abetz, it is not difficult to delineate the characteristics of the man who would hold court on the rue de Lille. The swagger, the unlimited self-belief and the ability to charm were ever-present in his make-up, allowing him to wheedle his way into fashionable Paris society and political circles. An examination of his pre-war dealings also reveals a man, fuelled with ambition, who was not afraid to act on his own initiative, and it was this tendency that resulted in his expulsion in 1939. Never a fully-fledged spy, Abetz had only himself to blame for his expulsion from France. Later in the Occupation, he would again attempt several independent démarches. While this behaviour underscored his ambition, it also highlighted the ambivalent and ever-shifting nature of Nazi bureaucracy where it was often difficult to define where responsibility lay. Accordingly, Abetz discovered that his efforts were not always appreciated by Berlin; and, as the demands of 'total war' took their toll on Hitler's mental health, he often became the scapegoat for reversals in Franco-German relations. Nevertheless, it is manifest that both before and during the Occupation he enjoyed considerable confidence on the part of the Nazi leadership. Although his marriage to a Frenchwoman raised eyebrows and although his links with Ribbentrop created rivalries, his commitment to the Nazi cause, his personal admiration for the Führer, and his ability to manipulate the French were never seriously in doubt. This is why Berlin pressed so hard for his reinstatement after the expulsion of 1939. Moreover, Hitler would never have consented to making him 'ambassador' in August 1940 if he had thought Abetz potentially disloyal. Abetz himself was eager to reciprocate this trust. Before 1939, adopting the cause of Franco-German friendship, a cause in which he might once have believed, he pursued largely diplomatic means. During the war, his National Socialist instincts came to the fore. Forever misleading Vichy about Hitler's true intentions, he oversaw the persecution of Jews, the deportation of workers, the plundering of Parisian art treasures and the corruption of French culture. The time to stop thinking of Abetz, the misguided Francophile, one of the forerunners of European union, has arrived. The moment to start thinking of Abetz, the diligent Nazi, has begun.

Acknowledgement

I am grateful to Benjamin Arnold, Michael Biddiss and Matthew Peaple, all of the University of Reading, for their advice in writing this chapter. I also owe a tremendous debt to Douglas Johnson whose lectures on twentieth-century France, when I was an undergraduate at London University, first ignited my interest in the Occupation.

Notes

1 On such figures, see especially M. Ferro, *Pétain* (Paris, 1987), J.-P. Cointet, *Pierre Laval* (Paris, 1993) and H. Coutau-Bégarie and C. Huan, *Darlan* (Paris, 1991). On the 'lesser lights', among many works, see L. Greilsamer and P. Schneidermann, *Un certain Monsieur Paul. L'affaire Touvier* (Paris, 1989); C. Moniquet, *Touvier, un milicien à l'ombre de l'église* (Paris, 1989); P. Froment, *René Bousquet* (Paris, 1994) and B. Lambert, *Bousquet, Papon, Touvier, inculpés de crimes contre l'humanité* (Paris, 1991).

2 The two seminal studies on local history are J. Sweets, *Choices in Vichy France* (New York, 1986) and P. Laborie, *L'Opinion publique sous Vichy* (Paris, 1990).

3 R. Tournoux, *Le Royaume d'Otto. France 1939–1945* (Paris, 1982), P. Ory, *Les Collaborateurs, 1940–1945* (Paris, 1976) and P. Burrin, *La France à l'heure allemande, 1940–1944* (Paris, 1995).

4 Specifically on Abetz, see J. Wallace, 'The Case of Otto Abetz', unpublished PhD thesis, University of Southern Mississippi, 1969, and his 'Otto Abetz and the Question of a Franco-German Reconciliation, 1919–1939', in *The Southern Quarterly*, vol. 13, no. 3, 1975, pp. 189–206; J. Fox, 'German Bureaucrat or Nazified Ideologue? Ambassador Otto Abetz and Hitler's Anti-Jewish Policies, 1940–1944', in M. Fry (ed.), *Power, Personalities and Policies. Essays in Honour of Donald Cameron Watt* (London, 1992); N. Atkin, 'France's Little Nuremberg: The Trial of Otto Abetz', in H. R. Kedward and N. Wood (eds), *The Liberation of France. Image and Event* (Oxford, 1995), pp. 197–208; and M. Peaple, 'Ribbentrop's Francophile: The Political Career of Otto Abetz in the 1930s', unpublished University of Reading MA thesis, 1998.

5 The proceedings are published as *D'une prison. Précédé du procès, vu par Jean Bernard-Derosne. Les quatres témoignages principaux, le réquisitoire et la plaidorie de Maître Floriot* (Paris, 1949).

6 Published by Greven Verlag, Cologne, in 1951. See, too, his *Pétain et les allemands. Memorandum d'Abetz sur les rapports franco-allemands* (Paris, 1948).

7 Wallace, 'Otto Abetz and the Question of a Franco-German Reconciliation', p. 190, for information on Abetz's funeral.

8 H. R. Kedward, *Occupied France. Collaboration and Resistance, 1940–1944* (Oxford, 1985), pp. 12–13.

9 Public Records Office (hereafter PRO), FO 371 31939 Z3005/81/17, Foreign Research and Press Service (Royal Institute of International Affairs), Balliol, Oxford, report of 8 April 1942, 'Otto Abetz: German Ambassador in France', drawn up by Professor Stewart. The same report is in PRO WO 208 4435.

10 Wallace, 'Otto Abetz and the Question of a Franco-German Reconciliation', p. 192.
11 PRO FO 371 31939 Z3005/81/17, report of 8 April 1942.
12 Wallace, 'Otto Abetz and the Question of a Franco-German Reconciliation', p. 194.
13 On Abetz and Catholicism, see A. Weil-Curiel, *Le Temps de la honte* (Paris, 1946, vol. 2), p. 123; A. Darlan, *Amiral Darlan parle* (Paris, 1952), pp. 281–2; *Documents on German Foreign Policy* (*DGFP*), Series D, vol. 11, telegram of Abetz to Berlin, 18 December 1940, pp. 891–8; and Peaple, 'Ribbentrop's Francophile'.
14 Burrin, *La France*, p. 60, pp. 99–100.
15 See B. Unteutsch, *Vom Sohlbergkreis zur Gruppe 'Collaboration'. Ein Beitrag zur Geschichte der deutsch–französischen Beziehungen anhand der Cahiers franco-allemands, 1931–1944* (Münster, 1990).
16 PRO FO 371 31939 Z3005/81/17, report of 8 April 1942.
17 *Ibid.*
18 Quoted in Wallace, 'Otto Abetz and the Question of a Franco-German Reconciliation', p. 193.
19 *Le Figaro* (14 July 1949).
20 Centre de Documentation Juive Contemporaine (CDJC) CII-2 Tribunal Militaire Permanent de Paris, 'Acte d'accusation dressé par Nous, Capitaine Flicoteaux, Substitut du Commissaire du Gouvernment près du Tribunal Militaire de Paris dans l'affaire du nommé, Abetz Otto Friedrich', Paris, 1949, p. 4.
21 For much of what follows, see A. Kupferman, 'Le Bureau Ribbentrop et les campagnes pour le rapprochement franco-allemand, 1934–1937', in CNRS, *Les Relations franco-allemandes, 1933–1939* (Paris, 1976), pp. 87–98.
22 *The Ribbentrop Memoirs* (London, 1954), pp. 68–9.
23 On the CFA, see Burrin, *La France*, pp. 60–3 and Ory, *Les Collaborateurs*, pp. 18–20.
24 Cited in Wallace, 'Otto Abetz and the Question of a Franco-German Reconciliation', p. 200. Details of the DFG are also to be found in *Documents Diplomatiques Français, 1932–1939*, 1ère série (1932–35), vol. 13, telegram of Arnal to Laval, 13 November 1935, pp. 330–3, and Kupferman, 'Le Bureau Ribbentrop'.
25 *The Goebbels Diaries* (London, 1948), entry for 22 January 1942, p. 4, and *Hitler's Table Talk, 1941–1944* (Oxford, 1988), p. 278. See Peaple, 'Ribbentrop's Francophile', for the SS enquiry.
26 Fox, 'German Bureaucrat', p. 180.
27 Burrin, *La France*, pp. 99–100. British sources suggest Abetz may have joined the NSDAP in 1931. See PRO FO 371 31939 Z3005/81/17, report of 8 April 1942.
28 *Ibid.*, CDJC CII-2, Tribunal Militaire, 'Acte d'accusation', p. 4.
29 J. Romains, *The Seven Mysteries of Europe* (London, 1941), p. 153.
30 *Ibid.*, p. 160.
31 *Ibid.*, p. 162.
32 CDJC CII-2, Tribunal Militaire, 'Acte d'accusation', pp. 6–7.
33 Wallace, 'Otto Abetz and the Question of a Franco-German Reconciliation', p. 203.
34 CDJC LXXI-113, procès-verbal of the Commissaire de Police à la Direction des Renseignements Généraux à résidence à Paris, 'Relations d'Otto Abetz avec certaines personnalités françaises', 21 November 1945.

35 A. Adamthwaite, *France and the Coming of the Second World War* (London, 1977), p. 332.
36 PRO FO 371 31939 Z3005/81/17, report of 8 April 1942.
37 *Ibid.*
38 CDJC CII-2, Tribunal Militaire, 'Acte d'accusation', pp. 6–7.
39 Wallace, 'Otto Abetz and the Question of a Franco-German Reconciliation', p. 204.
40 *Le Temps* (1 July 1939), cited in PRO FO 371 23038 C9238/94/18, telegram from Sir Eric Phipps, Paris, to the Foreign Office, 1 July 1939.
41 CDJC CII-2, Tribunal Militaire, 'Acte d'accusation', pp. 1–9.
42 *L'Humanité* (12 and 13 July 1949).
43 Adamthwaite, *France*, p. 332.
44 PRO FO 371 23039 C10340/94/18, letter from Ronald Campbell, Paris, to Sir O. Sargent, 21 July 1939.
45 D. Cameron Watt, *How War Came* (London, 1989), p. 324.
46 *Ibid. Documents on British Foreign Policy, 1919–1939 (DBFP)*, 3rd series, vol. 6, 1939, Sir Eric Phipps, Paris, to Viscount Halifax, 30 June 1939, pp. 212–13.
47 *Ibid.*
48 PRO FO 371 23038 C9238/94/18, telegram from Sir Eric Phipps, Paris, 1 July 1939, on which a London-based official has written the remarks about Ribbentrop.
49 PRO 371 22912 C10305/3630/4, letter from Aveling, Brussels embassy, to Halifax, 20 July 1939.
50 *DGFP*, Series D, vol. 6, telegram of Ribbentrop to Welczeck, 9 July 1939, pp. 886–7.
51 *DBFP*, 3rd series, vol. 6, Sir Eric Phipps, Paris, to Viscount Halifax, 30 June 1939, pp. 212–13.
52 *DGFP*, Series D, vol. 6, Welczeck to state secretary Weizsächer, 15 July 1939.
53 *DGFP*, Series D, vol. 6, telegram of Welczeck to Ribbentrop, 12 July 1939, pp. 907–8.
54 *DGFP*, Series D, vol. 6, telegram of Weizsächer to Welczeck, 13 July 1939, pp. 913–14.
55 *DGFP*, Series D, vol. 6, telegram of Welczeck to Foreign Ministry, 20 July 1939, pp. 946–7.
56 *DGFP*, Series D, vol. 6, telegram of Weizsächer to Welczeck, 2 August 1939, pp. 1043–4.
57 PRO FO 371 23039 C11129/94/18, telegram from Campell, Paris, 9 August 1939.
58 *DGFP*, Series D, vol. 6, telegram of Weizsächer to Welczeck, 4 August 1939, pp. 1062–3.
59 *Ibid.*
60 *DGFP*, Series D, vol. 7, telegram of Welczeck to Foreign Ministry, 11 August 1939, p. 22.
61 Details of the statement contained in *DGFP*, Series D, vol. 7, Bräuer, chargé d'affaires, Paris, to Foreign Ministry, 14 August 1939, p. 58 and PRO FO 371 23039 C11322/94/18, telegram from Campell, 14 August 1939.
62 *DGFP*, Series D, vol. 7, Memorandum of Weizsächer, 15 August 1939, p. 72.
63 *Daily Express* (18 December 1940), in PRO FO 371 31939 Z3005/81/17, report of 8 April 1942.

64   Burrin, *La France*, p. 98.

65   The reasons why Abetz believed he had been appointed ambassador are in CDJC LXXI–111, 'Note réduite par Otto Abetz, le 7 mai 1947', and a note of 25 August 1946. See, too, PRO FO 371 31939 Z3005/81/17, report of 8 April 1942.

66   Contained in PRO FO 371 31939 Z3005/81/17, report of 8 April 1942.

67   The responsibilities of Abetz are outlined in *DGFP*, Series D, vol. 10, Ribbentrop to Chief of the High Command of the Wehrmacht, 3 August, pp. 407–8, also contained in *Trial of the Major War Criminals before the International Military Tribunal (IMT)*, vol. XXXII, *Documents and other Material Evidence*, document 3614–PS, pp. 432–3. See, too, the proceedings of 27 March 1947 in *IMT*, vol. X, *Nuremberg 14 November 1945–1 October 1946*, pp. 124–7, which contains much on Abetz's duties.

12

Les Années noires? Clandestine Dancing in Occupied France

Robert Gildea

The phrase *les années noires* has served to conceptualize the Occupation from Jean Guéhenno's remarkable diary, published in 1947, to contemporary scholarly accounts. 'These were black years,' say Jean-Pierre Azéma and François Bédarida, 'for all those who were afflicted in their own flesh by so much bereavement, anxiety and suffering.'[1] The recurrent themes are those of the repression exercised by the German military administration, Gestapo and Vichy regime (with plentiful photographs of executions) and the hunger created by economic disruption and the demands of the German war effort. Markets were regulated by the French authorities to keep inflation under control and supplied by requisitioning goods from producers; everything from bread to bicycle tyres was rationed. Associative life was severely restricted by the Germans for security reasons and for moral reasons by the Vichy regime which preached that disaster had befallen the French people because of their decadence, and inaugurated the National Revolution to stiffen their sinews.

Survivors of the Occupation period are also inclined to stress the darkness of those years. They do so, it may be suggested, for one of two reasons. If they were involved in resistance activities, they want to stress the horror of the Nazi machine they were up against, while if they were not they feel bound to emphasize that all resistance was impossible. Thus Georges Audebert (b. 1913), who was section head of the prefecture of Tours, which was run by notorious collaborators, says that when he tried to challenge the Germans' demands for industrial materials, they replied, 'If that is how you feel, M. Audebert, you will be in Germany tomorrow.' At the Papon trial, he continues, it was suggested that 'you could resist the Gestapo. It is simply not true.'[2] In similar vein, but in their capacity as ordinary citizens, a group which experienced the Occupation in Chinon comes to the conclusion that 'it was four years of fascist dictatorship'.[3] And yet, against the grain, some survivors of the Occupation are prepared to admit that things were not that bad, after all. One of the participants in the Chinon discussion, Claude Bougreau, ventures that 'the Occupation at Chinon was very mild, in fact'. Michel Leterme (b. 1930), who was a schoolchild at Angers during the Occupation, argues that the 'presence of the German army was not disagreeable'. The military band used to play in the Jardin du Mail on a Sunday afternoon and French and Germans rubbed up against each other in cafés, the cinema and the opera. 'It was what people now call cohabitation,' he declares.[4] Abbé François Jamonneau (b. 1914), who was secretary to the very Pétainist bishop of Nantes during the Occupation, insists that 'it was not a Gulag, anyway'.[5]

*

The purpose of this essay is not to contest that for many French people the Occupation was a period of intense suffering and hardship. It is simply to suggest that this was not the whole picture and that, in some way, the degree of hardship also provoked a desire to throw cares away and pursue pleasure. In particular it will argue that many French people, especially the young, danced their way through the Occupation. Dancing was in fact banned, as the French authorities took the view that while 1.5 million French soldiers were in German POW camps, while French ports and cities were being bombed by Allied forces, and while French civilians were being executed or deported to Germany as political prisoners or forced labour, the rest of the population should act with modesty and decency. This reflected the opinion of a large section of the French population, who had absent ones to mourn. Others, however, particularly young people, saw no reason to sacrifice their desire for conviviality, music and intimacy with the opposite sex to the demands of France's new moral order. The issue of dancing thus became a battleground between the authorities and young people, and serves as a test case to refine enduring assumptions about the nature of Occupied France.

In the first place, the issue throws light on the balance of power and respective priorities of the French and German administrations. Under the Armistice a provisional state of affairs obtained in Occupied France whereby the French state remained sovereign and the French administration stayed in place but the German military reserved the right to take all measures they saw fit to protect the security of their armed forces. The matter of dancing reveals how far French and German authorities acted in concert, whether they had the same goals and whether it was possible for individuals to play one off against the other.

Second, although dancing, especially clandestine dancing, was not an organized form of associative life, it may nevertheless provide a pointer to the amount of associative activity that in fact continued under this period of 'fascist dictatorship'. Third, the case of clandestine dancing provides the opportunity to reassess the success or otherwise of Vichy's National Revolution. Pierre Laval said famously at Pétain's trial that he did not know what the National Revolution was, but the ideology of Travail, Famille and Patrie was certainly at the root of the regime's campaign to eliminate decadence and promote national regeneration, particularly in 1940–42.[6] It was and is easy to be cynical about the programme, but it is hoped here to be specific about the gap between official rhetoric and public response. Fourth, the issue of dancing offers some insight into the attitudes and motivations of young people in Occupied France and notably into how politicized they were. It may be argued that to dance in defiance of an official ban was in itself an act of resistance. Alternatively, dancing may be seen to express the opposite of political commitment, a frivolous hedonism that cannot be compared to transmitting secret messages or meeting parachute drops of weapons.

This chapter is based on the archival sources of three departments of Occupied France along the Loire valley: Loire-Inférieure, now Loire-Atlantique (southern Brittany), Maine-et-Loire (Anjou) and Indre-et-Loire (Touraine), a small part of which in fact lay in the Unoccupied Zone on the other side of the demarcation line. The most significant records are the reports drawn up by gendarmes who surprised clandestine dances and endeavoured to arrest and question the participants, who usually fled. It is clear, however, that the gendarmerie discovered only a tiny fraction of clandestine dances, and the written record has been supplemented by interviews conducted with two musicians of clandestine dances who survive in Touraine. These have much to say not only about the practice of clandestine dances, but also about what they represented to the young people of Occupied France.

The first measures banning public dances were taken not by the Vichy authorities at all but by prefects of the Third Republic during the fighting of May–June 1940. Once the hostilities were over, in July and August, owners of hotels, cafés and restaurants put pressure on the authorities for permission to reopen their dance floors. Owners were quick to point out that their rivals had restarted dancing, with or without permission, but while mayors were often sympathetic, prefects newly appointed by Vichy were intent on keeping the ban in force for reasons now dictated by the National Revolution. Here the ill-defined division of power between French and German authorities came in to play. Some individuals attempted to play one off against the other, appealing paradoxically to the Germans as the more liberal. Thus M. Bellan, manager of the 'Sporting Café' at Saint-Avertin outside Tours, reported by the gendarmerie for infringing the prefect-orial ban, wrote to the Feldkommandant of Tours to have the prefect overruled. The Germans were concerned above all to ensure the security of the occupying forces. Thus they had introduced a curfew to prevent hostile acts against the army under cover of darkness, banned all organizations which were uniformed or which might serve a paramilitary purpose, such as the Chantiers de Jeunesse, Compagnons de France and Boy Scouts, which flourished in the unoccupied zone, and prohibited all public commemoration of national festivals such as Armistice Day (11 November) or 14 July.[7] Dancing, so long as it did not violate the curfew, was not considered a security risk and was thus of no consequence to them. Feldkommandant Kloss of Tours accordingly told the prefect of Indre-et-Loire in June 1941 that 'in future responsibility for making decisions about permitting dances will be left to the French administration', and he confirmed the following month that this ruling had been made centrally by the German military command in Occupied France. The Vichy authorities did not however take this liberty to decide for themselves to mean licence for café-owners, and the prefect told Bellan that 'the instructions of my government are opposed to such goings-on'.[8] Even in the intensely Catholic Cholet area, where it might be assumed that the National Revolution would be widely acclaimed, hoteliers wrote to the German authorities for permission to prolong wedding feasts and other family occasions beyond the curfew. The Kreiskommandant of Cholet was prepared to be accommodating, but this provoked his French opposite number, the subprefect of Cholet, to reprimand him, pointing out that while the German author-ities were equipped to permit or refuse dispensations from the curfew, they were not to authorize dances, which was the responsibility of the French authorities alone and for himself he intended to take a tough line.[9]

One concession that the Vichy government was prepared to make was to permit the continuation of dance classes. But given that they were increasingly crashed by a wider public and came to serve as dance halls in disguise, they were subjected to close supervision. The Interior Minister Pierre Pucheu circulated prefects to the effect that dance classes would be approved if they admitted no more than fifteen individuals for a minimum of five sessions, served no drinks, were accompanied only by piano or gramophone music, not by an orchestra, and issued no publicity. The prefect of Loire-Inférieure decided to summon and interview all those who ran dancing classes in Nantes, to verify their 'morality' and exclude those who might have a 'bad influence' on young people. In most cases he need not have worried. One of the six teachers summoned, Mme Rauzières, told the prefect that her school had been open since 1914 and specialized in drawing-room dancing. 'I have never organized a "dance", except that once a year I put on a dancing matinée at the Hôtel de la Duchesse Anne, which is reserved for my pupils and their families. My lessons are attended by the families, are well run and enjoy an excellent reputation.'[10]

This highly restrictive provision was of no use for the youth of Occupied France, who wanted their own dances on their own terms. They assumed that the wartime ban of 1940 was temporary and in Touraine the prefect was bombarded by petitions of young people requesting the right to dance in public. Anticipating the National Revolution line, 'A group of young people aged from 18 to 20' and 'a young woman who adores dancing', both denied that the cafés where they had habitually danced were 'places of debauchery'. One petition of 'young people who want to make the most of being 20' affirmed that most girls were chaperoned at dances by their mothers, while another, signed by 50 young people, argued that 'we are working with all our heart and all our strength for the rapid recovery of our country. We hope that you will understand us and reward our labours by allowing the dance floors to reopen very soon.'[11]

The attitude of prefects, however, was becoming not more but less liberal, and began to extend from a ban on public dancing to a ban on private dancing as well. Thus in April 1941 the prefect of Indre-et-Loire was giving permission to dances at marriages on the grounds that 'if a dance takes place in a private house it is impossible to prohibit it'. Two months later, however, after a so-called private dance in the village of Reugny was crashed by 300 people, the prefect told the mayor that no more balls at all would be allowed. The following month, having received the green light from the German authorities to make all decisions about dances, he ruled that dancing would be allowed at weddings so long as they were confined to the family and invited guests. This far from satisfied young people, two of whom wrote to the prefect in July 1941 to argue that if the government wanted to encourage the return of town-dwellers to the soil it should not abolish all entertainment in the countryside. There were no cinemas in rural areas, they said, boules and card-games were for another generation, and dancing was the only pleasure to which they looked forward.[12]

The regional prefect of the Loire region, based at Angers, was even more repressive than the prefect of Indre-et-Loire, over whom (together with the prefects of Maine-et-Loire, Loire-Inférieure, Sarthe and Mayenne) he wielded ultimate authority. In July 1942 he published a regulation announcing that 'merry-making which is misplaced and flouted the country's state of mourning' could not be tolerated and therefore 'all public and private dances, noisy or rag processions, singing or licentious behaviour on the public highway, all music that could be heard outside, whether by loudspeaker, jazz band, gramophone or radio, are and remain forbidden'. This classic exposition of the mentality of the National Revolution was backed up by the General Secretariat for Youth which had a departmental delegate in each department whose task it was to coordinate the work of youth leaders in the battle to regenerate the youth of the country. Though their first concerns were for the young unemployed and delinquent youth, they were deeply troubled by the general demoralization of French youth, which they saw to be corrupted by alcohol, pornography and the cinema.[13]

Some measures were taken by the authorities to keep youth out of trouble. Thus the prefect of Indre-et-Loire closed cafés to youths under 18 who were not accompanied by their parents (27 March 1941) and placed cinemas out of bounds to under-18s on schoolday afternoons (4 April 1941), while the police in Tours enforced a municipal bylaw of 1932 under which shops and kiosks could not display pornographic material. But the Secretariat for Youth was fighting a losing battle. Its incursions were resented by the teaching establishment, the Catholic clergy and many municipalities, its attempts to recruit youth leaders as local organizers never got off the ground and it was powerless to reform a young population that had no taste for puritan moralizing and remained as ever thirsty for pleasure. In May 1943 the incoming delegate for youth of Indre-et-Loire simply reported that 'the National Renovation is becoming less and less influential each day'.[14]

Perhaps the most striking evidence of the failure of the National Revolution was the explosion of clandestine dances in the summer of 1943. The prefect of Maine-et-Loire informed the mayors of his department that even Marshal Pétain was exercised about 'the upsurge of clandestine dances, especially in the countryside'. He himself had observed the spate of dances, organized by café-owners, hoteliers or private individuals 'in isolated houses, barns, and even on the main squares of certain villages or hamlets. I will not dwell,' he added, 'on the shocking character of such festivities at a time when hardship afflicts so many families. Only too often women whose husbands are in POW camps are persuaded to dance, a scandalous state of affairs which demoralizes the prisoners who get wind of it.'[15] In the nature of things it is virtually impossible to quantify the amount of clandestine dancing that went on. Departmental archives contain thick dossiers of gendarmerie reports on dances that were raided. The gendarmerie of Vernantes on the Touraine border of Maine-et-Loire, for example, unearthed four clandestine dances of up to 80 people in that and the neighbouring commune in a single month in the Christmas and New Year period of 1943–44 alone.[16] But what they discovered represented only the tip of the iceberg. Confessions and denunciations give some clue. A peasant accordionist of Jarzé in Maine-et-Loire admitted to playing at about 20 dances, being paid 150 francs a time.[17] A somewhat illiterate denunciator asked the prefect of Indre-et-Loire to punish the young people who were involved in dances, adding that 'you will have a lot of them to deal with as they take place every Sunday, on one occasion at Jahan's at Saché, on another at Moreau's at Thibouze, and even in peasants' houses because there will be one next Sunday at la Goupillière, commune of Ballan-Miré, another at Brosseau's at les Brosses in the same commune, and in many other places like the Rideau's place at Druye'.[18] Persuasive evidence, finally, comes from those who were actually involved as musicians at clandestine dances. Abel Avenet (b. 1921), whose family had a small farm with vines and cattle at Dierre, in the Cher valley east of Tours, played the trombone at local dances which, he said, happened most Sunday evenings. The second time they organized a dance, in the spring of 1941, some youths arrived by train from Tours to join in. They let in the girls but not the boys. As a result they were denounced and fined 15 or 20 francs by the Justice of the Peace. But they started up again, confining the participants to locals, and were not bothered again. Sometimes they walked 6 or 7 kilometres to Bléré, on the demarcation line, and danced in an old mill where the thick walls muffled the sound. They returned home after the curfew, taking care to avoid German patrols.[19] André Roussel (b. 1926) had left school at 14 and served in his father's grocer's shop in Bléré. He played the accordion at dances in a large cabin in the forest of Amboise or else in a barn belonging to one family or another. They had look-outs posted on the road, and 'when there was the slightest alert we turned everything off, stopped everything, exactly as in the Indochinese war [in which he later fought], we didn't know each other any more, we scattered'. He remembers that they were more afraid of the French gendarmes, who 'would have caught us if they had been able to', than of the German customs officers on the demarcation line who 'couldn't care less, given that we weren't terrorists and weren't armed'.[20] One way or another there was a complete subculture of clandestine dances, which only rarely left a trace in the official records.

What did these clandestine dances represent? In the first place they were simply an expression of youth culture under the Occupation. In the countryside and small towns it was a youth that had left school at 14 or 16 early and begun to work, further study being ruled out by the demands of the family economy, the elitist nature of lycées and faculties and the special difficulties of the Occupation. They worked six days a week and wanted, come Sunday, in André Roussel's words, 'to give themselves a bit of a good time'. There were not many distractions at Bléré. The Pagoda cinema, open at weekends, was for family entertainment and attended by whole families. There was billiards at the café du Chalet and ping-pong at the café de Commerce, but the café de la Ville was full of elderly card-players. The youth of the little town separated into three age cohorts, each with its own den, where it met at the weekend to feast and play records. The older ones met at the old mill (which Abel Avenet discovered), the middle ones, including Roussel, in an old barn, the younger ones somewhere else. The older and middle ones each had their separate dances, Roussel playing the accordion for the middle ones with a drummer nicknamed Quick, although he was sometimes asked to play drums for his older friends. The middle ones were a tight-knit group of eight boys and eight girls; they all knew each other and there was no question of any wallflowers.

Sometimes the dances were part of the ritual observed by those 20-year-olds about to leave on military service or, since there was none under the Occupation, about to leave for Germany as labour conscripts under the Service du Travail Obligatoire (STO), as did the 'class of 1942' in June 1943. They regularly involved not only young people but also parents, who were called upon to provide facilities for the dancing. One acute if embittered observer in Touraine attributed the indulgence of parents to the demographic phenomenon of the only son who was destined to inherit the family farm, small but extremely viable if it was under vines. 'The child has no respect for old age, infirmity or suffering. Religion and the Family are for him vague notions which disappear under the weight of egoism and the increasing desire for comfort. From children who are never denied anything we get a youth which thinks only of pleasure and of every pleasure: new clothes, and clandestine dances which are more numerous than ever.'[21] The involvement of families in the organization of clandestine dances also had the effect of diminishing the deterrent effect of the fine, should the gendarmerie pounce. 'A fine of 120 francs,' said the commander of the Anjou gendarmerie, 'appears a derisory sum for the peasants of these regions who earn a great deal of money.'[22]

Clandestine dances were also expressions of the solidarity and identity of the local community. In Anjou, the *kermesse* or patronal festival of the village was generally rounded off by a ball which, complained the gendarmerie, was organized 'with the goodwill of mayors who turn a blind eye to the ban imposed by the prefecture'.[23] Of course, local communities were also riven by rivalries and jealousies, and were thus at the mercy of denunciation. An anonymous letter to the subprefect of Saumur complained that dances had taken place in the village of Chacé three nights in a row over Christmas 1944, which was an insult to those families still longing for the return of prisoners of war or deportees. The mayor was himself opposed to the dances but was told by one of the organizers, whose son was a local gendarme, 'You can let things go ahead, *Monsieur le maire*, everything will be all right, the gendarmes are in the picture.'[24] Even the gendarmerie could be persuaded not to do its job if subjected by family ties to the pressures of the local community. Given that the Catholic clergy was traditionally hostile to godless merrymaking, it is not surprising to find one curé who found that dancing under the bombs was in bad taste. Raymond Couella, curé of Nouzilly, reported that as the railway yards of Saint-Pierre des Corps, a suburb of Tours, were being subjected to aerial attack, the guests dancing at a wedding in his parish, 10 kilometres away, came out to watch the fireworks and then went back to dance until 5 a.m. 'While unfortunate victims were dying under the bombs,' he complained, 'these ignoble people continued their insanity. Where is this mentality going to lead us? To certain perdition.'[25]

Surviving musicians of the clandestine dances insist that they were all good clean fun. News of a forthcoming dance was communicated by word of mouth, says André Roussel, comparing it to the recent spate of raves, although on a smaller scale. 'We really didn't do any harm,' he explains, 'we didn't smash things up.' He asserts that drinking did not take place at the ball, but that afterwards the dancers sometimes retired to an isolated limestone cave to drink; on one occasion he had to take a legless friend home in a wheelbarrow. Other evidence suggests that dancing and drink mixed fairly well. As the gendarmerie of Longué, near Saumur, surprised a clandestine dance in a barn, 100 to 120 young people scattered, leaving a table groaning under unfinished bottles of wine and a 50-litre barrel of wine, drunk to the last 5 or 6 litres. The owner, Auguste Josselin, aged 70, confessed that he had organized the dance with his 31-year-old son Maurice, charging the boys an entry fee of 20 francs and the girls 10, and selling wine at 40 francs per bottle. The gendarmes were delighted to charge him not only with organizing a prohibited dance but with selling liquor without a licence.[26] In the nearby commune of Saint-Lambert-des-Levées, the dancers did not even bother to run when they were discovered in a barn celebrating the New Year, 1945, by the gendarmerie of Saumur. Stocks of wine and rillette sandwiches for sale were found and the gendarmes, noting the 'commercial turn' that clandestine dances were taking, charged the farmer, Henri Fefeu, aged 60, a recidivist in matters of clandestine dances, with a list of offences relating to the black market including illegal pig-killing for the rillettes, the purchase and resale of 60 loaves of rationed bread for which coupons had not been exchanged, and illicit price-raising.[27] Further north, at Cuon, the inhabitants complained that balls were simply being used as a cover for black market activities. Entry fees were exorbitant, wine, white bread and *charcuterie* were being sold by peasants at black market prices, and the profits were so enticing that recently traders had fought with knives for the right to supply the dances.[28]

A final question that has to be asked is whether, given the clandestine nature of the dancing, involving defiance of the Vichy authorities and, so far as the curfew was concerned, the German authorities, such dancing may be considered an act of resistance. Of course, resistance was not all about blowing up supply trains and ambushing German patrols. French youths found plenty of other ways of a symbolic nature to express their hostility both to the German occupation and also, on occasion, to Vichy. Public commemorations which were banned by the German authorities could be celebrated in covert ways. All commemoration of Armistice Day on 11 November 1940 was prohibited, for example, and the police and school authorities dragooned into enforcing it, but this did not prevent the school students of Nantes from trying to organize a school strike, attempting to lay wreaths at the war memorial and singing the *Marseillaise* on the steps of the Théâtre Graslin. Most daringly, a 16-year-old schoolboy, Christian Mondragon, and a student, Michel Dabat, climbed the tower of the cathedral under cover of darkness, and over the heads of the German sentries, attached a tricolour to the lightning conductor.[29] In March 1941 another symbolic gesture was in vogue, inspired by the BBC, to paint or draw 'V' signs on public buildings, preferably those occupied by the Germans. This severely antagonized the Germans who ordered the graffiti to be removed at once, while the French authorities published appeals to young people to refrain from vandalism which risked provoking the collective punishment of the population.[30] Ringing the changes, young people used Joan of Arc Day, celebrated each May, as another occasion for a patriotic demonstration. For this commemoration no parades were allowed but the authorities were permitted to lay wreaths at the statue to be found in many French towns, doubtless on the ground that Joan had booted the English out of France. But she was also fêted as the heroine of national revival whose patriotic orientation could be easily glossed. In Nantes, on 11 May 1941, a group of 50 female students processed to the cemetery where six British airmen had been interred the previous day; finding the gates locked they shouted 'Long live de Gaulle! Down with the Boches!' before being dispersed by the French police.[31] The previous night, at Bléré, the swastika flying over a German-occupied building was stolen and a tricolour hoisted on a nearby poplar tree. The German authorities were in no doubt that the perpetrators were young Gaullist sympathizers and noticed that the tricolour was surmounted by a cross of Lorraine, 'symbol of the Maid of Orleans' now adopted by de Gaulle. As a punishment they imposed a local curfew between 9 pm and 5 am until the perpetrators were found.[32]

French youth thus found symbolic ways, between passivity and armed resistance, of expressing their hostility to the German presence. The question at issue here is whether clandestine dancing constituted symbolic resistance. There are a number of reasons why it may be argued that this was not the case. First, it was an act of defiance not of the German military regime but of Vichy. The Germans, as we have seen, were interested only in protecting the security of their armed forces, not in preaching morality to the defeated. They specifically handed responsibility for the regulation of dancing to the French authorities and were concerned only about the curfew. Second, the penalties for discovery were extremely mild, a few francs and perhaps confiscation of the musician's instruments. Clearly if black market activity was involved the penalties were higher. But the genuine political activity of a student like Michel Dabat, who went on from hoisting tricolours to higher things, led directly to the firing squad as one of the hostages of Nantes-Châteaubriant.[33] Third, the motivations of those who took part in clandestine dances were only political in the sense that they defied the law; otherwise they were an expression of pure pleasure. 'Boys and girls just wanted to amuse themselves,' says Abel

Avenet. At the age of 18, echoes André Roussel, 'it's the time to live. The only thing that interested me was going out with girls and playing the accordion.' 'I don't know any partisans,' he says, as he insists on calling the maquisards, 'I never mixed with them, never associated with that lot.' Indeed, he says, once near the Liberation they decided to hold a dance on a village street and were obliged to scatter as 'partisans' arrived in the vicinity and started firing.

A fourth and conclusive reason why dancing did not constitute an act of resistance is that it continued to be banned by the French authorities after the Liberation and down to the end of the war in Europe in May 1945, so that the clandestine dancing that also continued was in defiance not of the Vichy regime but of the Republic. The Liberation, as everyone knows, was greeted by dancing in the streets, but the official line remained that so long as prisoners of war and deportees had not returned the ban was in force. Naturally most of the population did not see things in the same way. Maurice Lefebvre, who was collared by the gendarmerie in September 1944 for organizing a dance in his café, told them, 'I thought I was allowed to, given that Tours had been liberated.'[34] Albert Chevalier, who had his musical instruments worth 20,000 francs confiscated after a police raid on a dance at Saint-Lambert-des-Levées in January 1945, protested in the strongest possible terms:

> given that everywhere, I repeat EVERYWHERE, people are dancing. Here is proof. The same evening as we were caught there was a big dance in the cinema at Longué, for which posters had even been put up. A large orchestra that I know was playing. Recently there have been dances at Fontevrault, Montsoreau, Parnay, with the complicity of the gendarmerie of Fontevrault and, to cap it, gendarmes went to the dance that has just taken place at Turquant.

To reinforce his case he set out his resistance credentials, saying that he had played for maquisards who later became FFIs (members of the armed resistance) in the woods of Breuille and Vernantes, and refused to play for the Germans at Saumur. 'The Fourth Republic,' he concluded, 'has restored the motto Liberty, Equality, Fraternity to its place of honour. Are we to believe that it is only for the façades of public monuments?'[35]

It was not only organizers and musicians who were becoming impatient with the authorities. The dancers themselves were no longer in a mood to scatter when the gendarmerie appeared, and when FFIs or American soldiers were present at a dance, the situation could turn nasty. When the gendarmes turned up at a dance at Yzeures-sur-Creuse in April 1945 they were surrounded by a group of angry dancers, four of whom belonged to the Division Leclerc, who clearly felt that they had not liberated France for nothing and refused to go home.[36] On the front with the German forces holed up in the Saint-Nazaire pocket, the FFIs were still on active service in the spring of 1945 and reports about their brutal behaviour only too frequent.

When the gendarmerie raided a café in Coueron where a dance was going on, Lieutenant 'Pierre' as the landlord introduced him threatened them with a revolver and they beat an undignified retreat rather than risk an incident.[37] Back at Chacé, where the NCOs of the Cavalry School of Saumur, covered in glory since their last-ditch resistance to the Germans in 1940, had organized a dance, the meddling gendarmerie were asked by a sergeant, 'Why are dances allowed at Château-du-Loir and not here?' He answered his own question, 'because the prefect is an idiot (*un con*) and the subprefect too. If there is a fine to pay just send the bill to de Gaulle who is my boss. No one is going to stop us dancing.'[38]

Under the pressure of public opinion, the provisional government of the Republic was prepared to make some concessions. The Interior Minister Tixier told commissaires de la République and prefects in October 1944 that public dances were still banned, but later conceded that they might be permitted for three categories: charity balls for the benefit of POWs, deportees, civilian casualties and their families; traditional local and national festivals; and dances organized by resistance groups, army units and soldiers on leave. Permission in the first and third categories had to be requested from the prefect at least a fortnight in advance of the event and mayors were invited to assess whether the dance would 'shock public opinion'. This loophole provoked a rash of requests for dances that were charity balls only in name, and a significant section of public opinion continued to feel that while the fighting continued the ban on dancing should be respected. Thus a Resistance group of Saumur complained to the prefect that 'it is unacceptable that people are exploiting our unfortunate prisoners and deportees to organize celebrations uniquely to satisfy their personal pleasure and that young people neglect their sense of duty by forgetting the suffering of our dear ones who are still absent'.[39] This led the prefect of Maine-et-Loire appointed at the Liberation to reimpose a blanket ban on public dances on 14 March 1945 and freedom to dance was not finally announced by the government until 7 May 1945.

What general conclusions may be drawn about France under the German Occupation from the case of clandestine dancing? First, it is possible to reinforce the point, already well taken, that the National

Revolution was much rhetoric and propaganda which found a ready response among only the minority of French people who took its moralizing seriously. Most saw no reason to internalize the defeat as shame and to mend their ways and endeavoured to retrieve the maximum of relief from a sorry situation. Second, it must be recognized that though the sovereignty of the French state was limited in the occupied zone, the concerns of the German military administration were strictly limited to military security. The regulation of dancing was gratuitous and the responsibility of the French authorities alone. In this respect they demonstrated that they could be more repressive than the Germans, and a parallel might be drawn with the regulation of the cinema, where the Germans were exercised by catcalls during newsreels but not, as were the French authorities, by the dangers of pornography. Third, received ideas of 'fascist dictatorship' and *les années noires* should not reduce our appreciation of the Occupation to one of a police state characterized by fear and hunger. Despite the prohibition of paramilitary groups and the dissolution of masonic lodges by Vichy there was an active associational life of recreational, cultural and charitable organizations, even in the Occupied Zone, which deserves further study, especially given the current vogue of interest in civil society. Lastly, the phenomenon of clandestine dancing suggests that between the politicized sphere of resistance and collaboration and the *vie quotidienne* of ration cards and bread queues there is another story to be told, less about bread than about circuses, less about resistance than about hedonism. As the accordionist André Roussel said, 'those were the good old days, weren't they?'

Notes

1   J.-P. Azéma and F. Bédarida, *La France des années noires* (Paris, 1993), II, p. 514. See also H. Rousso, *Les années noires: vivre sous l'Occupation* (Paris, 1992) and J. Guéhenno, *Journal des années noires* (Paris, 1947).
2   Interview with Georges Audebert, Tours, 24 June 1998.
3   Interview with Les Amis du Vieux Chinon, 23 June 1998.
4   Interview with Michel Leterme, Angers, 13 May 1997.
5   Interview with Abbé François Jamonneau, Nantes, 26 May 1997.
6   See for example L. Gervereau and D. Peschanski (eds), *La Propagande sous Vichy, 1940–1944* (Paris, 1990), and D. Rossignol, *Histoire de la Propagande en France de 1940 à 1944* (Paris, 1991).
7   On youth organizations, see W. D. Halls, *The Youth of Vichy France* (Oxford, 1981).
8   Archives Départementales (referred to hereafter as AD) Indre-et-Loire (hereafter I-et-L) 10W48, Bellan to Feldkommandant, 12 June 1941; Feldkommandant to prefect, 20 June and 4 July 1941; prefect to Bellan, 1 July 1941.

9 AD Maine-et-Loire (hereafter M-et-L) 95W75, subprefect of Cholet to Kreiskommandant of Cholet, 7 July 1941.
10 AD Loire-Atlantique (hereafter L-A) 270W153, circular of Pucheu to prefects, 21 Oct. 1941; Mme Rauzières to prefect, 13 March 1942.
11 AD I-et-L 52W24, petitions of 4, 6 and 10 November 1940.
12 AD I-et-L 52W24, prefect to mayor of Sorigny, 16 April 1941; prefect to mayor of Dame-Marie-des-Bois, 13 June 1941; prefect to mayor of Semblançay, 24 July 1941; Pierre Renou and Maurice Rigoreau of Semblançay to prefect, 18 July 1941.
13 See, for example, the reports of the delegate of Indre-et-Loire, Marcel Bocquet, in AD I-et-L 46W87.
14 AD I-et-L 10W9, report of H. Biessy, 1 May 1943.
15 AD M-et-L 117W32, circular of prefect to mayors, 27 July 1943.
16 AD M-et-L 117W32 and 80W34, gendarmerie of Vernantes, reports of 5, 28 and 29 December 1943, 2 January 1944.
17 AD M-et-L 117W32, report of gendarmerie of Seiches, 20 November 1943.
18 AD I-et-L 10W38, letter signed 'un ami de la dance' to prefect, 12 May 1943.
19 Interview with Abel Avenet, Dierre (I-et-L), 25 June 1998.
20 Interview with André Roussel, Bléré (I-et-L), 25 June 1998.
21 AD I-et-L 218W17, report of cantonal representative of Neuillé Pont Pierre, April 1943.
22 AD M-et-L 18W50, report of 9th Gendarmerie Legion, 3 July 1943.
23 AD M-et-L 18W50, report of 9th Gendarmerie Legion, September 1943.
24 AD M-et-L 97W123, anonymous letter to subprefect of Saumur, 28 December 1944; subprefect to Captain Viala of Saumur gendarmerie, 30 December 1944.
25 AD I-et-L 10W38, letter of curé of Nouzilly, 24 April 1944.
26 AD M-et-L 80W34, report of gendarmerie of Longué, 18 May 1944.
27 AD M-et-L 117W2, report of gendarmerie of Saumur, 7 January 1945.
28 AD M-et-L 117W32, letter of 'un groupe d'habitants écœurés' to prefect of Maine-et-Loire, 24 July 1944.
29 AD L-A 13W25, official correspondence concerning 11 November 1940; E. Duméril (ed.), *Journal d'un honnête homme pendant l'Occupation* (L'Albaron, 1990), pp. 81–2; interview with Christian de Mondragon, Nantes, 29 April 1997.
30 AD M-et-L 97W33, Feldkommandant Kloss to prefect of Maine-et-Loire, 31 March 1941; AD L-I 13W8, Feldkommandant Hotz to prefect of Loire-Inférieure, 30 March 1941; AD L-I 51W21, commissaire central, Nantes, to prefect, 31 March 1941.
31 AD L-A 51W21, commissaire central de police, Nantes, to prefect of Loire-Inférieure, 15 May 1941.
32 AD I-et-L 10W84, report of gendarmerie of Bléré, 11 May 1941; AD I-et-L XV Z 23, Feldkommandant of Tours to Chef der Militärverwaltung Bezirks B, Angers, 13 May 1941.
33 The most recent account of this affair is L. Oury, *Rue du Roi Albert. Les otages de Nantes, Châteaubriant et Bordeaux* (Paris, 1997).
34 AD I-et-L 46W83, Maurice Lefebvre to gendarmerie of Châteaurenault, 3 September 1944.
35 AD M-et-L 97W123, Albert Chevalier to subprefect of Saumur, 31 January 1945.

36   AD I-et-L 46W83, report of gendarmerie of Preuilly-sur-Claise, 5 April 1945.

37   AD L-A 270W510, report of gendarmerie of Saint-Nazaire, 28 March 1945.

38   AD M-et-L 117W32, report of gendarmerie of Saumur, February 1945.

39   AD M-et-L 117W32, resistance group of Saumur to prefect of Maine-et-Loire, 12 March 1945.

13
Opening Pandora's Box: the Case of *Femmes tondues*

Claire Duchen

The Liberation period in France – the two years roughly between spring 1944 and the end of 1946 – was a 'complex amalgam of opening and closure'. Its abiding fascination lies at least partly in its ambiguity: it was no longer war but was not yet recognizably peace. The Liberation was a key moment of transition, its identity suspended between the past and the future and, as such, it represented a time when the tensions between the desires for a fresh start and for a return to 'normality' were played out.[1] These months were exhilarating in their celebration of freedom and vicious in their recriminations over wartime behaviour. In the chaos, people wanted order, closure, a clean slate from which to open a new era. To this end, all those who had been part of the collaborationist government of the Occupation (1940–44) at Vichy had to be purged from public life, punished for their involvement with Nazi Germany. Ordinary people were also targeted: those who had actively collaborated, supporting Nazi or Vichy ideology; those who had made money out of the Occupation, such as industrial manufacturers and owners of large corporations; those who had worked for Vichy or for the Germans and those who had enjoyed German company, benefiting from the material advantages denied the rest of the population. When compiling figures for the number of individuals involved in the purges, a breakdown by gender was not carried out.[2] Gender was not considered appropriate as a focus for attention, but I will argue that particular visions of gender roles and particular visual uses of the female form were a crucial part of France's attempt to settle its wartime score and look towards the future.

Like men, women were shot, imprisoned and suffered national degradation, depriving them of their recently conferred civil rights. Women were accused, often anonymously, of denouncing Resisters, Communists, Jews.

They were accused of black marketeering, or of settling personal scores by denunciation to the Gestapo, the Milice (a paramilitary police force) or the police. They were punished for working for the German authorities or for association with known collaborators. They were considered guilty by association and were punished for the collaborationist activities of their husbands or their children. Unlike men, however, women were also punished for sexual misconduct. They were accused of consorting with the Occupier, flaunting their associations with German soldiers, sleeping with one or more of the enemy, sometimes bearing a German's baby.

These forms of so-called 'horizontal' collaboration were not included in definitions of 'national indignity', punishable by national degradation, and most frequently did not reach the courts. The women were none the less punished. As part of the widespread unofficial retribution of 1944 and 1945, women were singled out for a particular kind of punishment: that of having their heads shaved in public. Theirs is a story told by others; the principal characters remain silent, neither punishers nor punished wishing to speak of the headshavings and their part in them. This silence contrasts strikingly with the apparent ubiquity of photographs of these shorn women, or *femmes tondues*, with the widespread knowledge and memory of the events and with their symbolic significance. This chapter will focus on *femmes tondues*. The first part will discuss the information and analyses available so far about these headshavings; the second part will focus specifically on photographs taken of the scenes. The principal questions addressed are the following: Why did France need to punish women for their wartime sexual conduct? Why have *femmes tondues* come to be emblematic of collaboration and guilt? How do photographs of *femmes tondues* contribute to a gendered analysis of the purges? How can these photographs serve the historian of the Liberation?

First, the facts

The function of the purges was to eliminate, to punish and to cleanse. There was also an undoubted element of revenge. Women who associated with Germans had most visibly, publicly, enjoyed benefits denied the majority; they were thus those who were primarily targeted for visible, public revenge. Fabrice Virgili has carried out research on the press reporting of the public headshaving of women: it took place in 1944, starting from the spring before the Allied landings, peaking in June and July and continuing until the autumn; a second wave followed in 1945, after the return of deportees from the concentration camps; it was reported in at least 77 of the then 90 departments of France;[3] it was done in towns, villages and rural communities; it was carried out for mostly sexual reasons, but also for denouncing others to the authorities or for 'keeping company' with Germans; a certain amount of score-settling was also undoubtedly involved; the number of women concerned varied widely, from one isolated case to that of 80 women in the town of Beauvais; other punishments (stripping, painting swastikas on the body, throwing things at the women or hitting them, jeering, parading them, often naked, through the town, making them sit in a public place with a placard around their neck proclaiming their crime) most usually accompanied the shavings; women were – although this was rare – shorn before being shot. The women subjected to the headshaving were mostly single (unmarried, divorced, widowed), usually young and poorly educated. They mostly worked for the Germans as cleaners or housekeepers, or had clerical jobs working for the local occupying administration. Some had higher level employment, working as translators or secretaries; some were prostitutes. The scenes occurred often at the end of the day, so that people could come and enjoy the sight on their way home from work; they took place at the woman's home, in the street, in a seemingly impromptu fashion; or they became a significant part of local Liberation ritual, taking place in sites of local power – at the town hall, in the town square, outside the main café, in front of the girls' school, at the police station. The headshavings were more or less tolerated by the new Resistance authorities, in spite of official condemnation which formally forbade Resistance involvement in any such event. Anecdotes suggest that Resistance fighters stopped the shaving in some places, turned a blind eye in others and were among the instigators in yet others – claiming a role as alternative authority as an alternative justice was enacted.

Writing the event

The press was on the whole gleeful in its reporting, although disapproval was occasionally expressed. Eye-witness accounts, reported mainly in memoirs, tend to demonstrate a certain unease with the situation. They describe the event generally from the perspective of one who witnessed but did not participate and did not condone what he or she saw.[4] This discomfiture is repeated, largely, in the historiography of the *femmes tondues*. Historians have had difficulty in finding an appropriate manner in which to discuss this aspect of the purges, wishing to condone neither the punishers nor the punished, and rejecting the apologetic mode for either side.[5] Many of the first historians of the Liberation and the purges were reticent about *femmes tondues*, or at first tried to excuse those who meted out the punishment. Peter Novick for instance, wrote: 'these acts of minor violence corresponded to a population's profound need to give free rein to their emotion, after four years of swallowing their resentment.'[6] He notes on the same page that these girls were probably the unsuspecting saviours of those members of the Milice and other collaborators who would otherwise have been on the receiving end of the people's wrath. Herbert Lottman wrote that the women were used to 'sop up the anger that would otherwise have ended in bloodshed'; Fred Kupferman mentioned the *femmes tondues* in passing and wrote: 'The chase began. It appeased the hatred and anguish of a starving, humiliated people . . . Yes, in 1944, the French were full of hatred.'[7] It is telling that many historians of the purges restricted themselves to reporting anecdotal evidence and left analysis well alone.[8]

Over the past five years, it has been recognized that the question of *femmes tondues* is rich in interpretive possibilities. The theoretical insights offered by Michel Foucault on the public spectacle of punishment, Mikhail Bakhtin on carnival and René Girard on the scapegoat[9] have all been particularly valuable for scholars who wish to develop a gender-based analysis of the purges.[10] For instance, in *Discipline and Punish*, Michel Foucault suggests that in a public execution, what was at stake was less the re-establishment of justice than the reactivation of power. The purpose was to 'make everybody aware, through the body of the criminal, of the unrestrained presence of the sovereign'.[11] The Liberation, and in particular the purges, can be considered as the ritual of a return to republican order. The *femmes tondues* were victims of this process.[12]

The most apposite of these theoretical works comes from René Girard. The 'episode' of the *femmes tondues* is frequently discussed in terms of a scapegoating. Girard has suggested that persecution – a scapegoating, a witchhunt, a lynching – takes place at times of social crisis, as a response to a sense of powerlessness. He points to the fact that the 'crimes' most frequently evoked during persecution are those which attack the foundations of cultural order, those which are the most taboo and which are often sexual, as tends to be assumed in this case.[13] These women are targeted because they allegedly transgressed codes of patriotism and also codes of feminine behaviour. It is the latter part of their transgression that is highlighted, or foregrounded.[14] Women are seen as body, and if women's bodies are the property of men and of the nation, then they must be punished in the body, by the nation. Ingrid Bauer makes a similar argument to explain the Austrian hatred of the postwar GI bride: an Austrian woman who married an American occupier after the war represented the 'loss of hereditary property rights' for Austrian men.[15] The women were involved in a public spectacle, in which the crowd took revenge, asserted and celebrated its own outraged innocence by designating the few as guilty. In a scapegoating, the weak are targeted as blame is transferred; the finger is pointed at another, in order to diminish one's own share of the blame.[16] French society pointed at its Other, thereby making the Self innocent. Via the *femmes tondues*, French society could reconstitute itself as a coherent whole, as a Resistant whole, morally and politically virtuous.[17] This meshed with the Gaullist construction of the notion that the true, eternal France had never accepted the armistice or the Occupation and had always retained its integrity; and that the people had risen up to liberate their own country. Excise the rotten elements of the nation and the whole may be exonerated, or indeed its guilt denied or forgotten.

Feminist approaches to the *femmes tondues* have been slow to gain recognition. Before the mid-1980s, it was never pointed out that there was a specifically gendered dimension to the headshaving scenes; that it was men who were doing it to women. It was primarily seen as resisters doing it to collaborators. The first hint of a gendered approach to the *femmes tondues* came from Marie-France Brive in the mid-1980s, at a conference on the Liberation in Toulouse. She suggested that by singling out some women for headshaving, all women were implicitly targeted and therefore more was at stake than righteous revenge or popular justice. This suggestion met with anger rather than interest and Brive's contentious argument was dismissed.[18]

Since Brive's intervention, it has been increasingly hard to ignore or sideline analyses of the events which have gender and sexuality at their core; male and female researchers all now emphasize the centrality of gender. The metaphoric significance of the headshaving has been favoured over the search for data concerning who did what to whom and why. For instance, a recent reading suggests that the women were scapegoated to compensate for the humiliation of the French male at the hands of the Germans in 1940.[19] It is possible to read the scene of the *femmes tondues* as a scene of sexual empowerment for men, by which they erase their failure of the defeat and the Occupation. The metaphor of France as a woman, invaded by the Germans while French men were unable to prevent it happening is too close to the stereotypical image of the *femmes tondues*, women giving themselves to Germans, to be ignored. At some unspoken level, the sight of women voluntarily choosing a German evokes the image of France (so often represented as a woman in the figure of Marianne) as a willing partner of Nazi Germany – an image which the French would like to forget. The *tondeurs* (headshavers), take the place of the powerful (and attractive?) male invader in a symbolic re-enactment and reworking of the scene. The French man was masterful once more, the French woman submissive; thus were 'proper' gender roles re-established and the normalization of French society encouraged.

Further insights are provided by feminist work on war and nationalism.[20] The headshaving scene can be considered in terms of the sexual violence of war. The headshaving can be thought of as one of the final acts of war, a symbolic rape. Rape, sexualized violence, has always been part of war, considered practically a soldier's right. It is about power, humiliation and abasement. The victim's suffering is transformed into the victimizer's display of power. Rape in war is crucial, according to sociologist Ruth Seifert, in that the suffering of the civilian population in war is central to the deconstruction of the culture one wishes to defeat.[21] Women represent, symbolically, the nation; literally, in times of war, women hold family and community together. Attack women and the nation, or the community, is destroyed. This can also work in reverse. As part of reconstructing a community, some women are defined as outsiders/Others. Shorn of their hair, they are raped symbolically by the newly powerful. The identity of outsider, the stigma of unworthiness, is thus clearly marked on the body.[22] It is marked in discourse too: in narrative accounts, the terms used to describe the *femmes tondues* distinguish between *les femmes à boches* ('Jerry's women') or a *tondue* and 'real' French women, 'les *Françaises*', as if a sexual involvement with a German meant loss of French identity.[23] *Femmes tondues*, then, had forfeited their right to be called French, and were discursively excised from the nation.

Feminists have long argued that control of women's sexuality was, and is, an integral part of patriarchal societies; that sexuality is political. Dorinda Outram posits a direct link between sexuality and politics in her work on the body and the French Revolution. She speaks of the 'continuum between virtue as chastity or female fidelity within marriage, and virtue as the upholding of the republic.'[24] The disruptive potential of unrestrained female sexuality endangered the revolutionary republic, because 'any deviation from chastity/virtue involves the collapse of republic/virtue'. This conflation of sexual and political virtue has rarely been more explicit than in the case of *femmes tondues*.

If women were punished for transgressing codes of femininity and also codes of patriotism, anecdotal evidence and photograph captions suggest that they were judged more for their 'deviant sexuality' rather than for acts of political betrayal. Luc Capdevila's research shows that the social profile of the women victims indicated that they were mainly single, employed, sexually active and without children, and in control of their own fertility. He argues that in the *femmes tondues*, France was punishing those who were perceived as a threat to social and moral order, undermining marriage, family, fertility and the nation.[25] The headshaving, then, was clearly not just about punishment; it was also about future agendas. The punishment of female sexuality was to safeguard the integrity of the new postwar Republic. As Ela Hornung and Ingrid Bauer have suggested in writing about Austria, a symbolic reconstruction of the 'family Austria' – or in this case, the 'family France' – was necessary in order to make reconstruction of the nation possible.[26]

Picturing the scene

The *femmes tondues*, then, have received significant attention from historians only in recent years. In the popular memory of the Liberation, however, they have been present since the events themselves. They have been evoked mostly via a well-known poem by Paul Eluard ('Comprenne qui pourra'), a well-known song by George Brassens ('La belle qui couchait avec le roi de Prusse'), by Alain Resnais's film *Hiroshima Mon Amour* and by a couple of much-circulated photographs by the celebrated photographers Robert Capa and Henri Cartier-Bresson. Photographs of *femmes tondues* have acquired an emblematic status and the same few pictures can be seen repeatedly on magazine and book covers every time an anniversary comes around.[27] The most familiar images show the public humiliation of two women, marched through the streets, surrounded by a jeering crowd (Figures 13.1 and 13.2). They are the most instantly recognizable visual signifiers of the purges, and by extension, of collaboration in France.

How can these photographs serve the historian of the *femmes tondues*? In spite of claims made on its behalf, the photograph is rarely an unambiguous record of an historical moment, and is a complex and unreliable form of testimony on a number of levels. The photograph, by definition the site of a multiple gaze, inevitably resists a single meaning; it is a visual medium but is none the less 'invaded by language'; it implies a story but simultaneously withholds that story; and without the implicit narrative made explicit, 'photographs . . . are inexhaustible

Figure 13.1 Capa 23 August 1944, First pictures of the capture of Chartres. Two French women collaborators, their hair shaved off, are led back to their home, while the population whistle loudly and jeer after them. (© *Magnum Photos*)

Figure 13.2 Capa 23 August 1944 Chartres. Two women collaborators, one with a baby by a German father, are led back to their homes after having their hair shaved off, amid jeers and whistles from the population. (© *Magnum Photos*)

invitations to deduction, speculation and fantasy'.[28] The complexity of the photograph as record, or testimony, is compounded by a number of imponderable aspects: photographic meaning can only too easily be altered, and another meaning imposed, by visual or verbal means (angle, cropping and framing, context, caption);[29] the identity of the photographer cannot always be ascertained;[30] the motivation of the photographers – a particular brief which pushed them towards one set of images or another – is even more elusive; pictures were always subject to censorship but the grounds on which decisions about individual pictures were taken cannot always be discovered; the way in which a picture ended up in one particular newspaper, newsreel, or archive, or illustrating one book or another, often seems to have been due to chance or accident as much as to deliberate policy.[31] The fundamental questions (some of which may have to remain frustratingly unanswered) are about why pictures of *femmes tondues* were taken at all, for whom they were intended and the function they have played in the memory and the historiography of the Liberation.

Allan Sekula has said that 'a photographic discourse is a system within which the culture harnesses photographs to various representational tasks.'[32] The photographs participate in the story (fantasy?) France wished to tell of itself, to itself and to others. At a first level, these photographs help the historian understand how the feminine was used politically, and how particular visions of the feminine are written into the mythology of the Liberation.

Gendered imagery occupies a significant place in the memory of the Liberation, and in its mythology. The female form provides some of the strongest, most evocative images of the time. In collections of images, in history textbooks and in monographs of the period, key aspects of the Liberation are encapsulated by images of women. The female form is identified with the euphoria of freedom: in paintings, drawings, posters, photographs and newsreels, we see young girls and women; they are waving, kissing soldiers on tanks, dancing, able to enjoy legitimately some frivolity at last.[33] The female body was used allegorically, triumphantly representing Liberty, or the Republic itself. At the Liberation, the preferred role of the female in political posters was maternal – represented usually in a mix of stylized semi-allegory and semi-realism – protecting the child in the family grouping which was to represent the harmonious and stable future of France. There were photographs of women Resisters, most obviously in Communist Party publications and at the meetings of the Communist-dominated women's organization, the Union des Femmes françaises (Union of French

Women), where huge portraits of women Resistance heroines hung behind the platform. But photographs of *femmes tondues* were (and are) far better known than these. They are the necessary opposite of the pictures of maternal or political virtue. They instantly signify the purges and collaboration, in a way that pictures of women Resisters do not instantly signify the Resistance, whose dominant image (visual and narrative) is masculine.[34] They draw on archetypes of female imagery from myth, history and fiction: Eve, woman-as-sex, woman-as-weak, both tempted and temptress; the witch as archetypal outsider; prostitutes and adulteresses, with hair shorn as punishment; aristocrats, waiting before the guillotine. The sexually and politically deviant woman has always been singled out, looked at, desired, punished. If the female body is ever-present in Liberation imagery, it is in this culpable, sexually marked way, linking the political with the erotic, that it has its greatest impact.[35]

Photographs of *femmes tondues* confirm the equation of male as Resister, female as collaborator,[36] male as heroic, female as guilty, male as combative, female as passive.[37] The attribution of collaborationist guilt to all women is achieved by the use made of the pictures at the time in the press and in newsreels, by the anonymity of the women captured on film, by the lack of precise information accompanying the photograph. Some of the captions are bland or terse ('Photograph of a *femme tondue*'); some confirm guilt ('Photograph of collaborators having their heads shaved'); some are explanatory ('The man on the left explains to the public what it means to have your head shaved'); some are intended to be jocular ('Doesn't she look awful without her hair'); others are triumphant ('This is what the French think of their collaborators').[38] The caption may give the name of the victim or the village or town where the shaving took place, but many of the pictures give no indication at all of specific individuals or locations.

The caption or lack of caption universalizes the scene and thereby universalizes the feminine associations with collaboration. It reinforces collective rather than individual responsibility; while certain individuals were targeted for headshaving, the women in the pictures could have been any women, anywhere in France. The way the photograph makes specific women anonymous and representative of all women confirms the mythic status of the image, whose symbolic significance far outweighs the actual incidence of headshaving. To misquote Roland Barthes, the photograph without context transforms history into myth; myth then turns history into nature.[39] These women become all women, their particular history becomes myth. Women, traditionally, have been represented as body, sentiment, moral and physical weakness; men as mind, intellect, strength. The *femme tondue* as 'woman' draws on these stereotypes, confirming the associations. The repeated circulation of the photographs served to implant the image in the public consciousness and, over time, in the collective memory of the French.

Opening Pandora's box

Fifty years later, the pictures tell a different story, or rather, tell the same story differently. Gone is a comforting story of France as told in 1944 or 1945 (the righteous men punishing the wicked women; the cleansing of France; the return of the morally and politically virtuous to authority) which might have emphasized the women's misdeeds and satisfied the viewer's sense of justice. Fifty years later, just looking at the pictures undermines this reading. Analyses of the headshaving, as we have seen, can suggest that it was an integral part of the return to a Republican patriarchal order; it can be described as part of a 'distasteful carnival'[40] or as a catharsis necessary for the reconstruction of the country. Maybe it was all these things. But the scene also shows a form of bullying, a public humiliation seemingly enjoyed by everyone but the women themselves. The pictures show ordinary women, young girls in dowdy coats, skimpy frocks, short socks and clumpy shoes, with faces as expressionless as possible (Figure 13.3). The crowd provides a compelling sight of collective sadism, taking vicious and petty pleasure in the proceedings (Figure 13.4). These images carry the resonances of the photograph used for surveillance purposes, or for anthropometric studies; they parody the accepted genres of photography (the portrait for the family album, the holiday snap, the pin-up girl); the woman is sometimes practically turned into an animal trophy – the fish, the deer's head – won by the crowd surrounding her (Figure 13.5). In some cases, the photographer seems to be in complicity with the crowd and the act of photographing itself becomes the centre of attention (Figure 13.6). The viewer of the 1990s is made profoundly uncomfortable by witnessing the scene through photography. Spectators are turned into voyeurs, looking at an image from a public yet intimate act.

The anonymity that had tainted all French women with the guilt of collaboration paradoxically encourages a rereading of the scene 50 years later, calling up the sympathy of those who were not there, while turning the tables on the punishers. From the standpoint of 50 or so years later, the waters of vice and virtue are muddied. The surrounding figures become as significant as the women themselves. If the viewer is

Figure 13.3   Axis collaborationists denounced in Cherbourg. A French Resistance Party member explains the shorn hair of young women of Cherbourg accused of violating rules set down by the French Resistance Party for conduct with the Germans. The women were rounded up on Bastille Day morning, 14 July, 1944, shorn, and paraded through the streets. (*Courtesy of the Imperial War Museum*)

to bring any judgement to bear on the events, it is more likely to be critical of the victimizers than of the victims. The photographs, over time, pass guilt from victim to victimizer; nobody occupies the moral high ground. The ambivalence that the spectator feels punctures the vision of the morally unassailable new France.

The silences and lacunae which make the photograph such an unhelpful witness in certain respects now become useful, and prompt the historian's questions which pour forth once the lid of this Pandora's box is opened: What was the relationship between the various actors in the headshaving – *tondeurs*, *tondues*, accusers, onlookers – were they neighbours, friends, colleagues? What does this tell us about sociability and community? What happened afterwards? Did they all live together again? The women involved have tended to deny any wrongdoing, but none the less continued to live with the shame of the accusation. Some women were driven to suicide; others were forced to leave town, as in

Figure 13.4   Members of the French Resistance Party mete out punishment to all known 'Axis collaborationists'. Housemaids, servants, etc. of the Germans were gathered together, shorn of all their hair, and paraded through the streets of Cherbourg, their bald heads the emblem of their violation of the rules of the Party. (*Courtesy of the Imperial War Museum*)

Figure 13.5 Collaborationist in Paris (no date). French civilians wave the victory sign after shaving the head of a French woman accused of collaborating with the Nazis. A swastika has been painted on her forehead. (*Courtesy of the Imperial War Museum*)

the film *Hiroshima, Mon Amour;* yet others faced their victimizers daily and simply carried on.

Why did sexual relationships constitute collaboration? How has collaboration been defined? Men bore the weight of political, economic,

Figure 13.6   Montereau (no date). (*Courtesy of the Imperial War Museum*)

ideological collaborationist activities because of the dominant role they played in public life; women seemed to carry the guilt of collaboration at the level of daily life and personal relationships. No distinction was made between different types of relationship between French woman and German man. For instance, if a woman was a prostitute, plying her trade with the available clients in the interests of making a living, she was usually (but not always) considered guilty. A young woman who had 'really fallen in love' with a German was equally guilty, if not more so. Her behaviour was considered a crime and not a tragic love story which might be how the participants viewed their situation themselves. Women were punished for failing to recognize that during war, private life was far from private and that the very personal had become very political. And whereas women who joined the Resistance were commonly said to have done this for personal rather than political reasons – following a boyfriend or husband – women who probably had little interest in politics had deeply political implications attached to their personal actions.

What was the intended message of the photograph? Who was the intended audience? Were the pictures of *femmes tondues* intended for French women, to tell them, unequivocally, how to behave in the future? Were they for foreigners, to demonstrate that the French had recognized their guilt and were punishing the guilty? Or were the French simply indulging in some self-flagellation, displaying their own sinners to the world? What is striking here is the emphasis on the visual. In most cases, the purges have been examined via court records, written accounts, memoirs. Where ordinary people are concerned (as opposed to the trials of politicians and intellectual celebrities or public figures – civil servants, lawyers, journalists), there has been little interest, and particularly little visual interest. In the case of *femmes tondues*, however, attention is paid less to the identity of the victims and the nature of the crimes than to the sight of the punishment itself. The photograph is an integral part of the punishment. The traditional role played by female imagery incarnating virtues and vices, together with prurient interest in the sexual transgressions – and in the downfall or misfortune – of others have ensured this prominence.

These photographs now invite the historian to rewrite the story, recast the *dramatis personae*, reassess the villains and heroes, shift the relationships visible in the pictures. Fifty years later, images of *femmes tondues* contribute to the destruction of the myth they originally helped to create. In one sense, the historian's duty is to reverse the semiotic alchemy and turn myth back into history, all women into some women. For this to be possible, the anonymity of the individuals and locations involved has to end. Only by learning more about who had done what to whom and why, individualizing and attributing some actions (sexual or not) to some women, can the association of women and collaboration be undone. This laborious process, undertaken via local and national police archives, prefects' reports and court reports is now underway as material becomes available to researchers.[41]

The photographs contribute to rewriting history in a broader sense too. Confronting the unease provoked by the photographs means recognizing the gendered dimension of the purges and examining the unacknowledged relation between sexuality, violence and politics that is present in the images. More than that, it means working towards new readings of the sexual politics of war and peace and reconsideration of the relationship between the image, society and the writing of history.

Notes

1 Cf. H. R. Kedward and N. Wood (eds), *The Liberation of France. Image and Event* (Oxford, 1995), p. 9, and C. Duchen, '1944–1946: Women's Liberation?', in D. Berry and A. G. Hargreaves (eds), *Women in 20th-Century French History and Culture* (Loughborough, 1993), p. 50.

2   Researchers are now trying to establish the gendered nature of the purges; see F. Leclerc and M. Weindling, 'La répression des femmes coupables de collaboration', in F. Thébaud (ed.), 'Résistance et Libérations'. France 1940–1945', *Clio. Histoire, femmes et sociétés*, no. 1 (1995).

3   Information is simply not available or research has not yet been carried out in the remaining departments; no known reported incidence of headshaving does not mean that none occurred. See F. Virgili, 'La tonte et les tondues à travers la presse de la Libération', unpublished thesis (Paris, 1992).

4   See A. Brossat, *Les Tondues. Un carnaval moche* (Paris, 1992), who provides the following anecdote related by Eugène de Brocard: 'Rue de Rennes, a car passes quickly. In the car, next to the driver, a French officer. Behind them, a German-general in full military dress. He is dignified, quite haughty in fact, with an air of disdain for the passers-by who are booing him. Further along, piercing cries. Down on the ground, a woman is defending herself from a wild crowd. 'Friend of the Germans,' someone says. A local hairdresser shears her head. He leaves her a long lock by which she is dragged along by some shrieking women. They tear her clothes off. Completely naked, she runs away. A charitable soul opens a door for her. That same night, I saw ten women like that, surrounded by guards, paraded through the streets, carrying a Nazi flag. Luckily, the next day, spectacles like this were categorically stopped' (p. 157).

5   That is apart from Robert Aron, whose many volumes on the history of the purges (1967–75) fed the 'bloodbath' version of events.

6   P. Novick, *The Resistance versus Vichy France* (New York, 1968), p. 78.

7   H. R. Lottman, *The People's Anger: Justice and Revenge in Post-Liberation France* (London, 1986), p. 68; F. Kupferman, *Les Premiers Beaux Jours* (Paris, 1985), p. 89. In the twentieth century, the French have not had the monopoly on headshaving: Republican women soldiers had their heads shaved by Nationalists during the Spanish Civil War; Irish girls risked tarring and feathering if they went out with British soldiers in Northern Ireland in the 1970s.

8   As well as those historians mentioned above, see H. Amouroux, *La Grande Histoire des Français après la Libération*, vol. 9, *Le Règlement des comptes* (Paris, 1991), and the work of journalists such as Catherine Gavin and Janet Flanner.

9   See respectively M. Foucault, *Surveiller et punir: naissance de la prison* (Paris, 1977); M. Bakhtin, *Rabelais and his World* (Cambridge, Mass., 1968); and R. Girard, *Le Bouc-émissaire* (Paris, 1982).

10   See Brossat, *Les Tondues*; C. Laurens, 'La femme au turban. Les femmes tondues', in Kedward and Wood, *The Liberation of France*, pp. 155–79; and Kedward in *ibid*.

11   Foucault, trs. Sheridan, 1977, p. 49.

12   Other avenues still to pursue which stress the metaphoric over the literal include the psychoanalytical approach to the question of hair and its sexual symbolism, from headshaving as castration to hair as fetish; the Biblical and historical echoes, linking headshaving with sexuality and transgression across cultures and through time; and the arguments, developed by many scholars, over the masculinity of the republican ideal.

13   Girard, *Le Bouc-émissaire*, pp. 24–6.

14   It has been suggested that the alleged sexual nature of the *femmes tondues'* offence, and indeed the attention paid to the *tondues*, has completely obscured other forms of female collaboration and other ways in which

women were punished; see Leclerc and Weindling, 'La répression des femmes coupables de collaboration'.

15 I. Bauer, 'The GI Bride: on the (de)construction of an Austrian postwar stereotype', in C. Duchen and I. Bandhauer-Schöffmann (eds), *When the War was Over: Women, War and Peace in Europe 1940–1952* (London, 1999).

16 See T. Douglas, *Scapegoats: transferring blame* (London, 1995), p. 195.

17 See also Lynn Hunt's arguments in *The Family Romance of the French Revolution* (Berkeley, Ca., 1993).

18 M.-F. Brive, 'L'Image des femmes à la Libération', in R. Trempé (ed.), *La Libération du Midi de la France* (Toulouse, 1986), p. 391.

19 See M. Kelly, 'The Reconstruction of Masculinity at the Liberation', in Kedward and Wood, *The Liberation of France*, pp. 117–28.

20 A. Stiglmayer, *Mass Rape. The War Against Women in Bosnia-Herzegovina* (Lincoln, Nebraska, 1994); B. Einhorn (ed.), 'Links across Differences: Gender, Ethnicity and Nationalism', *Women's Studies International Forum*, vol. 19, nos. 1/2 (1996); N. Yuval-Davis, *Gender and Nation* (London, 1997).

21 R. Seifert, 'The Second Front: the Logic of Sexual Violence in Wars', in *Women's Studies International Forum*, vol. 19, nos. 1/2 (1996), p. 41.

22 When deportees returned from concentration camps, the women survivors, mostly with no hair, wore turbans and scarves, as did women who had had their heads shaved. To avoid confusion between the women, some local Resistance authorities called for a law forbidding *femmes tondues* to wear turbans; Virgili, 'La tonte et les tondues', p. 30.

23 L. Capdevila, 'La collaboration "sentimentale": antipatriotisme ou sexualité hors-normes?', in F. Rouquet and D. Voldman (eds), 'Identités féminines et violences politiques (1936–1946)', *Cahiers de l'IHTP*, no. 31 (October 1995), p. 70.

24 D. Outram, *The Body and the French Revolution. Sex, Class and Political Culture* (New Haven, Conn., 1989), p. 126.

25 Capdevila, 'La collaboration "sentimentale"', pp. 81–2.

26 E. Hornung, 'The Myth of Penelope and Odysseus: an Austrian married couple relate their wartime and postwar experiences', in Duchen and Bandhauer-Schöffmann, *When the War was Over*.

27 For his thesis (referred to above) Fabrice Virgili consulted photographs at the Bibliothèque Nationale in Paris. He notes that of the 5,000 or so photographs of the period, only six represented *femmes tondues* (p. 3). For this chapter, the Photographic Collection at the Imperial War Museum, at the Musée d'Histoire contemporaine in Paris, the Agence Roger Viollet and the Agence de la Documentation française were consulted. The Imperial War Museum's photographic archive contained a large number of photographs of *femmes tondues*, while the collections in Paris had many fewer. Many of the photographs overlapped from one collection to another.

28 Cf. V. Burgin, 'Seeing Sense', in H. Davis and P. Walton (eds), *Language, Image, Media* (Oxford, 1983), p. 226, and S. Sontag, *On Photography* (New York, 1973), p. 23.

29 Caption writers were often probably quite unconnected to the scene, sitting in some newspaper office far away, after receiving a copy of the picture from the Allied War Pool – they could easily attach a more-or-less appropriate but invented caption to the picture.

30   Press photographers were most usually not given a credit for their picture before 1948.

31   I tried to pin down a few answers to some of these questions. Each service in the military had photographers attached to it; there were also press photographers such as Robert Capa working at that time for *Life* magazine, or freelancers like Henri Cartier-Bresson, Margaret Bourke-White and Lee Miller. The nature of the individual pictures partly depended on the objective of the photographer. Press photographers and freelancers hoping to sell pictures independently, tended to focus on 'human interest' stories; military photographers focused on images to be used for official recording purposes or for propaganda (Hilary Robert, Deputy Curator of Photographs, Imperial War Museum London, conversation with the author, 12 July 1994). All photographs were sent to the Allied headquarters for approval or refusal by the censors. It appears that the photographs were all put in the Allied War Pool. Some of the pictures I saw at the Imperial War Museum had 'Not for publication in the Western hemisphere' stamped on the back; others gave no indication of eventual publication, while yet others were clearly marked by the name and date of publication.

32   A. Sekula, 'On the Invention of Photographic Meaning', in V. Burgin (ed.), *Thinking Photography* (Basingstoke, 1982), p. 87.

33   Of particular interest are the images chosen for the catalogue of a Paris exhibition; see *La France et les Français de la Libération* (Paris, 1984).

34   S. Reynolds, 'The Sorrow and the Pity Revisited. Or Be Careful, One Train can Hide Another', *French Cultural Studies*, vol. 1 (1990).

35   See M. Pointon, *Naked Authority. The Body in Western Painting 1830–1908* (Cambridge, 1990) for a discussion of 'reading the body'.

36   The association of women with collaboration has been noted by Siân Reynolds in 'The Sorrow and the Pity Revisited' (note 34 above). Here she analyses the famous film in terms of its gender bias and shows how Marcel Ophuls makes an unspoken but obvious association between women and collaboration. See also Pierre Laborie, cited by Rouquet and Voldman: 'France, saved from Pétainism and collaboration by the spirit of struggle and "*insoumission*" of the Resisters, is assimilated to a symbolic form of political passivity, encapsulated by "horizontal" collaboration . . . The violence against women exalts the verticality of resistance opposing the horizontality of passive collaboration'; 'Identités féminines et violences politiques', p. 9.

37   Pictures of male collaborators that have had as much exposure as those of women have been pictures of named individuals, famous men whose activities were known; the women photographed are generally unnamed, their punishment unofficial, their crimes alleged.

38   This is a selection of captions from the collection at the Imperial War Museum.

39   R. Barthes, *Mythologies* (Paris, 1957; 1970 edition), p. 229.

40   Brossat, *Les Tondues. Un carnaval moche.*

41   See F. Virgili, 'Les "tondues" à la Libération: le corps des femmes, enjeu d'une réappropriation', in *Clio* (1995), p. 112, note 4.

14
Contemplating French Roots

H. R. Kedward

In June 1995 the international conference at Besançon on 'La Résistance et les Français' took as its theme the 'Armed Struggle and the Maquis' and discussed a number of studies on the relationship of the maquis to specific rural communities.[1] The year before at Rennes, the preceding conference in the same series had confronted the issue of 'the rural world and the Resistance', and it was noticeable that almost all speakers felt they had to start with the statistical under-representation of the peasantry in the Resistance and the image of a close relationship between Vichy and the peasantry, before going on to consider whether the general statistics and the special relationship were, in fact, accurate for the locality under consideration.[2] At Besançon this detailed testing of 'accepted truths' about rural France was continued and in the discussions it became obvious that these 'truths' constituted notions about ruralism in France which were not easily fractured by specific cases and were still dominant in historical practice.

It happened to be the centenary year of the birth of both Jean Giono and Marcel Pagnol, and exactly ten years before, in 1985, Claude Berri was remaking Pagnol's 1952 film of *Manon des sources* into his own more powerful *Jean de Florette* and *Manon des sources*. Sixty years before, in 1935, Giono began the first of his cultural gatherings and rural restoration on the plateau of Cantadour in north-eastern Provence, and 40 years before, in 1955, he and a group of graphic artists began the restoration of the village of Lurs. Giono was born in Manosque, in la Rue sans Nom, and in the 1960s a nuclear research station was aggressively imposed on the plateau where he had rebuilt the ruined mill and houses of Cantadour and created his own cult of rural life, poetry and pacifism. In 1995, shortly before the Besançon conference, France resumed atomic testing in the Pacific Ocean. The cult of anniversaries;

the historical pairing of anonymity and rurality; the national justification of nuclear testing, all gave particular thrust to the formal debates on national identities and rural France at the Besançon conference, and determined the central problematic, in which ruralism as image and myth was interwoven with an empirical re-creation of event and social context.

This essay emerged as a personal reaction to this ruralist theme. It is not about the French peasantry in itself, nor about agricultural activity. It is about the interface between rural identities and national politics, about attitudes to the countryside, and about ruralism, that sympathy and search for rural life which is a recurrent projection, alter ego, or some may say fantasy, of all industrializing or post-industrial states, nations and peoples. It is, finally, about certain moments in French history which had been largely lost and are only now beginning to be rediscovered. I begin and end with myths of very different kinds, but in the middle there is a historical argument about how images of ruralism in France between the 1920s and the 1950s became rigidified, with negative effects reaching into the very debates at Rennes and Besançon.

I am sure I would have been intolerant of Giono's stubborn romanticism and his egocentric disdain for realities other than his own, but he never ceases to surprise. Among the many revelations in his diaries, there is his unlikely wish to have made a film of Sartre's existentialist novel, *La Nausée*, published in 1938. He constantly asserted that a person's own existence was the most important material fact in life, and he must have applauded the decision of the novel's main character, Antoine Roquentin, to give up writing someone else's history and make his own creative gesture, perhaps by writing a novel, or, as he said, 'another kind of book, I don't quite know which kind . . . the sort of story, for example, which could never happen, an adventure. It would have to be beautiful and hard as steel and make people ashamed of their existence.' Giono must have approved of that, as I imagine he must have been responsive to Roquentin's moment of truth when confronting the root of a chestnut tree. Here is Roquentin, a lost urban intellectual, vaguely left-wing, full of modernist angst and the nauseating effect of too many objects and too much contingency, dying to get back to Paris, and he suddenly begins to work it all out. Where? In a provincial park. And he arrives at the primacy of existence thanks to the root and trunk of a chestnut tree: 'I got up, I went out. When I got to the gate, I turned round. Then the park smiled at me. I leaned against the gate and I looked at the park for a long time. The smile of the trees, of the clump of laurel bushes, *meant* something; that was the real secret of existence.'[3]

At that moment might Roquentin not have thought of the end of René Clair's film *À nous la liberté*, made in 1931, when the factory and the world of mass production are left behind for the open air and the countryside? Might he not have recalled Giono's *Regain* or Pagnol's film? Might he not have solved his own need to be creative by writing a ruralist novel? It is highly unlikely that Roquentin, alias Sartre, entertained any such thoughts.

Urban intellectuals of the time are assumed to have identified the concept of rurality with the doctrines of the nationalist Right which, since the time of the Dreyfus Affair, had idealized the image of a hierarchical nation composed of rural and regionalist traditions.

By the late 1930s this nationalism and its call for a regenerative 'retour à la terre' [return to the earth] had secured a definitive conceptual hold over the minds of more than its own supporters. In 1919 the phrase itself had made one of its popular literary appearances, when novelist Gilbert Stenger wrote *Le Retour à la terre* in a suffused heat of postwar anger at the evils of trade unions and the mechanical enslavement of urban life. Didactic and laboured, but with poetic evocations of the rural scene, Stenger's novel sees the young peasant Antoine Daudat lose his personality, his willpower and his freedom of action, in his search for urban wealth, and for good measure he is beaten up by trade unionists when he refuses to go on strike. Nearly dead, he is nursed back to life in his native village in the Bourbonnais, and he marries 'la petite Louise', strong, robust, happy and full of good common sense who is chambermaid at the château of the paternalist seigneur. This noble squire welcomes him back to the land and to the recovery of the good old days.[4]

The concept of 'le retour à la terre' as it was developed by the nationalist Right, and whatever the economic programme envisaged, drew on the types of argument deconstructed by Roland Barthes in *Mythologies* (1957). Despite the positivism to which Maurras was nominally wedded, the concept was essentialist and anti-historical. By the time it became a main pillar of the National Revolution of Vichy France, roughly two years after Roquentin's encounter with his chestnut tree, its apostrophe to eternal virtues was already a cliché. The concept had become a wide-ranging racist and gendered discourse of preferred values: country against town, provinces against Paris, peasant against industrial worker, family against decadence, tradition against modernity, and Christianity against Bolshevism and the Jews. The slogan itself heads a moral crusade, alongside, and interchangeable with, 'la femme au foyer' and pronatalist injunctions. Vichy in the end gave it rather weak *institutional* support and only a limited measure of financial investment, but it kept the wide-ranging *ideology* of the discourse on the front page of its propaganda. The Provençal President of the regional Chamber of Agriculture promoted it *con brio* on Radio Nîmes in October 1940: 'Life in the towns, stressful, irritating and false, leaves no time for reflection. We should not be opposed to progress, but the essential truths are eternal, and if our present circumstances can lead to a return to the land which is true and life-giving, then our health, morality, and the strength of our race, so endangered over the last 20 years, will rediscover their ancient vigour ... Vive la paysannerie française.'[5]

Personified in the peasant origins of Pétain himself, 'le retour à la terre' as ideology and cultural abstraction was promoted in Vichy France in a pageant of regional cultural events, all with their musical and poetic interludes detailed on embossed programmes for the Regional Prefect, which can still be found in the attics of provincial town houses: there is little record, if any, of peasant attendance. Charles Maurras and his disciples pressed strongly for the changes which would transform France into the likeness, if not the reality, of a traditional monarchy, in the belief that the ancien régime and the old provinces lay intact underneath the ruins of post-revolutionary France and could, to use an earthy phrase, be easily disinterred. The notion of a permanently 'underlying' folk reality is crucial, and accounts for much of the attraction which many regionalist poets, story-tellers and cultural anthropologists felt towards the Right.

The collective archetype of eternal peasant values, the fairy story and the child's imagination had been brought together in a small, but significant book by Georges Bernanos written in 1930, when he was in his flamboyantly royalist phase. Published as a sort of Christmas present for anti-Republican families and called *Noël à la maison de France*, it has all the mordant wit and irony for which Bernanos becomes famous, but it also has tenderness too. Its thesis is a simple one. He deplores the fact that republicanism has a stranglehold on the institutions of education, and he ridicules republican civic instruction and political science as lifeless, pedantic and unrelated to the truths of French history. 'Fortunately,' he argues, 'in this century of pedantry, these truths retain their vitality, not least in the Mother Goose folk and fairy stories of Georges Perrault, who must surely be Saint Perrault by now, such is his special relationship to paradise. Thanks to the eternal images of Mother Goose, the French child has nothing to fear from republican attempts to suffocate the imagination. The child *knows*. The adults are ignorant and misguided.' As an example he writes, 'I asked a young boy in Alsace "What is a king?" His reply was instant: "A man on horseback who has no fear". It does not matter that his parents are republicans. He knows.'[6] It follows from the main argument that the royalist and nationalist Right should look to youth and, still more, to children as the depositories of political wisdom, an argument to which Pétain was particularly responsive, with his litany of praise to the virtues of innocent and uncontaminated youth.

Judith Proud has shown how easy it was for Vichy to make the wolf in Perrault's 'Little Red Riding Hood' into the anti-Semitic stereotype of an evil, conniving Jew, and to allow her to be rescued by the youthful peasant hero, the woodcutter, 'a son of the new France'.[7] I would like to return to the wolf later.

Clearly, Bernanos was not directly responsible for the infantilism of Vichy propaganda, since he broke with right-wing nationalism during the Spanish Civil War. But infantilism was rampant in the back-to-the-land images of the nationalist Right in the 1930s. *Noël à la maison* is indicative of a populist ideology which sees the uneducated as the wise. The idealization of the peasantry follows the same course: the peasants do not need to be taught – they *know* already; they are 'eloquent in their silence'. What is vital to understand about this infantilism, and this abstraction of the peasant, is that they constituted such a powerful mythology of rural life under Vichy that any other interest in rural identities and concern for peasant values, rituals, traditions and behaviour found it almost impossible to establish serious credentials or create an area of independent thought, or sufficient space for effective political policy. The equation of rurality with the programme of the nationalist Right under Vichy obscured, and continued after the war to obscure, the nature of other ruralisms. These other rural programmes and ideas have had to be rescued from neglect, by research which has dragged them into the open from behind the all-pervading nationalist and Vichy discourse. They never constituted a comparable discourse or movement, but they were less mythological in the Barthesian sense, though eminently capable of using cultural myths and ritual, and they were not dependent on infantilism. Their tragedy was to be obscured and ignored at strategic historical points when they might have informed a multicultural pattern of French social and political development.

The whole subject has been fraught with difficulties. Any mention of peasant culture, regionalism or ruralism of the 1930s and 1940s still triggers such an immediate association with the mythology of Vichy that it may seem foolish to tamper with it. For example, the cinema, art and literature which could be used to assert the existence of an alternative ruralism, are surprisingly rich and well known, but if they are situated politically it is normally by reference to the dominant discourse of the Right. One example among many: the energetic and imaginative historian, Marc Ferro, has lectured on Jean Renoir's *La Grande Illusion* of 1936 and has illuminated the sequence when the French escapees from a German prisoner-of-war camp during the First World War are protected by a German peasant woman whose husband has already been killed in the war. It is an idyllic rural scene with cows, Christmas on the farm and the celebration of family values across the divides of nationality and religious culture, all against the backdrop of a dehumanizing war. Ferro suggests that, whatever Renoir's left-wing sympathies, this scene made the film ambivalent in its politics because it reflected the rural nostalgia of the Right and anticipated Vichy's peasant utopia, its family programme and even the collaboration of French and Germans. Not at all impossible; in fact when I first heard his analysis it seemed that the link was too obvious to miss.[8] But on subsequent consideration, why does the link forwards have to be with Vichy and the Right? Why not see an anticipation of the resistance actions of French peasant women in the Cévennes, for example, who took in foreign refugees and escapees during the Occupation without a question asked? Some of these refugees were German and many were Jewish, escaping from both the Nazi and Vichy police. Here is real comparability between film fiction and historical fact. Visually, many of the family snapshots which have survived of Jewish refugees in the farms and villages which protected them,[9] can stand close comparison with the images in *La Grande Illusion*, but no one looking back after the war made this alternative iconographic link, and it is still difficult in French historical circles to draw from it any substantive ideas about an alternative or lost moment of French history. The phenomenon of rural resistance under the German Occupation provided an opportunity for urban politics and historiography to respond to the plural identities of rural France. But it was an opportunity that was not taken. Before that there was the earlier opportunity provided by the Popular Front.

It is not hyperbole to state that the election of the antifascist Popular Front Government in 1936, with its humanist, Jewish Prime Minister Léon Blum, was one of the outstanding moments of hope and idealism in modern French history, whatever judgement one may go on to make about its successes or failures. The Magnum photographs of Robert Capa, David Seymour and Henri Cartier-Bresson, which captured the working-class gains of better conditions, a shorter working week and paid holidays, were themselves creative acts which surely could have inspired Roquentin or any other jaded and disillusioned historian. But for the nationalist Right the Popular Front was the embodiment of the industrial and decadent 'other', which, it claimed, threatened the life-blood of 'True France'. In reaction to the working-class triumph, the essentialist doctrines of Action Française were intensified, and used to justify the attempted assassination of Léon Blum. In this bitterly polarized situation it is extraordinary therefore to find evidence of an alternative ruralism, even if it is no surprise that its voice was largely lost in the clamour both for and against the ideology of the Right.

In 1936, close to Jean Giono's home at Manosque, a dedicated supporter of the Popular Front, François Morénas, set up a youth hostel which he called 'Regain' in homage to Giono's already classic novel. The young socialist men and women who came there over the next two years set about restoring a ruined village, and collaborated closely with Giono's community projects at Cantadour. It was not just a country outing of left-wing eccentrics nor did it cut itself off entirely from urban culture. At Cantadour itself, the gramophone played Mozart and Couperin, and Lucien Jacques taught locals the dance steps he had personally learnt from Isadora Duncan.[10] It seems from Giono's diary that in 1935 he was very close to accepting working links with the collectivist culture of the Popular Front, and he did indeed write for André Chamson's and Jean Guéhenno's weekly paper *Vendredi*. But in 1937 Giono issued a refusal. It was mainly a manifesto of pacifist intention, rejecting the whole move of the anti-fascist Left towards national defence, but it was received as a break with the whole ethos of the Popular Front.[11] His diary insists this was to protect his freedom of thought and expression. The break soon became open hostility and Giono retreated into a simplistic hatred of urban life and politics. During the war he was courted by Vichy and would have nothing to do with military resistance. But he did advise young peasants to reject labour service in Germany and he helped shelter two eminent Jews, one of whom, the musician Heinz Meyerowitz, cannot have been easy. Meyerowitz insisted on taking his piano everywhere with him and openly attended a German production of Wagner in Marseille.[12]

Historians who have alluded to the break in 1937 have pointed to communist intransigence within the Popular Front, particularly Louis Aragon, or to the growing fundamentalism of Giono's pacifism. Curiously enough, no one has researched it from the point of view of the ruralism which was at the centre of the Giono mystique and which an unknown number of young urban workers clearly found attractive, and there remains the hypothesis that a break might have been avoided had Giono's unique Cantadour experience been actively promoted by the

Popular Front in the interests of the humanism vaunted by the political leaders.[13] It is no longer a far-fetched research project to reconstruct the Popular Front's own interests in an alternative ruralism, and it has taken shape alongside the explosion of interest in the Popular Front's social-harmony aims which lay behind its promotion of sport and leisure activities and its science of popular holiday-making, when the sheer happiness generated by industrial workers going back to their rural roots during the first paid holidays raised the profile of rural values in much of the socialist and communist literature of the time. Undoubtedly the pages of *Vendredi* allow us glimpses of an alternative ruralism, and still more so the socialist weekly *La Terre libre*, and the communist *La Terre*. But mostly what was stressed in postwar studies of the Popular Front and rural France was that agricultural workers were not given the same right to paid holidays; that the Left identified most peasant discontent with the activity of the fascist-style greenshirts of Henri Dorgères; and that the creation of the Office national interprofessionel du blé (ONIB) by Georges Monnet, the Agricultural Minister, was an attempt both to stabilize wheat prices in the interests of the producers and keep food prices down for urban consumers, a dual aim which could only culminate in ambivalent results. Little, if anything, was written about the Popular Front's own ruralist ideas and initiatives.

In some ways this is surprising, given the rural gains of the Socialist Party in the 1936 elections and the accelerating implantation of the Communist Party (PCF) in rural areas throughout the 1930s, though the successes of the PCF with the rural electorate were not fully understood by the party until 50 years later. The Gascon communist Renaud Jean battled against the Party's urbanism, convinced that a shift of attitude would reap huge dividends in the countryside. Popular Frontism itself was a change in party policy and there was a resultant gesture in favour of peasant proprietors, but the doctrine of collectivization held its own as the party's panacea for agriculture. Within the SFIO, rural socialism and the agricultural cooperative movement had their own long-term roots and economic rationale, as Tony Judt has shown for the Var.[14] The consistent support of Jean Jaurès for 'the robust, democratic peasantry' was combined with a critique of their capacity for submitting to unacceptable conditions as immutable. His proposals for community ownership of rural productive processes were presented in the mid-1890s, alongside cheap credit for peasant proprietors, whom he never abandoned for a policy of collectivization.[15] His influence remained profound, particularly on Léon Blum and his socialist agricultural minister Georges Monnet, to the point where the Popular Front government attempted to create its own ruralist philosophy as a modernizing, republican antidote to the 'retour à la terre' nostalgia of the nationalist Right and the corporatist proposals of the Union nationale des syndicats agricoles (UNSA). In common with other aspects of Popular Front policy there was also a remobilization of the unitary republican ideology, which at the beginning of the century had presided over mainstream nationalism. This had permeated most areas of French life, with its climax in the *union sacrée* of 1914, and it also had its paradigm image of the patriotic peasant, whose protection had been guaranteed in the 1890s by the tariff reforms of Jules Méline who went on to give his agricultural programme the title *Retour à la terre* in 1905. The Popular Front's version of a 'New Deal' for rural France declared itself well within the unifying republicanism of this tradition, and although Monnet failed to carry many of his policies through a suspicious Senate, there is enough evidence of what Gordon Wright terms 'peasantism' in the Popular Front to encourage a much closer look at its ruralist credentials.[16]

This has now been done by Shanny Peer in her watershed study of the 1937 World Fair in Paris and its revelation of Popular Front attitudes to peasants, provincials and folklore, in which she deepens even more the analysis of cultural identities in the 1930s, perceptively pioneered by Herman Lebovics in *True France: The Wars over Cultural Identity 1900–1945*. She documents in absorbing detail the ways in which the Popular Front government set out to confer symbolic national, republican and unifying status on the diversity of peasant life and rural folklore, and to replace the romantic, archaic right-wing images of the peasantry with an imagery of rural democracy and modernization as the authentic tradition of the French countryside. Particularism was to be celebrated as evidence of cultural vitality in both the regional and local sphere, but more significantly as testimony to the richness of the whole. In a section headed 'Reinventing rural France' she concludes that the World Fair with its Rural Centre and model village gave new shape to a vision of the efficient, modern family farmer. At the same time, she adds, 'this image of a modernizing France retained a comfortingly picturesque veneer by reworking a familiar repertoire of images to reassure visitors that progress need not eliminate tradition in *la France profonde*.'[17] In the following section she further shows how 'the folklore of the rural French provinces was reinvented as a vital expression of France's national, popular culture.'[18] The book does not underestimate the conflictual presence of the right-wing peasantism and regionalism which opposed the Popular Front's programme as standardizing socialism, destructive of the peasant way of life: the Breton deputy Pierre Caziot, who later became Vichy's Minister of Agriculture, stated in the Chamber in February 1937, 'It is said that, in every society, there must be slaves to ensure the wealth of others. The Popular Front has also chosen its slaves: the peasants.'[19] Nothing in the Rural Centre's combination of tradition and modernity was able to outmanoeuvre Caziot's conservative scepticism, let alone the essentialism of the extreme Right's ruralist myth, but Peer allows us to see for the first time just how far the Popular Front went in its acculturation of rural France: it makes available material which had been ignored or even denied for decades after Vichy, and enables us to explore even more insistently the notion of a lost opportunity of multiculturalism within mid-century France.

From a different but related perspective Susan Carol Rogers had arrived at a similar conclusion on the symbiotic relationship of rural particularism and national identity in her influential analysis of Ste Foy in the Aveyron with its specific family structures of the *ostal* system. This example of a resilient rural culture in the 30 years after the Second World War, still defying the centralizing and levelling processes which were thought to be dominant in the first half of the century, does not lead the author into counter-generalizations about the dominance of diversity, but rather prompts her into subtle understandings of how such particularisms could flourish within the apparently unifying concept of 'being French'. Her theoretical solution to the great debate initiated by Eugen Weber's classic *Peasants into Frenchmen*, is that a centralized French State should not be seen as replacing a culturally diffuse society, but rather that the two should be 'understood as dialectically related, fulfilling and preserving each other'.[20]

Both of these superb studies break the mould of an urban-based historiography which took the polarity of 'ruralism' and 'modernity' as a given. Peer shows that ruralist ideas were not confined to the political Right, and Rogers shows that a particular form of rural particularism flourished in Ste Foy at a time when it was largely thought to be archaic. But neither study sets out to claim more for cultural diversity than the specific evidence allows. Neither the Rural Centre of 1937 nor the *ostal* system of Ste Foy constitutes by itself, or even together, an argument that France represented itself as a multicultural society in the mid-century. It could have done so but it did not. If there was diversity within unity, both studies acknowledge the constraining role of unity at a national level. It was indeed the notion of a unifying culture which predominated even when the evidence for diversity was available in the ways Rogers and Peer amply demonstrate. This dominant self-image was reinforced by the wholesale repudiation of Vichy, together with its 'retour à la terre' ideology, so that the diversity which was displayed in the Rural Centre, the specific rural identities which can be found in rural resistance, and any example of particularism such as the *ostal* system, all remained unacknowledged until the late 1960s at the earliest, on the basis that recognition of rural and regional plurality would play into the hands of the Vichy apologists and the surviving followers of the old Maurrassian Right. In short, the 'retour à la terre' ideology was allowed to act as a blocking mechanism over all other ruralist ideas and analysis.

The facts of rural resistance during the Occupation ought to have made this unnecessary. Like urban resistance it was a minority movement but even more so. To resist in the countryside meant refusing the seductive blandishments of Vichy's peasant propaganda and the primacy it gave to rural values. Rural resistance had its own rhythms and motivations, and in 1943 it revealed an alternative chronology. The liberation of Corsica in early October raised hopes of an imminent allied landing in France. As the winter approached and no landing came, the disillusionment throughout France was deep: Vichy seized the moment to step up its offers of an amnesty for workers who were hiding in the country from compulsory labour in Germany; hardship and hunger were widespread, and both the German troops and the Vichy police were confident of rounding up any lingering bands of poorly armed maquisards. The winter of 1943 did not, in this chronology, announce itself as an opportune moment to stabilize the fragile history of the maquis. But there was an alternative chronology operating, the chronology of the seasons. The maquisards were forced by winter to come down from the hills, but, like the sheep in transhumance, they were taken into certain villages and fed and protected. There developed in specific areas a community culture of the outlaw. In these rural places, the unpromising winter intensified and consolidated the existence of the maquis as a collective resistance force. The Germans only began to realize this in February 1944. Unable to liquidate the maquisards as they had expected, they relieved their frustration in an orgy of reprisals on villages in the Cévennes, the Limousin and elsewhere. Rural resistance, still a minority phenomenon, grew increasingly determined, successfully subverting Vichy's rural ideology. Its cultural resources lay often in ancient traditions of revolt and independence such as the eighteenth-century Camisards or even the medieval Cathars, and its local strengths should have alerted urban republicans to the cultural diversity of rural France and to traditions which were not inevitably reactionary, or atavistic in the negative sense used by most public administrators and a majority of politicians.[21]

Much of the lead in theorizing rural resistance as a cultural phenomenon has been taken by anthropologists and cultural historians of the Americas, India, the Far East and Africa. The concepts of James C. Scott and Ranajit Guha are immensely suggestive.[22] A new understanding of the subtlety of oral transmission in rural areas is also suggested by David Lan's anthropological study, *Guns and Rain. Guerrillas and Spirit Mediums in Zimbabwe.* With detailed insight Lan demonstrates how the spirit mediums in the Dande district of the Zambezi valley gave legitimacy to the guerrilla resistance against the British in the 1970s by providing them with access to the spirits of the chiefs of the past, the *mhondoro*, at a time when the chiefs of the present were widely felt to be too compromised with the colonial authority. It was not, argues Lan, the decision of the mediums or the guerrilla leaders to turn from the present to the past, but rather it was the ordinary people themselves who shifted their allegiance. This shift gave new and unexpected power to the mediums, and established a popular alternative to the authority of the colonial state.[23] An amazingly similar process of story-telling, memory and empowerment operated in parts of France during the Occupation, when generations of local history, rural knowledge and survivalism were transmitted by peasants and villagers to the maquis during the long evenings of the winter of 1943.

It is not easy to decide whether rural people did or did not transmit a specifically *regional* culture to the maquisards whom they fed and sheltered. Someone who thought they did was Max Allier, whose resistance poetry rejected Maurras and the Right in the name of a new Occitanism. He introduced me to the unwritten history of this radical new regionalism, which emerged fully in the late 1960s, but, he insisted, was already there in embryo during the late 1930s.[24] It was indeed. Charles Camproux, one of the best poets writing in the 'langue d'oc', animated a new Provençal Party to stand against the right-wing regionalism founded by the poet Frédéric Mistral in the late nineteenth century. Influenced by the degree of autonomy gained by Catalonia under the Second Spanish Republic, Camproux refloated Occitanism on a broad base of anti-fascism, and he was later to be found in the Resistance. There is still much to be done on the local history of the Institut d'Études occitanes which was launched on the wave of resistance across the Languedoc. It became a leading force in the regional support for the strike of the Decazeville miners in 1961–2, and the successful ten-year struggle to stop the expansion of the military camp on the plateau of Larzac.

If the roots of the new Occitanism clearly lie in the 1930s and resistance period, it is not so easy to show where the new Breton nationalism came from, since the prestigious Breton nationalists of the time were so heavily involved in collaboration with the Germans that this almost obliterated the tracks of others, notably in Basse-Bretagne, who were resisters and regionalists at the same time.[25] And yet a Breton geographer, Michel Phlipponneau, wrote in 1967 to argue that the virtually unknown committees of regional development which sprang up spontaneously after the Liberation, and with no allegiance at all to the old nationalist ideology, could have provided France with a fully democratic regionalist programme immediately after the war. As it was, the parliamentarianism of the Fourth Republic and the authoritarianism of the Fifth stood against this eminently rational solution.[26] De Gaulle's constitutional adviser, Michel Debré, would not even use the word 'Bretagne', referring to the region as 'l'extrême-ouest'. The Liberation Commissaires de la République, appointed on a regional basis, were soon abolished and were not replaced by Regional Prefects, on the basis that the very term smacked of Vichy. Surely the postwar governments might have acknowledged the debt that their own resistance credentials owed to many rural communities, which now sought greater regional identity, greater financial and cultural investment, a devolution of democracy and a new social charter? The opportunity to make this acknowledgement was not taken, and the field, so to speak, was left once again to right-wing leaders of peasant protest and to the perpetuation of ruralism within the syndrome of Vichy France.

It can, of course, be argued that the pressures of patriotism during the war had made it impossible to acknowledge the cultural diversity of rural France. If the Resistance was going to become representative of the nation it had to recuperate the notion of 'True France' which the Right had arrogated to itself. However disparate the springs of resistance at local level the image of resistance at national level had to be unitary, and this resulted in presenting the peasantry and rural France in ways which were remarkably essentialist. It was a rival essentialism, justified by the needs of resistance; a bold recuperation of images and language (such as 'le foyer') which had been monopolized by the Right. By these acts of recuperation the resistance legitimized itself and came to stand for the nation. But afterwards, after the Liberation, in the late 1940s and 1950s, was this essentialism so necessary?

Cultural diversity was manifest through the Popular Front's Rural Centre and creative in its contribution to local resistance, but postwar governments and political parties were basically unresponsive to cultural differences in rural and regional France. An unthinking unitary culture failed at crucial points to evaluate the *weakness* as well as the strength of its humanist claims. The multiculturalism which was so needed by France, both in the 1930s and particularly after the Second World War, was blocked by the monopoly which the old right-wing ruralist ideology established over any prolonged discussion of rural particularism. This was in turn compounded by the struggle for patriotism during the war. Together the two elements fused into a unitary cultural front which marginalized alternative ruralisms and refused to recognize the diversity and inversions which were so vital to resistance, and to the wider hopes of the Liberation. France entered the postwar world in active denial of its own internal cultural diversity. The readiness to fight repressive and inhumane colonial wars in Indo-China and Algeria cannot be held separate from this fact, but the precise relationship between the two areas of cultural intolerance still needs detailed research.

My own fantasy about the sequel to Roquentin's moment of truth is that he might eventually have been persuaded to explore the cultural diversities which were revealed in alternative rural actions during the resistance, actions which were not another aspect of Vichy's 'retour à la terre', but yet were full of traditions, of story-telling in the farms, of oral transmission in the villages, rituals of inversion and acts of transgression, all in the hard-edged modern context of mechanized warfare and economic rationalization. It is exactly Roquentin's sort of story, 'beautiful but as hard as steel'.

I will end by reclaiming not only Roquentin but the wolf. The southern paper, *Midi libre*, was full of stories of wolves in the Massif Central when I first began to read it in 1969. I thought little of it at the time, but later found several references to 'loups-garous' in the police reports from rural areas in 1942–4. In popular imagination the 'loup-garou' is a dangerous and evil spirit which haunts the woods and fields at night, and is said to feed on human flesh. Figuratively it signifies a surly and eccentric rural tramp or hermit who lives alone and shuns all company. On sighting such a figure children will call out 'garreloup' and run away. Did this popular imagery provide yet another camouflage for the first maquisards? I do not know. But it calls for a second look at the imagery of the wolf and rural society. This was already evident in the Occitan fête of 1982 at Marvejols in the Lozère. On 30 June 1764, Jeanne Boulet, a girl of 14, was devoured by 'a ferocious beast' in the hamlet of Ubas in the Vivarais. It was the beginning of the legend of the Bête de Gévaudan, an enormous wolf which the king's hunters were quite unable to find or destroy and the Church unable to exorcize. There are many versions of the story. But at Marvejols there was a theatrical event, a regional rite of passage. There was an entry into the mouth of a giant carnival wolf and an exit through its tail, completing a symbolic encounter with the culture of Occitanie. The very site in which it happened proclaimed inversion.

It was staged in the main square, once Place Royale but changed to Place Henri Cordesse, the tough, sensitive local schoolteacher who became Prefect at the Liberation in 1944, and personified the minority rural resistance of the Basse Lozère.

It is not guaranteed that all who go through a figurative Bête de Gévaudan will emerge with a new set of questions and a new understanding of different cultural identities. But if that is what Roquentin's creative act of adventure might have achieved, then contemplating French roots is indeed a moment of truth, and existentialism, finally, a humanism.

Notes

1 F. Marcot (ed.), *La Résistance et les Français. Lutte armée et Maquis* (Besançon, 1996).
2 J. Sainclivier and C. Bougeard (eds), *La Résistance et les Français. Enjeux stratégiques et environnement social* (Rennes, 1995).
3 J.-P. Sartre, *La Nausée* (Paris, 1938), trs. *Nausea* (Harmondsworth, 1965), pp. 193, 252.
4 G. Stenger, *Le Retour à la terre* (Paris, 1919).
5 Archives départementales du Gard, CA 1512 (Roger Rouvière, 'Causerie Radiophonique', October 1940).
6 G. Bernanos, *Noël à la maison de France* (Paris, 1930), pp. 16, 18.
7 J. K. Proud, *Children and Propaganda. Il était une fois . . . Fiction and Fairy Tale in Vichy France* (Oxford, 1995), pp. 30–5.
8 I first heard his lecture, entitled 'The Social Function of the Cinema', at the University of Sussex on 23 November 1976.
9 P. Joutard, J. Poujol and P. Cabanel (eds), *Cévennes terre de refuge 1940–1944* (Montpellier, 1987).
10 L. Heller-Goldenberg, *Le Cantadour 1935–1939* (Nice, 1972), pp. 123, 130.
11 J. Giono, *Refus d'obéissance* (Paris, 1937).
12 Heller-Goldenberg, *Le Cantadour*, p. 202.
13 The Cantadour experience is summed up by Heller-Goldenberg as follows: 'The Cantadour entirely defies definition; it goes beyond the framework of the youth hostels where one may trace its origins, it is equally detached from impracticable principles such as pacifism or the return to the earth which were only one aspect of its ideology. Nor is it fully explained in all those poetic, literary and social works to which it gave birth. Above all else it is a moment of delight'; *ibid.*, p. 249.
14 T. Judt, *Socialism in Provence, 1871–1914* (Cambridge, 1979).
15 H. Goldberg, *The Life of Jean Jaurès* (Madison, 1962), p. 181.
16 G. Wright, *Rural Revolution in France* (Stanford, 1964) pp. 58–74.

17   S. Peer, *France on Display. Peasants, Provincials, and Folklore in the 1937 Paris World's Fair* (Albany, NY, 1998), p. 133.
18   *Ibid.*, pp. 134 and 135–66.
19   *Ibid.*, p. 109.
20   S. C. Rogers, *Shaping Modern Times in Rural France. The Transformation and Reproduction of an Aveyronnais Community* (Princeton, 1991) p. 198.
21   See H. R. Kedward, *In Search of the Maquis* (Oxford, 1993).
22   J. C. Scott, *Weapons of the Weak. Everyday Forms of Peasant Resistance* (New Haven, 1985); R. Guha, *Elementary Aspects of Peasant Insurgency in Colonial India* (Delhi, 1983).
23   D. Lan, *Guns and Rain. Guerrillas and Spirit Mediums in Zimbabwe* (London, 1985).
24   Kedward, *In Search of the Maquis*, pp. 235–6.
25   See J. Sainclivier, *La Bretagne dans la Guerre, 1939–1945* (Rennes, 1994).
26   M. Phlipponneau, *La Gauche et les régions* (Paris, 1967). The book, like many others published in the 1960s on regional affairs, refers to the enormous importance of J.-F. Gravier's 1947 analysis of the neglect of the provinces, *Paris et le désert français*. Phlipponneau saw it as a voice of sanity crying in the wilderness of official policy and states that it became the bible of all those after the war who refused to accept regional decay.

15
From Greater France to Outer Suburbs: Postcolonial Minorities and the Republican Tradition

Alec G. Hargreaves

International migration is at one and the same time a major preoccupation in contemporary France and a blind-spot in constructions of the nation's past. As Gérard Noiriel has observed, immigration is a 'non-lieu de mémoire' (site of amnesia),[1] largely absent from both French historiography and public memorials to national identity. The prime symbolic site associated with today's immigrants is the *banlieue*, a marginal space on the outer rim of France's major conurbations. Just as immigrant minorities are located on the topographical periphery of contemporary society, so they have been relegated to the '*banlieues* (periphery) of French history'.[2] In exploring this marginality, I will begin by considering some of the main explanations which have been advanced to account for it and will suggest that the (post-) colonial origins of recent migratory flows merit particular attention. My argument will be that the incorporation of immigrant minorities is hampered less by a failure on their part to espouse majority ethnic values than by the failure of the majority population to apply fully the values associated with the discourse of republicanism. In the second half of my analysis, I will illustrate these tensions by a case study of the bicentenary celebrations of the French revolution, where the marginalization of minority groups offers revealing insights into majority ethnic attitudes.

Amnesia and repression

The absence of immigration from all but the final volumes of Pierre Nora's *Les Lieux de mémoire* is symptomatic of its non-status in most constructions of French national history. Immigrants are seldom mentioned in general histories of France and are entirely absent from the school syllabuses through which succeeding generations of citizens have been taught to think of national history. Only after immigration came to the forefront of political debate during the 1980s did a significant strand of historical research begin to emerge in this field.[3] Noiriel, a major figure in this re-evaluation, was eventually invited to contribute a chapter to the final instalment of *Les Lieux de mémoire*, published in 1992.[4] In it, he traced the 'historical amnesia'[5] of the French concerning immigration back to the Revolution of 1789, which established a framework of concepts and values which tended to marginalize or efface the significance of ethnic differences.

The founding myths of modern French identity were forged during the revolutionary period, well before the rise of mass inward migration. At that time, France was the most populous nation in Europe. In the United States, by contrast, nation-building and immigration were inextricably intertwined. The initially much smaller population of the fledgling American Republic originated almost entirely overseas, and as it grew with successive waves of immigrants the diverse ethnic origins of the nation were publicly recognized and celebrated. It was not until the second half of the nineteenth century that international migration began to gather pace in France. Immigrant workers came in to plug the gaps in a French labour market hampered by low rates of natural population growth. Although many immigrants settled permanently in France, they were seen as helping to sustain a nation with a long history behind it, rather than as partners in the construction of a new national community *à l'américaine*. Within the French assimilationist model, immigrants were to be absorbed into a pre-existing national mould which gave no recognition to ethnic particularities.

By abolishing privileges based on heredity or religious belief and making the nation the source of political sovereignty, the revolutionaries of 1789 had emphasized the primacy of individual freedom and denied any legitimacy to ethnic groups other than the national community, incarnated in the one and indivisible republic. While the distinction between French nationals and foreigners was to become a fundamental feature of official statistics and public policy under later republican regimes, the relative openness of French nationality laws made it easy for people of foreign origin to become French citizens. By the same token, their ethnic origins disappeared from the public domain. Official statistics are almost entirely silent about the foreign origins of the children and grandchildren of immigrants, who as French citizens are seen as part of a single, undifferentiated national community.

Republican values alone cannot, however, fully account for the absence of immigration from collective representations of French nationhood. It should not be forgotten that the United States has a longer and more continuous history of republican government than France. This has not prevented Americans from recognizing ethnic differences and taking account of them in public policy. Moreover, the core values of the French Revolution, encapsulated most obviously in the Declaration of the Rights of Man and of the Citizen, are more open to manifestations of cultural difference than is sometimes thought. Freedom of thought, opinion, religious belief and expression are all guaranteed by Articles 10 and 11 of the 1789 Declaration.[6] The current vogue for explaining majority ethnic aversion towards the recognition of minorities essentially in terms of republican values is not therefore wholly convincing.

Many of those who oppose the public recognition of ethnic differences are no doubt both sincere and to a certain degree justified in appealing to the republican tradition in support of their position. The sinister purposes for which ethnically-based categories were used by the anti-republican Vichy régime left a deep distrust in postwar France about the potential abuse of such categories by public authorities. By refusing to allow the state to collect ethnically-based statistics, many republicans believe they are protecting minorities from possible abuses of power.[7] As a quid pro quo, it may appear reasonable to expect minorities themselves to confine ethnic differences purely to the private sphere, thereby maintaining the apparent neutrality of public spaces. Yet that apparent neutrality is something of an illusion. The language and values of the public sphere in France are those dominant among the majority population. French alone is the language of the republic, and since the constitutional reform of 1992 this privileged status has been enshrined in the country's fundamental laws. As public spaces are *laïcs* (secular), religious expression is confined within carefully delimited parameters. Some self-styled republicans confuse the domination of majority ethnic norms in the public sphere with a mandate for the elimination of cultural differences on a wider scale. Not uncommonly, republicanism is implicitly equated with assimilation, and immigrant minorities who seek to retain cultural differences are branded as anti-republican. In this way, the moral legitimacy associated with the republican heritage is turned on its head, and serves as a cover for intolerance towards cultural pluralism.

Significantly, this reactionary brand of republicanism is selective in its intolerance. European immigrants, whose religious beliefs include a variety of Christian traditions as well as Judaism and other faiths (some of them relatively new to France), are seldom accused of challenging the republican order. The main target of republican zealots is Islam, by far the largest religion among non-Europeans in France. The discriminatory nature of this position was evident during the Islamic headscarf affairs of 1989 and 1994, when the sanctity of republican values was invoked to justify the exclusion from state schools of Muslim girls wearing headscarves while maintaining the right of Catholics and Jews to wear crucifixes and kippas. Leading intellectuals such as Régis Debray and Alain Finkielkraut, generally regarded as progressive in outlook, were horrified by the decision of Education Minister Lionel Jospin to accept the wearing of headscarves, which they described as 'the Munich of the republican education system,'[8] implying that Islam posed a threat to contemporary France comparable to that of Nazism half a century earlier. Although there was widespread support among politicians and the public at large for banning Islamic headscarves at school, the courts have repeatedly judged such a ban to be incompatible with the laws of the republic, which protect the rights of all individuals – including those who adhere to the Islamic faith – to freedom of religious belief and expression.[9] These court rulings amply demonstrate the inadequacy of the republican thesis as an explanation both for majority ethnic hostility towards certain minority groups and for the wider failure to recognize immigrants and their descendants as constituent features in the making of modern French society.

This historical blind-spot cannot in my view be fully understood without reference to the colonial origins of recent migratory inflows. Today's immigrants are yesterday's colonial 'natives'. The Africans and Asians who began settling in France during the 1960s and 1970s spent their formative years under French rule during the twilight of the colonial empire, which at its zenith had been popularly known as *la plus Grande France* (Greater France). Although subject to French authority, very few of the tens of millions of non-Europeans indigenous to the overseas empire were admitted to citizenship. The native populations of Greater France were excluded from the republican community not only in the sense that they were denied the political rights associated with citizenship, but also in the sense that the laws of the republic did not apply to them. The colonies were ruled through a system of executive orders untrammelled by parliamentary scrutiny, among them the notorious *code de l'indigénat* (native code), which gave authoritarian powers to French administrators and denied freedom of speech to the mass of the population. The notion of Greater France was thus highly ambiguous. While the empire served as symbol of French 'greatness', far from being incorporated into an enlarged French republic its inhabitants were held firmly beyond the pale.[10] During the First World War, the French sought to compensate for manpower shortages by mobilizing more than half a million colonial troops and bringing in almost a quarter of a million workers from the overseas empire. When the war ended, almost all were rapidly repatriated.[11] Until well after the Second World War, Algerians were the only significant but still relatively small group of peacetime migrants from the overseas empire. Far more than European migrant workers, Algerians remained separate from the majority ethnic population. Distrust engendered by decades of colonial domination led many Algerian migrants to seek security in minority ethnic enclaves, while the French authorities, fearful of the potentially corrosive effects of left-wing ideology vis-à-vis the colonial system, endeavoured as far as possible to keep colonial migrants away from the French labour movement.[12] For different reasons, there was tacit agreement on all sides that migration from Algeria was a temporary phenomenon. Instead of integrating Algerian migrants into metropolitan society, the authorities in France and the settler population in Algeria wanted to ensure that they remained tied to the colonial regime. Migrants themselves saw a period of employment in France as an economically necessary but short-term exile from home in North Africa. It was not until the postcolonial period that France became a locus of permanent family settlement for minorities originating in Algeria and other parts of the former empire.

In many ways, therefore, the psychological barriers which are today associated with the *banlieues* were presaged during the colonial period by exclusionary attitudes and practices which affected migrants originating in the overseas empire more deeply than those of European origin. Majority ethnic resistance towards the incorporation of minorities of colonial origin was sharpened by the violent and traumatic manner in which the empire was liquidated after the Second World War. Algerian immigrants, the largest national minority of non-European origin, played a key role in support of the struggle for Algerian independence, which was largely financed by money collected among the expatriate population in metropolitan France.[13] Unsuccessful French attempts at resisting the tide of decolonization – first in Indochina and then, most damagingly, in Algeria – left the colonial enterprise shrouded in a bitter legacy of humiliation and shame. As a result, there was a widespread desire to obliterate memories of the colonial débâcle and to reorient national energies in other directions, notably Europe.

Places and institutions associated with the former empire were renamed and shorn of their colonial labels. The Musée Colonial became the Musée des Arts Africains et Océaniens, the Académie des sciences coloniales turned into the Académie des sciences d'outre-mer, Nord-Africains and Arabes (words which resonated with the polarity between 'natives' and colonizers) were transformed into Maghrébins, the Colonial Ministry gave way to a Ministry of Cooperation, and *la francophonie* gained currency as a seemingly depoliticized label for the global space whose contours had once been those of the French colonial empire. French street names, a classic *lieu de mémoire* for public commemorations of the past, had been slow to absorb names from overseas during the colonial era and were slower still to incorporate traces of the struggle over decolonization. Campaigns by ex-servicemen's associations for a *rue du 19 mars 1962* (the date of the cease-fire which ended the Algerian war) to be created in every locality in France fell largely on deaf ears.[14] Most of the population, including many of those who had been conscripted during the war, preferred to repress memories of the Algerian conflict.[15]

Yet even as the French were struggling to filter out painful memories of their colonial past, the legacy of empire was brought ever more directly home by the rapid rise of immigrant minorities originating in former colonies. Until comparatively recently, Europeans were overwhelmingly dominant among immigrants to France. As late as 1962, when Algerian independence marked the end of the colonial period, 72 per cent of all foreigners in France were Europeans. Twenty years later, this figure had dropped to less than 50 per cent and by 1990 it stood at just 40 per cent, while Africans and Asians accounted for 57 per cent of the foreign population, with Maghrebis by far the largest group among them.[16]

During the Algerian war, Islam had served as a key emblem of national identity among those engaged in the struggle for independence from France. Small wonder, then, if Islam has appeared to the French as a religion fundamentally at odds with their own sense of national identity. Perceptions of Islam as aggressively anti-Western have been fuelled during the postcolonial era by a wide range of international developments including pro-Palestinian Arab terrorism, the revolutionary regime established by the Ayatollah Khomeini in Iran and the Gulf War of 1991. Although very few Muslims in France originate in or have any connections with countries under the sway of radical Islamic regimes, media coverage of those regimes has encouraged perceptions of Islam as an inherently fanatical faith. Consciously or unconsciously, therefore, the majority population is inclined to see in Maghrebis and other minorities originating in the Islamic world a threat to the continued domination of the secular values associated with the republican tradition.

The exclusion of immigrant minorities from constructions of French national identity has thus been strengthened by the legacy of the colonial period, which has helped to create an image of Muslims as fundamentally hostile to mainstream French values. It is true that naturalization rates are unusually low among Algerian immigrants, compared with other nationalities.[17] This undoubtedly reflects an enduring distrust among older Algerians vis-à-vis the French state, which waged a long and brutal war in an unsuccessful attempt to deny their separate sense of nationhood. Among the children of Algerian migrants, however, there is a much clearer sense of belonging to France. Although most are *de facto* binationals, with automatic French and Algerian citizenship as a consequence of the nationality laws of the two states, the overwhelming majority of second-generation Algerians choose French rather than Algerian identity papers. Until the recently announced abolition of conscription in France, male descendants of Algerian immigrants could choose to do their military service in either country. The vast majority have opted for France.[18]

There are of course practical advantages in French citizenship, such as access to public sector jobs and easier foreign travel. These pragmatic factors may well have carried considerable weight alongside or in some cases in place of an affective identification with the French republic. Yet there is ample evidence to show that France's postcolonial minorities have to a large extent internalized the legacy of 1789, with the younger generation in particular strongly committed to secular republican values. Numerous surveys have shown that most Muslims in France recognize and accept the dividing line between the private and public spheres. Indeed, they take a stronger view on this than is required by law, since only a small minority believe that Islamic headscarves should be allowed in state schools.[19] This trend is firmly entrenched among the second generation, most of whom are, as Jocelyne Cesari puts it, *Musulmans et républicains*[20] (Muslims and republicans).

The commitment of France's postcolonial minorities to republican values was clearly visible – though largely ignored – during the commemoration of the bicentenary of the French Revolution. By examining the bicentenary celebrations in greater detail, we can see both how the values of republicanism have impregnated immigrant minorities, and how, in the incorporation of those groups, majority ethnic distrust represents a far greater threat than minority separatism.

Commemoration and exclusion

In 1986, President Mitterrand created a Mission du bicentenaire de la Révolution française (MCRF), with responsibility for organizing a year-long programme of events in 1989 marking the 200th anniversary of the French Revolution. Towards the end of the commemorations, an opinion poll was commissioned on behalf of the MCRF in order to measure public attitudes towards the revolution and its legacy. Asked whether they thought the ideals of the revolution made it important to help East European countries to democratize following the collapse of the Soviet bloc, 66 per cent of those interviewed answered affirmatively. While 59 per cent inferred from those ideals that it was important to help the poorest countries in the Third World, only 36 per cent drew a similar inference concerning the importance of treating immigrants well; 45 per cent could see no connection between the ideals of the revolution and the need to treat immigrants in France decently.[21]

As the phrasing of the question was somewhat convoluted,[22] the responses need to be interpreted with caution. Read in their most negative light, the data could be taken to imply a general refusal to treat immigrants in accordance with the values of 1789. On a more charitable reading, the figures could be taken to mean that whereas Third World and East European countries had yet to benefit from those ideals, the treatment accorded to immigrants in France was already felt to be in tune with them. Bearing in mind that the survey was carried out at the height of national hysteria over the wearing of Islamic headscarves and that only a few weeks later an anti-immigrant Front National *député* was elected in the town of Dreux with 61 per cent of the vote, the latter interpretation rather lacks plausibility. In some way, however imprecise, the poll does seem to indicate the existence of a mental block among the majority ethnic population vis-à-vis the relationship between immigrant minorities and the founding myths of the republican tradition.

The blatant displays of racism seen in the political arena and elsewhere during the 1980s had led the MCRF President, Jean-Noël Jeanneney, to express the hope that the bicentenary celebrations could serve to improve the treatment of France's immigrant minorities. In 1988, Jeanneney had encouraged those planning bicentenary events to include the abbé Grégoire and Toussaint Louverture among the key figures highlighted in the celebrations. Toussaint, leader of the slaves' revolt in Haiti and servant of the First Republic following the abolition of slavery, offered obvious lessons in the struggle against racism, while the abbé Grégoire was highlighted by Jeanneney 'because of his struggle for the integration of Jews, Protestants and Blacks into the national community'.[23] A parallel with the situation in contemporary France was clearly signalled by Jeanneney's use of the word 'integration', which had recently become a buzz-word in public policy towards immigrants: just as Jews and Protestants had been incorporated into the nation during the revolution after centuries of intermittent persecution, so by implication the way must now be cleared for the integration of France's new Muslim minorities.

However, in selecting events to be supported by the MCRF, members of Jeanneney's team were highly suspicious of proposals emanating from groups with Islamic connections. Before his death in 1988, Jeanneney's predecessor as President of the MCRF, Edgar Faure, had invited Roger Garaudy, a former Marxist who had converted to Islam, to prepare a report on ways of involving Muslim immigrants in the bicentenary celebrations. Garaudy's report, submitted shortly before Faure's death, was shelved by the MCRF, which failed even to acknowledge its receipt.[24] Another proposal submitted by the Association Droits de l'Homme en Islam (Association for Human Rights in Islam) was rejected out of hand by the MCRF, a member of which wrote in the margin of the proposal: 'Project rejected because of suspected [Islamic] fundamentalism!!'[25]

While that particular proposal may well have been a front for Islamic activities which had little to do with the republican conception of human rights, similar suspicions appear to have been far less well grounded in relation to another project connected with the Islamic world rejected by the MCRF. This was a proposal from the Association de Culture Berbère (Berber Cultural Association – ACB), run by members of the Berber-speaking minority in Paris. Since independence in 1962, the Algerian state has pursued a policy of Arabization, systematically repressing the use of Berber and those campaigning for its acceptance. The ACB wanted to organize a concert featuring Ferhat Mehenni, a Berber singer who had helped to found Algeria's first Human Rights League in 1985, as a result of which he had been imprisoned. Jean-Jacques Lubrina, a senior official of the MCRF, rejected the proposal, commenting in a hand-written note: 'This proposal raises a lot of questions! And seems far removed from the main concerns of J.N.J. [Jean-Noël Jeanneney], as does everything to do with immigration!!'[26]

Although this remark may appear at first sight to mean that Jeanneney was not very interested in immigration, that is probably not what Lubrina had in mind. It is more likely that his remark was intended to indicate that many of the proposals submitted to the MCRF by immigrant-based groups appeared to be quite tangential in relation to the main objectives set by Jeanneney for the bicentenary celebrations. Yet in inviting Mehenni, the ACB had clearly set out the objective of demonstrating 'the contemporary relevance of the Declaration [of the Rights of Man] for men and women originating in a distant land where those "immortal principles" are more often than not trampled underfoot and who, in France, find themselves blocked by a wall of exclusion.'[27] Ironically, Lubrina's peremptory dismissal of the ACB's proposal was itself an illustration of these exclusionary processes.

The ACB eventually succeeded in gaining support for its proposal by resubmitting it via the Fonds d'Action Sociale (FAS), a government agency responsible for the integration of immigrant minorities, which in conjunction with the MCRF sponsored a range of bicentenary events involving minority ethnic groups. This time the ACB project was submitted as part of a wider proposal from a publicly funded organization called Images, Spectacles, Musiques du Monde (ISSM), responsible for fostering the cultural integration of Third World minorities.[28] The ISSM emphasized Mehenni's role in the Algerian Human Rights League, whereas the ACB had put the main emphasis on his defence of Berber culture. Although these were two sides of the same coin, they were evidently perceived differently by the MCRF. While the defence of human rights fitted readily with the MCRF's agenda, the defence of minority cultures was apparently judged to be less in tune with the myth of the one and indivisible republic inherited from the revolution. The unspoken phobia underlying Lubrina's reflexes was almost certainly an aversion to what the French call *communautarisme*, a spirit of ethnic separatism often attributed – with very little justification – to immigrant minorities.

The climax of the bicentenary celebrations was a Bastille Day parade masterminded by Jean-Paul Goude. A central feature of the parade was its cosmopolitan character, with colourfully clad participants representing virtually every corner of the globe as well as regional minorities within France. Wholly absent were representations of the nation's immigrant minorities. There was an unmistakable whiff of neocolonialism in the manner in which foreign participants were enlisted in the celebration of a historical revolution through which France was perceived as having transmitted to the world values of universal significance. The only immigrants in the parade were African migrant workers hired to play the role of *tirailleurs sénégalais*, 'native' troops who served in the French armed forces during the colonial period.[29] The implicit assumption was that while France had a glorious history of transmitting enlightened values to the colonial world, minorities originating there could contribute nothing of significance to the representation of republican values within the former colonial motherland. Yet the bicentenary celebrations had shown exactly the opposite to be true. One of the main axes of activity sponsored by the MCRF was organized around the theme of Human Rights. Three hundred of the 500 proposals brought forward under this umbrella were submitted to the FAS, which had earmarked 2 million francs to support events 'linked to the incorporation of immigrants into the community'.[30] The committee responsible for allocating these funds was instructed to award support on the basis of three main criteria: local historical links with the theme of the bicentenary, partnership with other organizations and the involvement of foreigners as participants and/or subjects of the proposed events. The FAS supported a total of 172 projects. About half of these were put forward by teachers in schools with high concentrations of minority ethnic children, who were to be encouraged to learn more about the concept of human rights inherited from the revolution. The other successful projects were submitted by municipal or voluntary organizations, often involving significant numbers of minority ethnic participants.

Projects designed or supported by activists of sub-Saharan or Caribbean origin frequently highlighted the role of Toussaint Louverture and the wider question of slavery during the revolutionary period.[31] Associations run by immigrants of West African origin organized guided tours to the small town of Champagney, in the Haute-Saône, whose inhabitants had called for the abolition of slavery in the early stages of the revolution, and to Joux, in eastern France, where Toussaint had been imprisoned after turning against Napoleon.[32] Toussaint was the central figure in a rock concert organized by SOS-Racisme under the direction of Harlem Désir, the son of a West Indian immigrant, with major funding from the MCRF, which officially designated the concert as one of the seven high points of the bicentenary celebrations.[33]

In events of this kind, minorities of African and Caribbean origin publicly commemorated their ancestral connections with the revolutionary period while the majority ethnic population basked in the warm glow of the abolition of slavery by the First Republic in 1794, glossing over the centuries of profit extracted by France from the slave trade, as well as its reinstitution by Napoleon only a few years later. Maghrebis, by far the largest of France's postcolonial minorities, have no comparable historical connections with the First Republic. While France had been present in West Africa and the Caribbean since well before the revolution, the colonization of Algeria did not begin until 1830. Moreover, the bloody manner in which Algerians were forced to fight for their independence made it virtually impossible for most of them to conceive of commemorating the historical foundations of a republican state which for over a century had denied them the human and civic rights proclaimed by the revolutionaries of 1789.

It would be wrong to infer from this that the republican heritage was devoid of any impact on Maghrebis. On the contrary, the small 'native' elites who were given access to the French educational system quickly assimilated the values of Liberty, Equality and Fraternity, turning them into battering rams against the colonial system.[34] As part of the bicentenary celebrations the Noroit Cultural Centre in Arras decided to commission a new play in memory of Robespierre, who was born in the town. Significantly, the commission was given to an Algerian author, Kateb Yacine, a veteran of the nationalist struggle for independence. 'I couldn't miss the chance to stage France's revolutionary past, since it is also my own past as an Algerian revolutionary,' Kateb declared.[35] In the final scene of the play, *Le Bourgeois sans culotte*, the actors are transformed from eighteenth-century revolutionaries into twentieth-century immigrant workers, clearly signalling the continuing relevance of the revolutionary heritage for today's postcolonial minorities.

The children of immigrants have all been educated within the French school system, through which a deep attachment to the values of the revolution has been widely absorbed. This attachment is evident in the projects submitted by young Maghrebis to the FAS bicentenary committee.[36] Many of these proposals were rejected on the grounds that they did not contain sufficiently explicit references to historical events. It was on these grounds, for example, that the largest national organization of young Maghrebis, France-Plus, was refused funding in support of a project for releasing into the air thousands of bicentenary balloons.[37] The message on the balloons – 'All human beings are born free and with equal dignity and rights'[38] – could hardly have been more republican, and was entirely in tune with France-Plus's primary aim of promoting active citizenship among second-generation Maghrebis. The leader of France-Plus, Arezki Dahmani, had summed up his organization's philosophy as follows: 'Our values are those of the French revolution. Our values are the values of *laïcité* [secularism]. Our values are the values of democracy. We are totally committed to them.'[39] In pursuit of this agenda – far removed from the *communautarisme* frequently imputed to immigrant minorities – France-Plus campaigned for the inclusion of minority ethnic candidates on the lists of major parties during the municipal elections of 1989 and opposed the wearing of Islamic headscarves in state schools.

One of the most important historical projects supported by the MCRF was the work of an association called Génériques, led by young Maghrebis committed to documenting the history of immigrant minorities in France. As part of the bicentenary celebrations, Génériques organized a major exhibition tracing the history of France's minority ethnic press, stretching almost as far back as the revolution itself. Many of the newspapers and magazines featured in the exhibition and accompanying book[40] were produced by dissidents from authoritarian regimes for whom France represented a place of asylum thanks to its revolutionary heritage. Here, as in events commemorating the abolition of slavery, there was an opportunity for the majority ethnic population to bask in a warm, self-satisfied glow while contemplating the shining example set by France to other, less enlightened countries. Yet an important strand of the Génériques exhibition documented the anti-colonial publications produced by 'native' activists originating in the French empire, where freedom of speech was non-existent. Not uncommonly, anti-colonial newspapers produced in France were closed down and prison sentences were meted out to their editors. Today's postcolonial minorities implicitly remind the French of a period in their history which is far from exemplary when judged by republican values. The work of organizations such as Génériques bears testimony both to the diffusion of those values among immigrant minorities and to the continuing sensitivity of the historical experiences conditioning postcolonial migration. They suggest that if France has difficulty in commemorating the role of immigrants in the construction of the nation, this is due less to any resistance to republicanism among minority groups than to the selective application of republican values by the majority ethnic population itself.

Conclusion

The incorporation of immigrant minorities is one of the most difficult issues clouding the immediate horizon in contemporary France. The political atmosphere has been poisoned by the electoral successes of the racist Front National, which has capitalized on widespread antipathy towards postcolonial minorities. Neither the historical background in which those minorities are rooted nor the values to which they currently adhere support the contention that they are fundamentally out of step with the republican tradition dominant among the majority ethnic population. The history of the empire shows that the French have a far from unblemished record in the field of human rights. If they are to resolve current problems relating to immigration successfully, a more lucid awareness of the colonial past will help to establish a better framework for mutual understanding across ethnic boundaries.

Notes

1   G. Noiriel, *Le Creuset français: histoire de l'immigration: histoire de l'immigration XIXᵉ–XXᵉ siècles* (Paris, 1988), pp. 13–67.
2   C. Liauzu, 'L'Immigré dans les banlieues de l'histoire de France', in *Horizons maghrébins*, nos 18–19 (1992), pp. 176–85.
3   Noiriel, *Le Creuset français*; Y. Lequin (ed.), *La Mosaïque France: Histoire de l'immigration et des étrangers en France* (Paris, 1988), P. Milza, *Les Français devant l'immigration* (Brussels, 1988).
4   G. Noiriel, 'Français et étrangers', in P. Nora (ed.), *Les Lieux de mémoire*, part iii: *Les France*, vol. 1, *Conflits et partages* (Paris, 1992), pp. 275–319. The earliest volume in this collection had been published in 1984.
5   *Ibid.*, p. 282.
6   A. Fenet and G. Soulier (eds), *Les Minorités et leurs droits depuis 1789* (Paris, 1989).
7   N. Weill, 'Un rapport s'interroge sur le rôle de la statistique sous Vichy', in *Le Monde* (8 September 1998).
8   Elisabeth Badinter, Régis Debray, Alain Finkielkraut, Elisabeth de Fontenay and Catherine Kintzler, 'Profs, ne capitulons pas!', in *Le Nouvel Observateur* (2 November 1989).
9   A. G. Hargreaves, *Immigration, 'Race' and Ethnicity in Contemporary France* (London and New York, 1995), pp. 125–31.
10 R. F. Betts, *Assimilation and Association in French Colonial Theory and Practice* (New York, 1961); R. Aldrich, *Greater France: A Short History of French Overseas Expansion* (Basingstoke, 1996).
11 N. MacMaster, *Colonial Migrants and Racism: Algerians in France, 1900–62* (Basingstoke, 1997), p. 58.
12 MacMaster, *passim*.
13 B. Stora, *Ils venaient d'Algérie: l'immigration algérienne en France, 1912–1992* (Paris, 1992), p. 322.
14 F. Rouyard, 'La bataille du 19 mars', in J.-P. Rioux (ed.), *La Guerre d'Algérie et les Français* (Paris, 1990), pp. 545–52.
15 B. Stora, *La Gangrène et l'oubli* (Paris, 1991).
16 *Recensement de la population de 1990: Nationalités, résultats du sondage au quart* (Paris, 1992), table R6.
17 *Recensement de la population*, tables 11 and 12; M. Tribalat, *De l'immigration à l'assimilation: enquête sur les populations d'origine étrangère en France*, (Paris, 1996), pp. 151–3.
18 Y. Biville, 'Les Jeunes d'origine maghrébine et le service national', in *Hommes et migrations*, no. 1138 (December 1990), pp. 7–18.

19 IFOP poll in *Le Monde* (30 November 1989); IFOP poll in *Le Monde* (13 October 1994); SOFRES poll in *Le Nouvel Observateur* (15 January 1998).

20 J. Cesari, *Musulmans et républicains: les jeunes, l'islam et la France* (Brussels, 1998).

21 CSA poll conducted in November 1989 in J.-N. Jeanneney, *Le Bicentenaire de la Révolution française: rapport au Président de la République* (Paris, 1990), p. 370.

22 'In your opinion, with reference to the ideals of the French revolution, do you think it is important or not, or without any relevance, to treat immigrants in France well?'; *ibid.*

23 'Un entretien avec M. Jean-Noël Jeanneney, Président de la Mission du Bicentenaire. Toussaint Louverture, l'abbé Grégoire et les "enjeux de la mémoire collective". Propos recueillis par Michel Kajman', in *Le Monde* (7 October 1988).

24 Archives Nationales (hereafter AN) 900502/888; unpublished letter from Roger Garaudy to A. G. Hargreaves (23 April 1993).

25 AN 900506/897; the two exclamation marks are part of the original punctuation.

26 AN 900506/891; again punctuated as in the original.

27 *Ibid.*

28 AN 900506/881.

29 S. L. Kaplan, *Adieu 89* (Paris, 1993), p. 385.

30 AN 900506/871.

31 AN 900506/873, 875, 876, 891.

32 Archives of the Fonds d'Action Sociale (FAS) 75894 89/74 00 N 01.

33 AN 900506/898.

34 B. Stora, 'L'Effet "89" dans les milieux immigrés algériens en France (1920–1960)', in *La Revue des études sur le monde musulman et méditerranéen*, nos 52–3 (1989), pp. 229–39.

35 Interview in *La Voix du Nord* (28 February 1989).

36 AN 900506/873–878, 881, 891, 899.

37 FAS 75555 89/74 00 N 02.

38 *Ibid.*

39 Statement by Dahmani in *Être Français aujourd'hui et demain: Rapport de la Commission de la nationalité présenté par M. Marceau Long* (Paris, 1988), vol. 1, p. 474.

40 *Presse et mémoire: France des étrangers, France des libertés* (Paris, 1990); cf. AN 900506/892.

16

Angels on the Point of a Needle: Counting Catholics in France and Spain

Maurice Larkin

'How many divisions has the Pope?'
Stalin

'Treat the Pope as if he had 200,000 men.'
Napoleon

If, as popularly alleged, the medieval schoolmen discussed how many angels could dance on the point of a needle, they had at least the modesty to confine the debate to how many *could* dance, rather than *did* dance, in these somewhat cramped conditions. Their twentieth-century counterparts are not so cautious, and there is consequently much lively argument and uncertainty about how best to calculate the membership and the strength of the Catholic Church in contemporary society. The discussion is particularly animated in southwestern Europe, where Catholicism has traditionally been the dominant religion of France, Italy, Spain and Portugal. In a collection of essays devoted to France, it would be inappropriate to try to illuminate the particularities of the French situation by comparing it with conditions in all three of her Latin neighbours. Indeed, some Francophiles may resent the loss of space to comparisons with even one foreign neighbour, albeit one so close as Spain with so much that is a striking mixture of resemblance and difference. A comparative approach, however, may often be a valuable means of discerning new significance in familiar material. Human beings perceive and gauge the importance of what surrounds them – size, three-dimensional shape and distance – by virtue of their bodies having two stand-points from which everything is sensed: two eyes, two ears and two nostrils, which permit them to locate and assess sights, sounds and smells with a precision that would always elude a single perspective. And in a not dissimilar way, the examination of a historical or contemporary problem is given depth and dimension by approaching it through two national experiences.[1]

The pattern of history

At first glance, it might seem that much of the politico-religious history of twentieth-century Spain is a re-enactment of what happened in France some 30 years earlier. French chauvinists might be tempted to call it a grotesque parody in which the Spanish experience, like some court dwarf, limps along behind the king, miming his gestures. But even if such a jaundiced view of Spanish history were accepted, it might be added that the court dwarf served a serious purpose, as did the tame intellectuals of John F. Kennedy's Camelot or François Mitterrand's Elysée. His parody of the king's mannerisms and attitudes warned his master – and the court in general – of where his quirks might lead him, if allowed to develop in an unrestrained way. In like fashion, the politico-religious history of Spain warns Francophiles where French history might have gone, had not indigenous circumstances and external factors interacted in a different manner.

In 1931–3, Republican Spain underwent a compressed and hence more strident version of the anticlericalism that France had witnessed in 1900–6. The Church was disestablished, Catholic schools were closed and there was sporadic street-level violence against priests and church buildings of a kind not seen in France since the Commune of 1871. Indeed this grass-roots violence became more virulent in 1936 as a result of the temporary suspension of the government's anticlerical programme, following the right-wing electoral victory of November 1933. The opening months of the Civil War then triggered an epidemic of church-burning and priest-killing which made it progressively easier for conservative propaganda to demonize the Republic as a communistic anti-Christ – much as the right-wing press had done in France when faced with the popular violence of 1871. The victory of Franco's Nationalist forces and the 36 years of dictatorship that followed created a situation that suggested some of the features of Vichy. If chronologically its inception preceded Vichy, its longevity restored the 30-year *décalage* which broadly separated the political histories of the French and Spanish Churches. Admittedly the Spanish bishops' endorsement of the Nationalist cause as a Christian 'crusade' in July 1937 had already predated by three years the papal nuncio's verdict on Vichy as 'the Pétain miracle'. Yet the close but chequered relationship between the Church and Franco that characterized the next three decades had much in common with the sweet-and-sour dealings of the Church and Vichy in France.

In similar fashion the fortunes of the Spanish Church during the post-Franco transition of 1975–82 echoed in many respects those of the French Church at the Liberation and during the early Fourth Republic. The mutual support that Church and government had given each other during the greater part of the lives of the previous authoritarian regimes led many foreign observers to assume that the return of democracy would see massive anticlerical retribution, as had happened several times in France in the nineteenth century and even more sharply in Spain in the 1930s. But in both countries the Church was the beneficiary of its cooling relations with dictatorship in the last quarter of the Vichy and Franco regimes. And of equal importance significant elements among the junior clergy and laity had been increasingly supportive of the regimes' critics and opponents. The Church was also an indirect beneficiary of the inward consciousness of a large section of the population in both countries that they themselves had passively worked within the regime and had done little or nothing to weaken it. This disinclined many from throwing the first stone at the Church – even if both Vatican and the nation's episcopacy, as self-accredited moral leaders, bore the greater responsibility.

Apart from their differing life-span, one of the major distinctions between the Franco and the Vichy regimes was the German factor in France; not only did the German victory in 1940 lead directly to the Vichy regime, but the German Occupation blurred the issue of loyalty in a variety of ways. On the one hand, the dual nature of authority in Occupied France encouraged a spectrum of part-genuine, part-specious claims that obedience to Vichy was a method of containing the risk of yet further German control, while on the other it left all forms of obedience to the regime open to the charge of indirectly helping the occupying forces. German aid was admittedly a major factor in ensuring Franco's victory in the 1930s, but Franco was careful not to display his gratitude in the Second World War. Support or opposition to his rule was therefore a stark choice between indigenous alternatives, in which the grimly inhibiting factor was not a German occupier but the overwhelming military strength of the *de facto* government. Moreover the onset of the Cold War disposed the Western powers to close their eyes to the repressive nature of the regime, much as fear of Russia in the 1930s had disposed their predecessors to feel more indulgent to Franco rather than to the Republic, despite their professions of neutrality and belief in democracy; this was admittedly at a time when the full nature of the Franco regime had yet to be revealed. Indigenous Spanish opposition to Franco was therefore unlikely to get much practical encouragement or sympathy from neighbouring governments; and therefore passive acceptance seemed to many Spanish democrats the only realistic attitude to adopt in the foreseeable future.

If, for the first three decades of Francoism, the attitude of Church leaders in Spain had gone much further than passive acceptance, the last ten years found them increasingly critical – even if this criticism was principally directed against state interference in ecclesiastical matters. The unaccustomed firmness of this criticism owed much to the impulse given by the Vatican; and in this respect the situation was markedly different from that in France in 1943–4. In France, overt episcopal criticism of the government had largely centred on Vichy's compliance with Nazi treatment of the Jews and on the conscription of French youth for compulsory labour service in Germany. It had also been mainly voiced by a few courageous bishops speaking in their own right rather than on the collective behalf of their colleagues. If the dangers of outspokenness were far greater in Occupied France than they were in Francoist Spain in the 1960s and 1970s, this was also an inhibiting factor when it came to plain-speaking by the Vatican. Despite the undercover aid that Rome was giving to the Jewish refugees, the notoriously cautious tone of official papal pronouncements gave little indication of the depth of the Pope's abhorrence of Nazi policies, as expressed in private conversations. With little public support from the pope, it was not surprising that few bishops in France were prepared to show their heads above the parapet.

The role of Rome in the 1960s, regarding Spain, was very different. The Vatican was defending what it saw as Church rights against a head of state who still broadly regarded the Church as an ally and who was unlikely to resort to punitive measures. Moreover the Second Vatican Council (1962–5) had theoretically committed the Church to a set of liberal principles which sat uneasily with the current nature of Church–state relations in Spain. Indeed the Spanish bishops returned from the Vatican Council in some bewilderment, feeling that the traditions which they had so tenaciously defended since the 1930s were now abandoned not only by their episcopal colleagues in other countries, but also by their supreme leader in Rome. Belatedly, they started to talk the language that progressive French clergy had been using since the 1930s – without being all that sure that they had got it right. Traditionalists in the Spanish Church accused them of losing confidence in themselves and thereby losing the confidence of the junior clergy and laity whom they were supposed to lead. Moreover, it was not long before conservatives were blaming them for the fall in Church attendance that hit Spain in the 1960s and 1970s, just as it affected other Western countries.

The time-lag between French and Spanish religious history was nowhere more clearly evident than in the contrasting attitudes of the progressive elements in the French and Spanish clergy towards the Vatican Council and its outcome. Oppressed by the heavy weight of Francoist and episcopal conservatism, the Spanish liberal clergy saw the Vatican Council as a momentous and unexpected liberation from past and present burdens. However much the outcome might fall short of the implications of the agenda, the fact that the council had been called on the Vatican's initiative seemed itself to be an invigorating indication that the defensive caution of Pius XII had died with him, and that, to use the new pope's phraseology, a window on the world was about to be opened. To progressive clergy in France and northern Europe, the actual changes the council brought about in Church structures and teaching were disappointingly meagre. But in Spanish eyes a strikingly new terminology was now in use which seemed to imply a transformation of attitudes in the long term, if not necessarily in the short. Even the modest short-term changes triggered by the council were seen as a vital signal to both episcopate and government that their prevailing attitudes were outdated and had to be changed, if Spain was not to be marginalized in the world that was coming into being. As for the Spanish bishops themselves, their traditionalism was centred on a strong ethic of hierarchy and obedience, which at least seemed to augur a loyal effort on their part to conform with what appeared to be the new instructions from Rome and their international colleagues, whatever their own uncertainty and misgivings. Franco and his entourage regarded the council's *newspeak* as a potential threat to the values of order and discipline that had made the Church such a respected ally. But they could not afford to alienate the Vatican by saying so too openly. Moreover the Spanish government harboured growing aspirations to be accepted as part of the new Europe that was arising from the prosperity of the 1960s and from the economic initiatives of the various bodies for European co-operation; reticence on the council's vagaries therefore seemed doubly advisable. It was admittedly the potential economic advantages of alignment with the new

Europe that inspired this prudence, rather than any desire to share its political ideals; but the new generation of Franco's advisers were well aware that Spain's recent record of repression was an embarrassing encumbrance in the pursuit of their hopes.

In France, on the other hand, the ideals of the Vatican Council seemed to be a belated recognition of what progressive French Catholics had been urging for decades; and indeed there were some who were inclined to attribute the welcome openness of the new pope, John XXIII, to his salutary immersion in French Catholic life during his brief period as papal nuncio during the heady euphoria of the Liberation era, when a shamefaced episcopacy and a distrusted pope were keeping a low profile and doing little to restrain the bold speculations of the Catholic *avant-garde*. Admittedly the onset of the Cold War and the strength of Christian Democrats in western European governments had rapidly restored the Vatican's nerve and traditional readiness to discipline adventurous theologians. But with the death of Pius XII and the advent of a new pope who had witnessed without necessarily sharing all the enthusiastic hopes of the Liberation era, there seemed a possibility of better things; and the calling of the Vatican Council appeared to confirm this. The language of the Council, and the encyclicals and documents that followed in its wake, seemed to embody a new spirit of collegiality and free choice – albeit subject to the pope's ultimate authority, and with periodic reminders of the spiritual consequences of making the *wrong* free choice. But what the death of one pope might give, the death of his successor could equally take away.

How different developments might have been, had John XXIII lived longer, is anyone's guess; and the subsequent record of John Paul II (1978–) indicates that the election of Paul VI (1963–78) as John's immediate successor was not necessarily the prime reason for the limited impact of the council on the Church's subsequent development. Certainly in France the cautious conservatism of Paul VI was seen as a democratically disguised recrudescence of the traditional attitudes of his former master and mentor, Pius XII, despite his extensive visits abroad and other displays of attentiveness to grass-roots opinion. And in most of the developed western countries, the encyclical, *Humanae vitae* (25 July 1968), reaffirming the Vatican's traditional condemnation of modern means of contraception, was regarded as a bitterly dispiriting rejection of the recommendations of the committee which the Vatican had appointed to consider the matter.

If Paul VI was widely disparaged in France as having betrayed the promise of the Second Vatican Council, his image in Spain was very different. Although in retrospect Spaniards tend to divide his pontificate into two parts – one progressive, and the other more conservative – he is nevertheless regarded by many liberal as well as traditional Catholics as a positive factor in the development of the Spanish Church. This largely stems from the firm line that he took in his dealings with the Spanish government – even if cynics claimed that what was at issue was merely whether the Church was to be ruled by the autocrat in Madrid or the autocrat in Rome. And however small the post-conciliar reforms appeared to the French, they were heartening to a Spanish Catholic population starved of progress in both religious and political life. In this way the chronological *décalage* between Spain and France in the development of religious thought and attitudes ensured that what seemed like shattered hopes to the French appeared to the Spanish as bright promise.

The pattern of religious practice

This *décalage*, however, started to narrow very rapidly in the years that followed, particularly in respect of levels of religious practice in the two countries. It is arguable that regular mass attendance in both nations had been uncharacteristically high in the 1940s. The issue is hard to resolve in that it is conversely also arguable that church attendance had been artificially low in the 1930s. In the case of France, men with ambitions in the more politically sensitive branches of public employment had often felt it to be inadvisable to be widely known as a regular churchgoer; and although these fears had only been justified in the more militant periods of government anticlericalism, notably before the First World War, the uneasiness remained. It was probably no coincidence that the *détente* between Church and government that characterized the interwar years saw a modest rise in male church attendance in what had traditionally been regions of strong anticlericalism – despite the socio-economic factors that continued to erode religious practice in the developed western economies. Even so, this rise represented only a small recovery of the numbers lost during half a century of periodic distrust between the Church and the various layers of French government.

In Spain, levels of church attendance were even more contingent on the locality and who was in power at the time. As in France, maps of religious practice in the twentieth century showed a significant affinity with maps of what are thought to have been the socio-political circumstances of earlier centuries – even after allowing for the fragmentary and tentative nature of the evidence from the past. In broad terms, Catholic religious practice, however measured, was greater in much of the north than it was in most of the south. There were major exceptions to this pattern, Catalonia being the most obvious. Some historians claim, with varying degrees of seriousness, that the south was never fully rechristianized after the reconquest from Islam. Poverty, ignorance and poor communications made evangelization difficult, and the south remained short of priests throughout the post-Islamic period. Priestly vocations among the indigenous population were fewer than in the north, and the tight diocesan structures of the Church inhibited the easy movement of priests from elsewhere to fill the gaps, even supposing enough priests had been ready to leave the familiarity of their own native dioceses for the arid sparseness of the south. Ironically, there had been greater readiness and effort invested in the evangelization of the overseas empire, where the prospect of millions of souls to be rescued and the self-sacrifice involved provided more of an exciting challenge to the idealism of young men with a firm belief in the rewards and punishments of the hereafter.

There is more general agreement that the differing structures and traditions of land tenure in north and south affected attitudes to the Church. Primogeniture and analogous systems of heredity, where land passed undivided to the next generation, tended to sustain a system of middling-sized farms, where large estates were the exception rather than the rule. This was the case in many of the northern dioceses until the much later changes, whereas in large parts of the south – notably Andalusia and the southwest – the division of property among heirs resulted in the creation of progressively unviable holdings, made all the more difficult by the inhospitable nature of the terrain. Peasants had increasingly been obliged to sell their tiny farms and become wage labourers on the *latifundia*, many of which owed their size and power to the government's misguided sale of common land in the mid-nineteenth century. An oppressed workforce felt little sympathy for a church which was widely regarded as an ally of the large landowners, and which until the early nineteenth century had been itself a large landlord. However deep the emotive religiosity of the south, with its devotion to local saints and extravagant displays of communal worship, it was often accompanied by a bitter distrust of the Church as a secular institution. This partly accounts for the widespread anticlericalism of much of the south and the lower levels of regular church practice – in paradoxical contrast with the massive participation of the same regions in the annual celebrations of patron saints' days and the like. In the southeast and central Spain, the problem primarily took the form of estate-owners renting their land to small farmers on very short leases which left tenants vulnerable and with little incentive to improve.

In the north, it might have been expected that the younger siblings, who did not receive a share of their father's farm, would form a rural proletariat, analogous to if smaller than that of the south. That this did not happen owed something to the northern custom of younger sons retaining close links with the family farm. In many provinces they had a customary right to board and lodging and very commonly continued to work there, as in their fathers' lifetime. This feature was itself a factor that helped to preserve family traditions and loyalties, including religious allegiances. At the same time the parallel custom of long leases for rented farms likewise created stability and family cohesion. In some northern provinces, notably the Basque provinces, it had also been the custom until the early nineteenth century for parishioners to elect their parish priests, before formal appointment by the local bishop. This tradition, notably absent on the French side of the Pyrenees and equally absent in southern Spain, may well have been a factor creating stronger links of sympathy between priest and laity, and contributing to higher levels of religious observance. If these laws and customs had long since disappeared, their influence on religious behaviour and attitudes still lingered, if much diluted by subsequent developments.

France under the *ancien régime* had also had its differing patterns of inheritance – partible inheritance in parts of the north and undivided inheritance in most regions of the south. But the nationalizing of inheritance law, under the Revolution and Napoleon, came earlier than in Spain, as did the sale of Church lands. And if the Church continued to be popularly distrusted as an ally of the large landowners in the nineteenth century, this was a significant contributor to low religious observance mainly in those regions where seigneurial rule had been harsh and where large landowners continued to be widely seen as exploitative employers and rent-collectors. The French map of religious practice tended to correspond much more closely to what have become the familiar textbook explanations of secularization. Observance was highest in the remote pastoral areas of France where there was less sustained contact with the main arteries of economic change and development; and if the industrial regions that bordered on Belgium and the German Rhineland seemed to belie the generalization by also having large congregations, demographers pointed to the deep-rooted religious traditions that these areas shared with the populations on the other side of the frontier. In Spain, however, as already seen, remoteness and high church attendance were not necessarily companions, even if highly developed regions like Catalonia seemed to conform to the classical theories of secularization.

The Second Republic of the 1930s bipolarized and intensified existing tensions in Spanish society, including those between the Church and anticlericals. As in France 30 years earlier, liberals widely identified the Church with obscurantism in education and reaction in politics, while socialists saw it as a buttress of the inequalities and injustices of the existing social order. In areas where republicanism had become strong, lukewarm Catholics who had previously gone to mass out of habit or family loyalty felt more inclined to stay away, especially in localities that had seen anticlerical violence. On the other hand, the anticlerical phases of the Second Republic – 1931–3 and 1936 – were too short-lived to see much appreciable impact on the composition of the civil service, of the kind experienced in France before the First World War. The Civil War, however, in a matter of days intensified politico-religious hatreds to a level unknown in France since the Revolution. With church services officially banned in Republican areas during the first half of the war, religious practice became a private, clandestine activity, in which only the highly committed were likely to risk themselves. In the strongly Catholic Basque areas of Spain, which preferred the decentralist-minded Republicans to the highly centralist Nationalists, the prohibition was widely circumvented in practice; and the Madrid government, needing all the friends it could get, was not inclined to enforce it. For the same reasons the prohibition was progressively softened elsewhere, as the hope of a Republican victory diminished. Even so, the image of the Republic as the perpetrator of anti-Catholic persecution was not easily dispelled, and Franco throughout the dictatorship was still regarded by many as the saviour of Catholicism in Spain, whatever his increasing disagreements with the Church on administrative issues. In Francoist Spain – far more than in Vichyite France – being known as a practising Catholic became an asset to entry and advancement in public employment. In both countries this may have helped artificially to raise the level of church attendance among men. What is certainly true is that the Civil War in Spain and the Second World War in France created widespread feelings of fear and uncertainty which brought back many indifferent Catholics to the regular churchgoing that they had largely abandoned in adolescence. And when the wars were over, a mixture of euphoric gratitude and apprehension as to what the future might still hold helped to keep the churches full in the immediate postwar period. When gratitude evaporated, the Cold War helped to keep apprehension alive;

and the Church in both countries was seen as a valued element in the Western bloc.

The crisis of numbers

By the 1960s, however, Western Europe was experiencing the coming-of-age of the 'baby-boomers' of the immediate postwar years, a generation that had no personal memories of the fears and privations of war and whose acquisition of personal independence coincided with a period of unprecedented economic prosperity when everything seemed possible. Neither they nor their elders imagined that the early 1970s would rapidly demonstrate the limitations of what lay within their reach. Although far behind France in terms of the size, potential and distribution of wealth, Spain too shared in this prosperity, as government policy became more outward-looking in the last decade of Franco's rule. It was in France, however, that the prospect of this impending heaven-on-earth had the more immediate effect on religious practice. Not only was there more money in the back-pockets of the new swinging generation, but the youth culture on which to spend it was far more developed than in Spain. At the same time the contraceptive pill, sexual permissiveness and the greater equality that the pill established between men and women affected France much more rapidly than Spain, where a repressive regime, strongly imbued with traditional Catholic values, sought, however ineffectually, to monitor the tide of magazines, films and clubs that proclaimed the new age of liberation in neighbouring countries. But a government seeking closer relations with the advanced economies – and which relied so heavily on the tourist trade – became progressively aware of the difficulties of enforcing the morality laws, despite their continued existence. As in France, condoms had long been available to those who knew where to buy them. But from 1964 the pill could be legally prescribed by doctors for the treatment of hormonal disorders; and in practice it was being increasingly supplied for contraceptive purposes, with half a million women allegedly using it by the time of Franco's death.

Sexual liberation – in an age yet ignorant of AIDS – and a seemingly limitless horizon of material prosperity made the here-and-now appear to youthful eyes as a much more likely source of happiness than the problematical uncertainties of the Christian afterlife. Their parents too, for all their sobering experience of wartime terror and hardship, cast wistful looks at what their children enjoyed without as yet any apparent ill-effects. Unbuttoned theologians of the *avant-garde* suggested that premarital cohabitation was a commendable trial run which saved many couples from making unhappy marriages, while the general expectation that the Church would modify its hard line on contraception made many hope that they could have all this and heaven too. *Humanae vitae*'s reaffirmation of the hard line, and the growing feeling that the Second Vatican Council had resulted in little change except in rhetoric, dashed the remaining hopes of those restless Catholics who had lingered in the club in the hope that relaxed rules and a new management would make for a more comfortable existence, in which corporate effort could be directed single-mindedly to the misfortunes of humanity rather than to mortifying the flesh. It is true that many committed Catholics both in France and Spain had long since practised contraceptive routines that were traditionally condemned by the Church – particularly *coitus reservatus* and all the other Latinisms of the marriage bed – while others had started to use modern contraception in the mid-1960s on the assumption that a softening of the prohibition was imminent. Although *Humanae vitae* precipitated the departure of only a minority of practising Catholics, many who stayed behind chose to ignore its prescriptions, on the grounds that the encyclical ran counter to the prevailing opinion of the pope's advisory committee on birth control. As an eminent Spanish Jesuit privately predicted at the time, *Humanae vitae* greatly accelerated the growing tendency for committed Catholics to be selective in their respect for papal pronouncements; and the Church thereby not only lost members but also lost much of its authority over those who remained.

Lest too much importance be given to the issue of papal authority in sexual matters, it has to be remembered that religious practice was also dropping in other denominations in Western Europe during the 1960s. Indeed, as demonstrated in so many secular sectors of life, traditional authority in all its forms was being scrutinized and challenged by the young in the 1960s – and the Christian Churches were but part of the target.

In France weekly attendance at mass fell from well over 20 per cent of the adult population in the early 1960s to less than 15 per cent by the end of the decade. By 1986 it had dropped to 11 per cent, falling to about 8 per cent in 1994.[2] Statistically, this overall drop was principally caused by the substantial departure of the younger generations of adults who were most affected by the ethos of the 1960s and what it engendered. And, significantly, the slowing down of the departure rate among subsequent generations of young adults was an important factor in reducing the overall speed of decline in the 1980s and 1990s, even if the direction was ever downwards.

In Spain the less developed nature of the urban economy and culture, buttressed by the restrictions of the Franco regime, delayed the full force of these features; but even during the last years of the dictator, weekly mass attendance among adults was falling from what one sample survey put at 64 per cent in 1970 to 56 per cent in the year following his death.[3] The end of the Franco era and the dismantling of the tight alliance between state and Church unleashed a plethora of pent-up defections which almost certainly would have occurred earlier, had they not been artificially delayed by the ethos and favours of the regime. This is strikingly reflected in the surveys of 1976 and 1983, which indicate a drop from 56 to 31 per cent.[4] Contrary to the predictions of conservatives, the advent of 14 years of Socialist government in 1982 did not accelerate the fall; indeed this period, as in France, saw a slackening of the fall, with levels hovering in the upper 20s between 1989 and 1995 – the surveys for both these years registering 27 per cent.[5] Survey figures in the intermediate years fluctuated between 26 and 30 per cent, but these differences were perhaps more a result of the sampling methods employed rather than the reflection of a genuine oscillation in religious practice.

Quantifying the crisis

This raises the issue of how reliable these various statistics are, and of how safe it is to make comparisons between them. Not only are there differences of methodology between France and Spain, but there are more serious differences between the methods used by the Church in both countries during the postwar decades and the secular sample surveys which provide most of the subsequent data. The quantitative problems raised by these statistics are further complicated by the different interests of those who compile them and by the yet more chequered concerns of those who chose to make use of them.

The figures of church attendance in the 1960s and thereabouts were mainly collected by the parish priests at the behest of the French and Spanish episcopal authorities, who were then particularly anxious to assess the proportion of baptized Catholics who could still be regarded as practising members of the Church. The Catholic Church believed that regular attendance at weekly mass was the most reliable measurable criterion of practising membership. The mass was not only regarded by Catholics as the re-enactment of Christ's redemption of mankind, but the accompanying sermon kept parishioners in regular contact with the moral exhortation and teaching of the Church; and on both counts

Sunday Mass was regarded as an essential channel of personal salvation. Indeed deliberate non-attendance was considered a matter of grave sin. Although good works and a virtuous life were equally necessary to salvation, these could not be quantified and the Church did not attempt to do so. Church historians, for their part, were grateful for this tangible criterion of membership, since it enabled them to make comparisons between the present and the fragmentary records of the past, when individual bishops had sporadically encouraged their clergy to keep statistical records of church attendance and Easter communion. Political historians and political scientists were similarly well disposed towards this method of measurement, since it was a guide to the strength of the Church as an influential force in society, even if assiduous church attendance was no sure guarantee of how an individual would vote, even on issues such as private education and abortion. It was also a useful indicator of regional differences in attitudes and behaviour – and, being based on parishes, was of relevance to even the lowest level of local politics.[6]

Counting people at church, however, was a laborious exercise, more arduous than quantifying Easter communicants, where it was a simple matter of counting the number of wafers left in the ciborium at the end of mass and subtracting the figure from the initial total. Even simpler for the Church was to leave the matter to the secular sample survey institutes and to use their figures as a rough-and-ready guide to current trends. This was what increasingly happened in the 1970s and after. The last significant episcopal count of church attendance in France took place in 1970 and covered only a few dioceses, while that in Spain occurred in 1982–3 and covered about two-thirds of the total.[7] The subsequent surveys were no substitute for head-counts. The responsible institutes had only a tiny fraction of the workforce available to the Church and largely depended on oral questions put to periodic samples of 1,200–2,000 people on a nationwide basis. Not only were the results contingent on the veracity of the replies, but the small numbers polled could give little indication of regional differences. These shortcomings were of less importance to the principal secular users of these surveys, who were mainly interested in the direction of national trends rather than in local detail. But historians, political scientists and the Church itself were poorer for the disappearance of these geographical distinctions.

An effort was made to compensate for these deficiencies by three organizations: the Observatoire interrégional du politique in France; the Fundación Santa Maria, working jointly with OYCOS, S.R., in Spain; and the international Estudio Mundial de Valores – all of whom widened the quantitative base of their surveys to include a representative number of respondents from each French Region and each Spanish Autonomous Community. They also offered to make available their databases to qualified researchers who wished to study the information on the more detailed level of the French department and the Spanish province. But with an overall base of about 14,000 (or 700 per region), the French OIP figures for some of the smaller departments were as little as 50 each, which scarcely gives a viable sample for study. The Santa Maria-OYCOS base of 4,300 poses even greater problems, with little chance of meaningful comparisons at provincial level. This is partially compensated by the Estudio Mundial de Valores which, with a base of 8,000 in Spain, can offer upward of 100 replies for the smaller provinces and substantially more for the larger. But this is still tiny compared with the 1,879,000 *fiches* gathered during the Spanish Church's last great survey in 1983, organized by the Comisión Nacional de Liturgia. Even so, interesting if very imperfect comparisons can still be made between past and present regional patterns in France and Spain.

As one would expect, the broad differences between regions remain, but the degree of difference has narrowed, especially between rural areas that in the past had been characterized by sharply contrasting levels of religious observance. Regions where there had traditionally been strong peasant hostility against the landlords and the Church still have very low levels of churchgoing, and these continue to decline. But if, as in the Extremadura, there is little urban growth and considerable emigration of young people, such regions may now be left higher in the national league table of religious observance than they were before, despite this numerical decline. Conversely rural areas of traditionally high observance have in many cases had these levels heavily eroded by growing urbanization and the other familiar factors of secularization. Even so, a Spanish sample survey of 1989 in the six largest Autonomous Communities showed levels of regular Sunday mass attendance that broadly replicated, albeit in reduced numbers, the pattern that diocesan surveys had shown in the 1960s and 1970s, with the pattern in Castilla-León (45.8 per cent) nearly four times higher than in Catalonia (12.5 per cent).[8] In the same year a similar *sondage* in the French regions likewise showed a pattern with overall affinities with the diocesan surveys of the 1960s – but with a much steeper drop than in Spain and with a greater levelling between the regions. It nevertheless still left churchgoing in the Pays de la Loire (the old Vendée and its surroundings) three and a half times higher (13.3 per cent) than in the old stronghold of truculent peasant anticlericalism, the Limousin (3.8 per cent).[9]

In parallel with the softening of regional differences, these sample surveys also show that the traditional differentials between men and women, though still significant, became less marked during the course of the century. This first became noticeable in France after the First World War, when the *détente* between Church and government made churchgoing among men less of a social embarrassment in left-wing localities and less of a liability in certain branches of public service. But more importantly it probably reflected the increasing secularization of women's lives, as more acquired higher educational qualifications and moved into the expanding professions and senior administration. In Spain both factors were present, but the chronology differed from that in France. It was essentially the Franco era that made churchgoing a positive asset for men in public employment, while the secularization of women's attitudes after entering former male preserves in the workplace became particularly significant during the democratic transition, when outward lip-service to *bien-pensant* attitudes ceased to be thought of as an advisable precaution. In both countries the sexual and social liberation that the contraceptive pill brought to women was also a strong element in detaching them from traditional Church teaching; but once again the chronology was different, being several years slower in coming to Spain. Yet even in France a sample survey of 1994 showed regular church attendance among women as half as high again as that among men, and belief in God a quarter higher.[10] In Spain in the same year, there were twice as many women as men claiming to be regular churchgoers, but the difference was much less marked among younger adults (13 per cent of women in their twenties, as against 11 per cent of men).[11]

While Church historians and political scientists remain nostalgic for the old diocesan surveys that indicated, however crudely, the potential influence of the Church in each locality, sociologists are less perturbed by the deficiencies of the nation-based sample surveys. They are more concerned with religious factors as a source of personal motivation and social behaviour, irrespective of the institutional allegiance, if any, of the individual. A major positive outcome of this different perspective is that the sample survey questionnaires range far beyond the simple issue of church attendance and are themselves just part of much broader surveys of social attitudes and political preferences – thereby enabling tentative correlations to be made between these various elements. Ironically the Church itself has tended in some measure to fall into line with the sociologists' way of doing things, if not necessarily with their preoccupations. Having abandoned the costly and time-consuming practice of counting heads at mass, French and Spanish bishops make a virtue of necessity and say that the Church should endeavour to measure the quality rather than the quantity of Church allegiance. But, as the religious demographers of 40 years ago fully recognized, quality is impossible to quantify; and just as those pioneers had to fall back on the inadequate but consistently measurable criterion of mass attendance, so their successors now compile figures of membership of the various Catholic charitable, devotional and fund-raising organizations – despite the ambiguities and uncertainties that surround the significance of such membership. Ironically, church-going Catholics who are also participating members of Catholic associations and activities are remarkably few in France and Spain – only 5 per cent of the adult population in each country in 1990, as compared with 23 per cent in Holland.[12] This probably reflects the state-centred, socio-cultural traditions of many Latin countries, as against the grass-roots co-operative initiatives that are more characteristic of northern Europe.

Clerics endeavoured to take some heart in the fact that a 1994 survey of 1,000 French adults showed a slightly higher level of weekly mass attendance in the 18–24 age-group (roughly 3 per cent) than in either the 25–34 group or that of 35–49 (both just over 2 per cent). Thereafter the level steadily rose to some 28 per cent among the over-64s.[13] Spanish figures for 1993 showed the 18–25 age-group as having a lower rate of church attendance than their elders, although the different criteria of the survey preclude direct comparison with the French situation.[14] However, both French and Spanish surveys suggest a greater propensity among this age-group to believe in the likelihood that there is a God and a life of sorts after death; these young adults are also more open to the appeal of what are often termed 'parallel religions' and *parasciences* – astrology, telepathy and the like. But, like alternative medicine, enthusiasm for them often wanes with age and their impact on ethical choices tends to be limited. It is often suggested that this greater readiness of the younger generation to entertain the possibility of non-material forces shaping life may reflect bewilderment and disillusion with the failure of governments and economic experts to find convincing solutions to unemployment and the other major problems that currently confront society; and in tandem with this, it may also reflect a disillusion with the failure of governments and scientists to protect the world from the destructive side-effects of scientific progress, such as chemical and nuclear pollution. Symptomatically, the 'parallel religions' contain many close sympathizers with the Green Movement.

What these figures have to say about the future is endlessly debated in the Catholic and secular press. There is broad agreement among sociologists that the predictable future is likely to see the continuing 'disinstitutionalizing' of religious sentiments and convictions. *Bricolage*, or 'do-it-yourself', has long been the fashionable way to describe the future; *à la carte* is another. Yet both these phrases suggest a degree of considered choice that perhaps belies the reality of the broad and often ill-defined principles and instincts that commonly influence attitudes and ethical behaviour. For many people these principles owe much to the faith or values of their upbringing. But what adults most willingly retain from this inheritance are generally those elements that accord easily with the life they currently lead – or aspire to lead – and which fit comfortably with the prevailing attitudes of the society they live in. If what they have discarded from their past leaves gaps in their system of values, they tend to fill these – when the need arises – with whatever is currently well-viewed in the milieux in which they live and work. Only a very small minority have the knowledge or inclination to scan the menus of the other great religions and systems of belief to find suitable fillers.

Catholics who contest the vision of a disintegrating Church point to the findings of a European sample survey of 1990 which show 85 per cent of Spanish respondents choosing to call themselves 'Catholic' rather than 'without religion', and 58 per cent of French respondents choosing likewise.[15] Others argued that in many of these cases 'Catholic' merely signified a broad socio-cultural background or upbringing, rather than any religious or institutional commitment; and critics made much of the fact that in the same survey only 48 per cent of Spanish and 20 per cent of French replies declared a belief in a personal God. On the matter of belief in life-after-death, the Spanish and French replies were much closer – 42 per cent in Spain and 38 per cent in France.

Viewed from the perspective of the political scientist, a continuing 'disinstitutionalization' of religious practice would undoubtedly weaken the Catholic Church still further as a force in French and Spanish politics. Its traditional defence of Catholic schools would increasingly depend on the support of all those other sections of society that are suspicious of a unified state system of education: either because it denies them choice, or because their children may be graded and streamed in a way that they feel underestimates their potential ability. This dependence was already evident in the huge demonstration of 1984 against the Savary Bill on French private schools in receipt of state funding. Similarly the Church's objections to a further widening of the existing legislation on abortion, divorce and other related issues would be entirely dependent on a currently shrinking body of voters who as individuals happen to share the Church's misgivings on these matters. On the other hand, to claim, as many do, that the erosion of Church membership will *ipso facto* erode an important buttress of the moderate right-wing parties in France and Spain is to beg the question of whether the many practising Catholics who support these parties do so primarily because they are Catholics. It could also be argued that their voting preferences spring principally from the fact that they come from socio-economic milieux which favour both conservatism and Catholicism – two distinct entities, whatever their alleged affinities. Former Catholics with right-of-centre inclinations are unlikely to modify their political stance as a result of leaving the Church. And, however much history has made these perceived affinities between Catholicism and conservatism a reality in so many countries, especially in Spain, conservative leanings and support for the Church are better understood as twin products of a given social situation, rather than as a parent–offspring relationship, with Catholicism as the parent of conservatism, as is so often suggested in analyses of voting behaviour. Admittedly, the Church continues to insist that moral principles have immutable foundations in God's purpose for humanity; and this unquestionably makes it hostile towards any legislation seen as threatening to these principles. Not only is this an inbuilt factor favouring conservatism on such issues, but it often results in ecclesiastical support for conservative parties committed to these principles – even if on other matters, such as social justice and the relief of poverty, churchmen may be highly critical of their programmes.

If continuing loss of numbers is the immediate prospect for the Church in France and Spain, Rome continues to draw comfort from its millions of enthusiastic supporters in other continents, not least in postcolonial French Africa and Spanish America. On the other hand, matters may look less comforting for the Church, if westernization increasingly imbues these enthusiasts with the current doubts and uncertainties of their former European masters, the conquerors who made them Catholic in the first place. But beleaguered optimists in the Vatican like to point out that Western history has been characterized by alternating stages of doubt and affirmation. The self-styled certainties of mid-twentieth-century economics, politics and culture have given way to the diffidence and relativism of postmodernity; and in all probability the pendulum will eventually swing back once more to affirmation, even if this takes the familiar form of a clamouring marketplace of competing affirmations. The clamour, say the hard-pressed optimists in Rome, need not be a cause for worry, since affirmation and assertion, by their very nature, crave discipleship and mutual support. That in turn will require organization and some semblance of disciplined structures – at which point the age-old churches may find themselves back in business. Sceptics reply that if this hypothesis were to be anything more than mere whistling in the dark, it would require enormous changes in current concepts of ecclesiastical authority, as well as much greater flexibility in the understanding of Christian 'doctrine', itself a highly unpopular word. In the words of an eminent Spanish priest, 'We can at least unite behind what Christ told us about living *this* life. It's God that's the problem.'

Notes

1 Although Douglas Johnson is best known for his outstanding work on French history, his publications and his teaching have always been informed by a sharp comparative sense of the resemblances and differences between French and British politics. Colleagues and students who have relished and profited from his company know how important the comparative dimension of history is to him. As well as paying tribute to a generous mentor and friend, I should also like to thank the British Academy for its award of a research travel grant, which although used for a much wider project, indirectly resulted in these preliminary thoughts on Franco-Spanish religious history.

2 SOFRES sample survey, 1986; Institut CSA survey, 1994. For a general analysis, see *L'Actualité religieuse dans le monde*, no. 122 (15 May 1994).

3 FOESSA sample survey, 1970; DATA survey, 1976. For a discussion of the implications, see A. de Miguel, *La Sociedad Española, 1995–96* (Madrid, 1996), pp. 233–345.

4 DATA survey, 1983.

5 DATA survey, 1989; TABULA-V survey, 1995.

6 Some subsequent surveys by secular institutions have confused matters by including in their regular mass-going totals the increasing number of Catholics, notably in France, who go just once or twice a month. Since Church teaching remains unchanged in its insistence on weekly attendance, these casual attenders are not included in the figures of 'hard-core' Catholics used in this essay.

7 For details of the 1970 and earlier episcopal surveys in France, see F. Boulard, Y.-M. Hilaire and G. Cholvy, *Matériaux pour l'histoire religieuse du peuple français*, 3 vols (Paris, 1982–92) and F.-A. Isambert and J.-P. Terrenoire, *Atlas de la pratique religieuse des catholiques en France* (Paris, 1980). For the Spanish episcopal survey of 1982–3, see F. Azcona San Martin, *Asistencia a la Misa Dominica: encuesta* (Madrid, 1985).

8 Fundación Santa-Maria–OYCOS, S.R. survey, 1989, fully analysed in P. González Blasco and J. González-Anleo, *Religión y Sociedad en la España de los 90* (Madrid, 1992).

9 Unpublished data kindly supplied by the Observatoire interrégional du politique.

10 Institut CSA survey, 1994. See note 2 above.
11 De Miguel, *La Sociedad Española*, pp. 242–3.
12 J. González-Anleo, 'Análisis del hecho religioso español: hacia un pluralismo centrífugo', *Sociedad y Utopia*, no. 8 (October 1996), p. 167.
13 Institut CSA survey, 1994. See note 2 above.
14 González-Anleo, *Religión y Sociedad*, p. 173.
15 'Enquête sur les valeurs, 1990', analysed in Y. Lambert, 'Religion et modernité: une définition plurielle pour une réalité en mutation', *Religion et société* (*Cahiers français*, no. 273), p. 7.

17
The Rule of Memory and the Historian's Craft in Contemporary France*

François Bédarida

The historian and public space

Some while ago, just after the Second World War, people said that France 'vivait à l'heure de son clocher', meaning that it was living a narrowly parochial existence measured out by the chimes of the village church-tower.[1] Today one might well argue that France is living 'à l'heure de son passé', meaning that it is experiencing the present with its attention fixed on the past. Indeed at times one wonders whether France is not in the process of becoming one gigantic 'lieu de mémoire' (or 'site of memory'). Sometimes there emerge from this site of memory moralizing and media-driven commemorations, at other times indictments, even attempted justifications of past actions. And sometimes a mixture of all three.

We are assailed on all sides by the past. This situation calls into question the responsibilities of the historian whose function it is to know and understand the past and whose mission it is to provide his or her contemporaries with the reference points and explanations which allow them to establish a fixed identity and relate their being to the chain of events. The historian mediates between past and present, works with time, attributes meaning to time. In other words, the historian's mission is to organize the past.

In order to define the historian we can make use of the words that Péguy used when he recalled the fragile historical character of our spaces of freedom, truth and culture: 'We have inherited a domain which is always under threat, a domain for which we are responsible and accountable.'[2] This is what constitutes the foundation of responsibility in historical research. Every historian is accountable for the history that he produces. We have the right to ascribe to him the vision of the past that he elaborates and transmits through his writings: in other words, he is responsible for the interpretation of time and of events that he lays before his contemporaries.

What is novel today, however, is that historians in their professional capacity find themselves asked by the public to respond to a growing number of varied requests relating to social, civic and moral issues. As a consequence we find that historians, especially historians of contemporary society, are repeatedly engaged in debates concerning the nature of their responsibility. Indeed, at the close of the twentieth century, we find that under pressure from the *Zeitgeist* the condition of the historian is undergoing change: there is a need to reformulate, if not to restructure, the way in which history is produced, written and communicated to the public. Half a century ago Lucien Febvre spoke pertinently of 'the social function of the historian'.[3] But in the present context, given the expectations and desire for meaning which characterize our 'age of extremes', it appears to be necessary to go a stage further. On the one hand, we need to adopt a new scientific and ethical position when we act in our role as professional historians; on the other, we need to redefine the threefold responsibility of the historian: intellectual, social, moral.[4]

The omnipresence of memory and the spirit of the times

One social phenomenon may be said to characterize the last quarter of the twentieth century: that is, the investment of public space by the paradigm of memory. Until this recent period the term 'mémoire' (memory) remained restricted to certain specialized uses, for example in psychology or in sociology. In normal parlance people used rather the term 'souvenir' (in the sense of memories, recollections, remembrances). Suddenly, towards the middle of the 1970s, the word 'mémoire' became omnipresent, and this was an international and not a purely French phenomenon. In a short space of time the value attached to the concept of 'mémoire' grew exponentially and it became the central structuring factor in our societies. Not only were the virtues of memory praised and encouraged, memory was also promoted as a categorical imperative. The 'devoir de mémoire', meaning the duty of memory, or the obligation to remember, was deemed absolute, universal, mandatory. Around it there developed a new form of worship: the cult of memory.

It is to be noted that the vogue for memory was accompanied by two related notions which likewise have been highly successful. First, there is the notion of 'patrimoine' (heritage), originally a juridical notion arising from Roman law. This has taken on an ever-wider meaning and has come to encompass the network of bonds which the inhabitants of a nation, a region, a town or a village hold with their past. Second, there is the notion of 'racines' (roots). This corresponds to the socio-cultural and historical reference point that people search for in the past in order to discover a sense of origins, continuity and belonging. The unprecedented craze for genealogy shared by people of all ages and all social milieux testifies to this need to find roots.[5] In reality the term 'mémoire' carries with it a range of ambiguities: semantic, psychological, political, affective. Its ambivalence arises from its two distinct functions: in the first instance memory restores and preserves the past; in the second it relates to the present time and to the process of transmission.

The first process is that of 'remémoration' (remembering again, or recommitting to memory). In this case memory acts to recover and revivify the past in order to safeguard it. Thus the past is reinvoked, recalled in the present and becomes present. The consequence is a corpus of more or less structured representations and images (national memory, working-class memory, women's memory, and so on). These are located within a tradition itself half-real, half-imaginary. From this arise the first premises of an historical consciousness, but of a consciousness as vague as it is spontaneous, unclear and often confused, a consciousness within which the legendary mingles with the real while it awaits validation (or invalidation) by historical knowledge.

The insertion of the past into the present leads therefore to an 'usage de la mémoire' (a use of memory, or its instrumentalization): this is the second process within the system of remembering. This act of transmission generates a new process of selection: the first selection had been undertaken as part of the act of 'remémorisation' (recommitting to memory), the second selection concerns the utilization of the corpus of memory (or memories). At a stroke, the ambiguities, interferences and the confusions of memory are multiplied, opening up the possibility of material undergoing all manner of distortion, or even manipulation. At the same time, however, this process provides the historian with a vast field of investigation and interpretation.

This shift in favour of 'mémoire' (memory) marks a turning point for history – and also for historiography.[6] The shift occurred in the mid-1970s – it can be dated with reference to two publishing successes, *Le Cheval d'Orgueil* by Pierre-Jakez Hélias, and *Montaillou village occitan* by Emmanuel Le Roy Ladurie[7] – and it may be located within a broader historical context of worldwide significance: it was felt as a shock-wave.

Among the constituent elements of this phenomenon were: the economic crisis after 1973, which generated a mood of pessimism and doubt following the euphoria of the 'Trente Glorieuses', the 30 years of postwar economic growth in the West; the retreat of the philosophies of progress and the simultaneous loss of confidence in the philosophies of time – that is, of Christianity and Marxism; the end of the utopian dream and the withering of those revolutionary doctrines and hopes which were by definition oriented towards the future and which rarely turned to the past, and then only as a foil to their aspirations; the anti-totalitarian movement. The new realization which arose in the 1970s (the Solzhenitsyn revelations, the Vietnamese 'boat people', the process of 'de-Maoization' in China) also contributed to the valorization and the extolling of memory. To the extent that totalitarianism wished to control the past in order better to control the present and the future, as Orwell famously put it, one of the first duties of the anti-totalitarian struggle consisted in denouncing the will to obliterate the past, characterized as the obliteration of memory. In Orwell's future society ('Angsoc'), time was abolished, as his hero Winston Smith noted. The occultation of the past is systematically organized by suitably designed machines which are installed in offices and in public places so as to produce gaps in memory. Events are suppressed, human beings are 'vaporized'. Hence the new imperative: at all costs conserve, remember again (*remémorer*), for we must transmit to the future the memory of the past. In this fashion memory was central to the anti-totalitarian project, providing it with the highest form of legitimation.

In the eyes of many, the future seemed so dark as to appear opaque and menacing. This led to a crisis in the way in which the relationship between the past, the present and the future was perceived. This in turn produced a tendency to return to the past, for the passionate concern for 'memory' is closely allied with the search for a sense of identity. When the *Zeitgeist* represents the world as gloomy, shifting and uncertain, memory offers an anchorage in the storm, or perhaps a fixed point which resists the undertow. In a manner different from national heritage (*patrimoine*), which by its very nature is multiple, national memory expresses a unifying will, a search for belonging and cohesion, a way of integrating the individual into a group, be it large or small. Memory is an agent of social bonding. It protects individuals, prevents them from becoming 'sans domicile fixe' (with no fixed abode), and saves them from becoming like so many travellers with no baggage, wandering about without hope in a desert peopled with the shades of the absent and the dead.

This explains the appeals constantly made in public debate to the goddess Memory. During a period at the close of the 1970s and at the start of the 1980s, the spectre of a nation without memory was raised: happily we have now moved beyond the resulting controversy – at once superficial and polemical – concerning the teaching of history in French schools. On the other hand, there are some who, drawing on a misunderstanding of the work of Halbwachs, continue publicly to champion an opposition between historical memory understood as conceptual and constructed, and collective memory understood as a reflection of lived time, inscribed within a social framework. Two illustrations of this attitude of mind are the vogue for oral history and the craze for commemorations. The former has lessened somewhat but the latter remains very much alive. Some have gone so far as to talk of a French disease called 'commemorationitis', with its accompanying liturgies and rituals, its sacred communion with the past, its fetishistic veneration of anniversaries, ranging from great acts of commemoration (the Bicentenary of the French Revolution, the Holocaust) to very small commemorations of a local or sectional character. Small or large, such commemorations have the capacity to become controversial on account of their symbolic power. Such events do not possess a unitary meaning and are open to forms of political recuperation. Recent examples of this phenomenon included the celebration of the quatercentenary of the Edict of Nantes or, earlier, the anniversaries of Hugh Capet and of the baptism of Clovis, an even more extreme case which produced what became a sometimes pointless controversy. Hence the question: by dint of all these references to a more or less mythical past, is the nation we call France perhaps turning into a 'museum' (*nation-musée*)? Or is France becoming the devoted follower of a new secular religion, the religion of commemoration (*religion commémorative*)? That is, unless Marianne, France's allegorical personification, is going through an identity crisis.

This is dangerous on three counts. First, it mixes together elements of anachronism and determinism. Unlike classical mechanics our world is not governed by laws which are indifferent to time – in Newtonian physics future and past time fulfil the same role. Time is not 'an illusion' as Einstein liked to put it. Given that we are not automata, time, while it is irreversible (and also because it is irreversible) leads us to entertain an evolutive vision of the world, one with a future and a past. According to the physicist Arthur Eddington, entropy is the arrow of time. This is why the historical universe, contrary to the version of things that memory tends to prescribe, is one actualization among other possible actualizations which have not taken place. According to Raymond Aron – in his great thesis of 1938 – the task of the historian is to liberate the past from determinism (*'défataliser' le passé*). Aron urged the historian to put himself in the situation of past historical agents who had a future but who were ignorant of what would subsequently take place, and who, with regard to their expectations, their fears or their hopes, lived in complete uncertainty with regard to the future, finding themselves in the midst of a plurality of possible outcomes.[8]

The second danger arises from the manner in which memory has no fixed shape. Memory flows, it can be channelled in certain directions as it can be moulded, twisted into new forms. This renders memory vulnerable since it can serve all kinds of orthodoxy. Collective memory contains deep but contrasting feelings and emotions – veneration and execration, respect and rejection. It privileges faith over reason and there is a real risk that it will lead to 'official' history. Whereas the historian exercises properly his mission of critical vigilance, collective memory may seek to legitimate itself with the seal of scientific authority.

Finally there is the third danger: based on the current frenzy for commemorations and amid the threatening noises we hear that we must commemorate, experience has shown the extent to which there was room for all sorts of manipulation on the part of those in charge of organizing celebrations, be it on the part of the State, community institutions, pressure groups, local authorities or micro-groups of all types.

The misadventures of history and memory: discord and discordance

In the space of around twenty years memory has invaded all public space and this has led to an intellectual confusion which threatens the discipline of history. The dividing line which separates history's territory from the domain of memory has become blurred. It has become necessary to resist the dictatorship of memory which the 'memory militants' threaten to establish in France. They dream impossible dreams of absolute and immediate transparency and are past-masters when it comes to suspecting the motives of others. In these circumstances it is worthwhile recalling the demands made by the analytical and rational approach and which are part and parcel of the practice of the historian as he constructs a body of knowledge. He dispels distortions born of partisan emotion, dismisses attempts to telescope different time-scales. And he does this regardless of the expectations which society at large has of him, for he knows that society often remains a prisoner of prejudice and myth.

In this sense history and memory stand opposed to each other as contrasting terms. This has been eloquently described by Pierre Nora:

> Memory is life, and is always borne by existing social groups. . . .
> Memory remains unaware of successive distortions, vulnerable to all
> sorts of uses and manipulations, and is prey to long silences and
> sudden reappearances. History is the ever-problematic and incom-
> plete reconstruction of what is no longer . . . because it amounts to an
> intellectual and secular (*laïcisante*) operation. As such it calls for ana-
> lysis and critical inspection. Memory invests remembrance with
> sacred meanings, whilst history strips them away. Memory always
> has a dubious status in relation to history, whose true mission is to
> destroy and repress memory.[9]

Charles Péguy, for his part, had already emphasized in *Clio* that history
and memory correspond to two systems which intersect 'at right
angles':

> History is essentially longitudinal, memory is essentially vertical.
> Essentially history consists in moving alongside the event, above all
> in not departing from the event but remaining inside the event,
> exploring it as it unfolds. Memory and history form a right angle.
> History is parallel to the event, memory is central and axial.[10]

We are now in a position to measure the risks – and the pitfalls – arising
from public expectations which are not only fed by memory but which
are also enthusiastic manifestations of the religion of memory. The
historian, on the other hand, should remain true to his critical method.
He must set the record straight, demythologize, argue a case in order to
ensure that the form of history which prevails frees humans from their
past and educates them for the present and the future.

Of course each social group, each nation, has a need for symbolic
figures. These may be either models of what the group valorizes or
embodiments of what it rejects. Of course it is necessary to fight against
historical amnesia, to take responsibility, to transmit knowledge of the
past. But it is also necessary to be aware of the ambiguities and the blind
alleys which are associated with the notion of a 'duty of memory' (*devoir
de mémoire*). Today this duty of memory, this injunction to remember,
has been transformed into a categorical imperative. It is running out of
control and cannot be reined back. It is the doctrine which zealous new
crusaders preach to the infidels. We especially need to take care since

the duty of memory contains an ethical requirement which, while it springs from motives which are undeniably and initially welcome, rapidly turns into a timeless morality, a morality disengaged from the context and defining conditions which obtained when an action took place in the past. Sometimes this goes so far as to take the form of an abstract and complacent moralism, extravagant and intolerant, ready to offer summary condemnations without seeking to analyse or understand. This produces a society caught up in lawsuits and trials, a society focused on the past, bogged down in anachronisms, a society which is unaware of the nature of its dependence on a past which imprisons the present. In order to escape from such mirages and in order to rediscover the true meaning of time, I suggest that we use the expressions 'travail de mémoire' (labour of memory) and 'devoir d'histoire' (the 'duty', or 'responsibility', of history, for the past). For in reality, as Saint Augustine forcefully demonstrated, historical temporality includes not simply the past but also and equally both the present and the expectation of the future.

For this reason, instead of finding ourselves torn between extremes, between an excess of memory and an insufficiency of memory, we need to practise the art of forgetting, which, as Nietzsche showed, is a force which allows scars to heal and which, in turn, lets us understand and assimilate the past by rebuilding that which has been broken. But here we are dealing with a form of forgetting which is active and selective, which Paul Ricoeur has called 'forgetting the debt owed, not the facts behind the debt'.[11]

Transforming memory into history

We are now in a position to evaluate the size of the challenge which has been thrown down to historians: memory is omnipresent among our contemporaries, but how may historians historicize it? For the conflictual relationship between the notions of history and memory can be expressed in two ways. The first way of representing the history/memory relation is to view it as dialectical, as the opposition between thesis and antithesis in the expectation that a hypothetical synthesis will be produced. The second is to use the image of the mirror to express a relationship of reciprocity (or of retroprojection) between history and memory, a relationship which reflects their similarity, each one illuminating the other. But that would be to run the risk of trapping ourselves within a play of reflections, of distorting mirror images. Why should we not foster a higher ambition and aspire to pass through the mirror and explore that which lies beyond?

The first point to make concerns the relationship between historical knowledge on the one hand, and time, action and the future on the other. As a consequence of the contemporary withdrawal into the mode of memory, society now tends to seek a legitimating agent in memory rather than in a representation of the future. In the words of Olivier Mongin: 'The resurgence of memory has brutally replaced the future as a way of legitimating action in the present.'[12]

This produces a paradox. Between, on the one hand, a future more unpredictable than ever (and considered all the more threatening) and a past conceived as a heritage museum, historiography, that is the production of history, is in quite good health. People even talk of too much history-writing and point to an abundance of books and articles as well as to the (relative) authority attributed by society to historians. But that does not much help the predicament of the historical consciousness caught, on the one hand, between an excess of memory and, on the other, a deficit when it comes to the act of history. Here we encounter the conclusion drawn by Jacques Rancière in *Les Noms de l'histoire*: the act of 'writing history' (*faire de l'histoire*) is accompanied by the refusal to 'make history' (*faire l'histoire*). To quote Olivier Mongin once more: 'Now we find that the present, which has become more and more "historiographical" and less and less "historical", contemplates its reflection in a memory which rapidly weighs heavily upon it.'[13] At heart, are we not dependent upon a society which 'loses itself in memory because it lacks a sufficient sense of the present?'[14]

This brings us to the problems arising from the uses made of memory and from the tendency to carry too far the reliance placed upon memory.[15] In the face of our history, and when confronted with the tragedies which have marked the twentieth century, it is essential certainly that we commit to memory the horrific deeds to which human beings have resorted (the Holocaust, the world of concentration camps, totalitarianism of all colours). It is imperative that both the memory of human bestiality and that of the conditions in which it took shape be preserved.

But this 'labour of memory' (*travail de mémoire*) which I have discussed could never lead to the representation of the past which is current today and which tends more and more to express itself in inquisitorial language and take on a judicial air. The healing power of memory may be immense but we should be careful not to become unconditional worshippers at the shrine of memory. All the more so since, as we have seen, our collective memory remains a prisoner of the past and lacks a project for the future. These factors mean that collective memory is susceptible to over-determination by ideology and that 'le mémoriel', or modes of storing memory, lends itself to recuperation and manipulation since it can be directed in the desired direction. The conclusion is the forging of a distorted historical consciousness which has diverged from historical reality.

Let us be frank: between lived experience stored by memory (*le vécu mémoriel*) and critical reason exercised by the historian (*la raison historienne*), the task facing us is difficult because it means we have to tread a perilous but necessary path. For, when dealing with extreme cases it is necessary, on the one hand, to beware of the emotions, the passions, even of anger and indignation, while on the other to assert the rights of rational knowledge, in order to explain that which is apparently inexplicable: these are the conditions which will determine the transmission of memory.

In truth, the historian who constructs a reasoned argument and follows a scientific approach which maintains a distance with regard to the material under investigation is not led to reject lived experience; he contributes to the situating of lived experience in its rightful place. Following in the tradition of Michelet we should take care not to diminish the significance of concrete lived experience merely for the sake of a scientific and rational approach. This is why the obligation of the historian – to maintain a hold over both of these at the same time – means that his task is as perilous as it is stimulating. Put another way, between the past and the future, 'scientific memory' (*mémoire savante*) must function as a mediating or enabling agent.

For historians of the contemporary (but also for those who study older periods), the present implies the presence of the world. This is the case whether one accepts the relationship of reciprocity illustrated so well by Reinhard Kosselleck (the past considered as a 'field of experience' and the future as a 'horizon of expectation'),[16] or whether one prefers Emmanuel Lévinas's conception of time ('drawing time together into a continuum of memory and aspiration').[17]

Thus, between the opposite poles of memory and history, and to defend against the pitfalls of amnesia or polarized memory, there are two imperatives to be proclaimed loud and clear. First, we should affirm and promote the continuity – or solidarity – between the past and the future through the mediation of the present. Second, we should remember that history was based – and should remain based – upon one simple axiom: that is, there exists an external reality and that external reality is capable of being rendered intelligible. Marcel Proust defined literature as 'the joy of the real regained'. In my view, the phrase may be applied with equal relevance to history.

Notes

*Translation by Ceri Crossley and Martyn Cornick.

1 'A l'heure de son clocher' is a phrase used by the Swiss journalist Herbert Lüthy in a work published in the 1950s; see *A l'heure de son clocher: essai sur la France* (Paris, 1955).

2 C. Péguy, *Par ce demi-clair matin* [1905], in *Œuvres en prose complètes*, II (Paris, 1988), p. 96.

3 L. Febvre, 'Vers une autre histoire [1949]', in *Combats pour l'histoire* (Paris, 1953), p. 438.

4 For further discussion, see the special issue of the journal *Diogène* no. 168 (October–December 1994), 'La responsabilité sociale de l'historien', containing my article 'Praxis historienne et responsabilité'. Cf. also F. Bédarida, 'Les responsabilités de l'historien "expert"', in P. Boutier and D. Julia (eds), 'Passés recomposés: champs et chantiers de l'Histoire', *Autrement* (January 1995), pp. 136–44, and D. Krüger, 'La responsabilità degli storici e degli archivisti: il caso tedesco', *Passato e presente*, XV, 40 (1997), pp. 121–31.

5 A recent report by the Archives of France showed that a third of the people using the National Archives in Paris worked on the 'minutier central des notaires' [the equivalent of birth and death registers]; in other words they are amateur genealogists.

6 As contrary evidence we might mention the three volumes of *Faire de l'histoire*, published under the direction of J. Le Goff and P. Nora (Paris, 1974), *the* work of reference on the methodology of historiography: in some 550 pages there is no mention of memory.

7 P.-J. Hélias, *Le Cheval d'orgueil: mémoires d'un Breton du pays bigouden* (Paris, 1975); E. Le Roy Ladurie, *Montaillou village occitan* (Paris, 1976). The English translations of these works are, respectively, *The Horse of Pride: Life in a Breton Village* (New Haven and London, 1978) and *Montaillou: Cathars and Catholics in a French Village, 1294–1324* (London, 1978).

8 R. Aron, *Introduction à la philosophie de l'histoire; essai sur les limites de l'objectivité historique* (Paris, 1938), esp. pp. 181–7.

9 P. Nora, 'Entre mémoire et histoire', in *Les Lieux de mémoire*, 1, *La République* (Paris, 1984), pp. xix–xx.

10 C. Péguy, *Clio* (Paris, 1932), p. 230.

11 P. Ricoeur, *La Critique et la conviction* (Paris, 1995), p. 190. See also M. Augé, *Les Formes de l'oubli* (Paris, 1998).

12 O. Mongin, 'Mémoire sans histoire: vers une nouvelle relation au temps', *Esprit* (March–April 1993), p. 108.

13 *Ibid.*

14 P. Chaunu and F. Dosse, *L'Instant éclaté* (Paris, 1994).

15 On this point it is worth consulting Tzvetan Todorov's little book *Les Abus de la mémoire* (Paris, 1995).

16 R. Koselleck, *Vergangene Zukunft. Zur Semantik geschichtlicher Zeiten* (Frankfurt, 1979), part III, chapter 5.

17 E. Lévinas, *Dieu, la mort et le temps* (Paris, 1987), p. 134.

Index

Abetz, Otto, 180–96 passim
Académie française, the, xiv
Academy of Nantes, 66
Action Française, the, 112, 116, 141, 239
Agincourt, battle of, 142
agriculture, 41, 42, 76, 150, 240
Agulhon, Maurice, xiv, 46, 58, 82
Aide-toi, le ciel t'aidera, 68
Algeria, xiv, 64, 137, 246, 253, 257, 259
Algerians, 253, 255, 260
Allier, Max, 244
Alsace, 129, 186, 236
Alsace-Lorraine, 109, 121, 181
Amboise, 3, 8, 203
anarchism, 172
ancien régime, 19, 20, 24, 44, 53, 85, 95, 236, 272
Andler, Charles, 172
Angers, 72, 198, 202
Anglais coués, les, 127
Anglophilia, 128
Anglophobia, 128, 130, 133, 140
Anglo-Saxon 'superiority', 134, 141
animals, 76, 81–103 passim
Annual Register, The, 143
anticlericalism, 265, 270, 271, 278
anti-Semitism, 113, 115–18, 128, 186
Arago, 69
Argueil, 38, 39
Aron, Raymond, 290
Arras, 260
Assembly of the Hundred Days, 64
Association de Culture Berbère (ACB), 257
Assolant, Alfred, 133
Aube, 56
Austerlitz, battle of, 68
Aveyron, 17–29 passim, 242
Azéma, Jean-Pierre, 197

Balzac, Honoré de, 5, 175, 183
banlieue, the (suburbs), 249–63 passim
Barrès, Maurice, 116, 149, 171
Barthes, Roland, 127, 223, 235
Bas-Languedoc, 21
Bas-Rouergue, 17, 20, 21
Basse-Bretagne, 245
Bauer, Ingrid, 217, 219
Bédarida, François, xiv, xv, 197, 283
Bernanos, Georges, 236, 237
Berry, Georges, 130
Berthelot, Marcellin, 173
Besançon, 233, 234
Beugnot, 31, 32, 33, 35, 38, 41
Bicentenary of the French Revolution, the, 289
Birmingham, The University of, xiii–xvi
Black Legend, 2, 5, 7, 13
Blanc, Louis, 72
Blangy, 37, 38, 39
Blanqui, 67
Blum, Léon, 172, 176, 238, 239, 240
Bodley, J. E. C., 134
Boer War, the, 131, 137, 139, 140, 142, 160, 161, 162
Bois, Paul, 40
Bonald, Louis de, 23, 85
Bonaparte, 28, 31, 33, 39, 163
Bonapartism, 104
Bonhomme, Jacques, 130
Bonnejoy, Ernest, 94
Boulanger Affair, the, 109, 115, 118
Boulanger, General, 106, 108, 109, 113, 115, 118, 128
Boulogne, 144, 153
Bourgeois, Léon, 156, 162
Bourgeon, Jean-Louis, 2, 6, 11, 12
Bourzès, Louis de, 23
Brasillach, Robert, 183
Brassens, George, 219
Braudel, Fernand, xiii

Bray, Pays de, 32, 40–6
Brest, 72
Brinon, Fernand de, 184
Britain, xv, 77, 82, 94, 100, 109,
    125–48 passim, 150, 152,
    157, 158, 162, 164, 184
Britannia, 139
Brittany, 70, 74, 122, 199
Brive, Marie-France, 217
Brunetière, Ferdinand, 141, 173, 176
Brusque, 26
Bryce, James, 156, 161
Buchez, 65, 67
Burns, John, 134
Burrin, Philippe, 180, 182, 185

Cabet, 67
Calas, Jean, 20, 21
Calvinism, Calvinists, 8, 19, 20, 21,
    22, 24
Camarès, 18, 19, 20, 21, 22, 25, 27
Camproux, Charles, 244
Cantal, 56
Capa, Robert, 219, 238
Capdevila, Luc, 219
Capitaine Corcoran, 133
Carcopino, Jérôme, xiv
Carlyle, Thomas, 129
Carthage, 127, 128, 132
Cartier-Bresson, Henri, 219, 238
Catalonia, 244, 271, 273, 278
Catholicism, 105, 108, 111–16, 119,
    120, 182, 264–84 passim
Catholics, 4, 5, 7, 8, 9, 12, 13, 17,
    19, 21, 22, 23, 25, 27, 112,
    115, 116, 120, 121, 176,
    252, 264–84 passim
Caux, Pays de, 31, 32, 36, 37, 40, 42–6
Caziot, Pierre, 242
Céline, Louis-Ferdinand, 183
Cesari, Jocelyne, 255
Cévennes, 238, 243
Chabot, François, 26
Chahaignes, 53
Chamberlain, Joseph, 140
Champagney, 259
Chaper, Achille, 70, 71
Charente-Inférieure, 56
Charle, Christophe, 160

Charles IX, 3, 4, 6, 7, 8, 9, 11, 12
Chateaubriand, François René de, 6, 70
chef-lieu, 38, 55, 56, 60
Chesney, George, 140
Chevalier, Michel, 67, 72
Chevrillon, André, 142
Chinon, 197, 198
Cholet, 200
Christianity, 73, 88, 90, 97, 235, 288
Citron, Suzanne, 109
civilization, 22, 134, 149, 186
Clair, René, 235
Clark, Stephen, 81
Clemenceau, Georges, 91, 92,
    106, 168–79 passim
Clement VII (Pope), 3
Clergy, the, 24, 37
Cloulas, Ivan, 11
Clovis, 289
Club de l'Oratoire, 71
clubs, political clubs, 23, 24, 52,
    55, 67–71, 72, 188, 274
Coligny, Admiral, 2, 6–12, 14
colonial rivalry, 126–8
Combes, Emile, 112
Comité France-Allemagne, 184
commemoration, 65, 184, 200, 206,
    207, 255, 289
Committee of Public Safety, 18, 26, 27
commune, 24, 25, 58, 60, 70, 73,
    76, 203, 206
communism, 73, 182
Condé-sur-Vesgre, 70
Condorcet, marquis de, 64
Constituent Assembly, 39, 55
Constitution of 1793, 53
Constitution of 1802, 34
Consulate, 28
Coppolani, J.-Y., 33, 36, 38, 39
Corrèze, 59
Corsica, 243
Council of State, 36
Croix, La, 115

Dahmani, Arezi, 260
Daily Star, 125
Daladier, Edouard, 182, 189, 190, 191
dancing, clandestine, 197–212 passim
Darwinism, 91, 98, 134

Daudet, Léon, 170
Debray, Régis, 252
Debû-Bridel, Jacques, 128
Declaration of the Rights of Man,
    the, 17, 22, 68, 251
decolonization, 253, 254
Degrelle, Léon, 189
Delcassé, Théophile, 163, 164
Delors, Jacques, 125
Demolins, Edmond, 134, 151
Dermigny, Louis, 20, 22
Déroulède, Paul, 106
Descartes, 75, 83
Deutsche-Französische Gesellschaft,
    184
Dienststelle Ribbentrop, 180, 184, 185
Dieppe, 34
Dimnet, Ernest, 176, 177
Dine, Philip, 132
Directory, 28, 54, 96
Dorgères, Henri, 240
Doubs, 60
Douglas, Mary, 81
Dreux, 256
Dreyfus, Alfred, xv, 93, 105, 108,
    110, 112, 113, 115, 116, 117,
    119, 120, 123, 129, 137, 139,
    140, 141, 149, 150, 153, 156,
    157, 160, 163, 170, 172, 174,
    176, 177, 235
  Dreyfusards, 113, 161, 162, 163
  the Dreyfus Affair, xv, 93, 105,
    108, 110, 112, 113, 115, 116,
    117, 119, 120, 123, 129, 137,
    141, 157, 160, 163, 170, 176,
    177, 235
Drieu La Rochelle, Pierre, 184
Drumont, Edouard, 115, 116, 163
Duclaux, Emile, 156, 158, 163
Dufour, Arlès, 69
Dujardin-Beaumetz, Georges, 94
Dumas, Alexandre, 5
Dupin, Charles, 69
Duroselle, Jean-Baptiste, 150, 170, 171
duty of memory, 286, 291, 292

Ecole normale supérieure, xiv, xv, 172
ecology, politics and, 31–49 passim
Eddington, Arthur, 289

Edict of Nantes, 17, 289
Edict of Toleration, 20
Edinburgh, xiv, 68, 135, 150, 151,
    152, 153, 156, 157
Edinburgh Review, the, 135
electoral participation, 50–63 passim
Ellis, Havelock, 137
Enfantin, Prosper, 67, 69–71
England, 1, 3, 13, 41, 42, 83, 92, 97,
    128, 129, 130, 131, 132, 136,
    137, 138, 161
English Civil War, 42
Entente Cordiale, 127, 139, 143, 162
Esquiros, Alphonse, 88
Estates General, 53
Estudio Mundial de Valores, 277, 278
ethnic differences, 250, 251
ethology, 66, 73
Extremadura, 278

Farrère, Claude, 132
Fashoda, 134, 137, 139, 140, 143,
    150, 164
Faucher, Léon, 130, 131
Faure, Edgar, 257
Febvre, Lucien, 286
Fécamp, 34
femmes tondues, 213–32 passim
Ferdonnet, Paul, 184
Ferro, Marc, 238
Ferry, Jules, 82, 112, 114, 122
Ferry, Luc, 82
Féval, Paul, 132
Figaro, Le, 163, 188
Finkielkraut, Alain, 252
Finot, Jean, 138, 139
First Republic, the, 128, 256, 259
First World War, 110
Florence, 2, 3, 82
Fourierism, 70, 73
Fourierists, 66, 70, 71, 73
France-Plus, 260
Franco, General, 265, 266–9, 273,
    274, 276, 279
Franco-British Council, xviii
Franco-British relations, 125–48
    passim, 150, 160, 164
Franco-German relations, 180, 185,
    192

François-Poncet, André, 185
Francophonie, 254
Franco-Prussian War, the, 106, 113,
    114, 128, 140
Franco-Scottish relations, 152, 156,
    162, 163
Fraternelle universelle, La, 72
French Revolution, xv, xvi, 17, 50, 52,
    88, 90, 104, 116, 122, 129, 150,
    218, 251, 255, 256, 289
Front National, the, 256, 261
Fundación Santa Maria, 277

Gall, 68
Garaudy, Roger, 257
Gascony, 19
Gaulle, General Charles de, xv, 122,
    187, 207, 209, 245
Geddes, Patrick, 149–67 passim
Génériques, 261
Germany, 13, 94, 106, 107, 109,
    143, 149, 150, 158, 182, 183,
    184, 186, 188, 189, 190, 191,
    197, 198, 204, 213, 218, 239,
    243, 267
Gestapo, 197, 214
Giono, Jean, 233, 234, 235, 239
Girard, René, 216
Girardin, Madame de, 74
Girondins, 64
Gleizes, Jean-Antoine, 87–9, 96–9
Globe, Le, 67
Goncourt, Edmond de, 170, 173
Goude, Jean-Paul, 258
Gournay, 37, 46
Grande Illusion, La, 238
Great Mutiny, the, 133
Great Universal Dictionary, 130
Greater France, 249–63 passim
Guardian, The, xvi, 126
Guéhenno, Jean, 197, 239
Guépin, Ange, 64–80 passim
Guesde, Jules, 120
Guha, Ranajit, 244
Guise, Dukes of, 3, 7, 10, 11, 12, 14
Guyenne, 20

Hancock, Sir Keith, xiv
Harlem Désir, 259

Haute-Loire, 56
Haute-Saône, 259
Hautes-Pyrénées, 60
headshaving, 214–25
Hélias, Pierre-Jakez, 287
Henri IV, 19
Henry II, 3, 4, 6, 8
Henry of Navarre, 4, 9, 10, 13
Henty, G. A., 143
Hérold-Paquis, Jean, 128
historians' craft, 285–95 passim
historiography, 6, 112, 215, 222, 238,
    242, 249, 285–95 passim
Hobsbawm, Eric, 105, 111
Howard, Michael, 133
Hugo, Victor, 69, 91
Huguenots, 2, 4, 5, 6, 7, 8, 9, 11, 12, 13
Humanae vitae (encyclical), 269, 275
Hundred Days, the, 34, 36, 37
Hyde Park, 140

identity, 104, 108, 110, 112, 122, 129,
    130, 137, 150, 205, 213, 218, 222,
    229, 233–48 passim, 249, 250, 254,
    255, 283, 288, 289
immigration, 249–263 passim
imperialism, 126, 127
Indochina, 253
Indre-et-Loire, 199, 200–3
inferiority, 23, 126, 137
invasion, problem of, 140
Islam, 252, 254, 257, 271
'Island Story', the, 142

Jacobinism, 17–28 passim
Jacobins, 18, 24, 26, 29
Jacques, Lucien, 239
Jarnac, battle of, 8
Jaurès, Jean, 110, 111, 119, 120,
    168, 240
Jean de Florette, 233
Jean, Renaud, 240
Jeanneney, Jean-Noël, 171, 256,
    257, 258
jingoism, 133, 141, 142
Joan of Arc, 1, 207
John Bull, 130, 131
John Paul II (Pope), 269
John XXIII (Pope), 269

Johnson, Douglas, xiii–xvi, 32, 125, 128, 163
Joly, Henri, 91
Jospin, Lionel, 252
journalism, 70, 126, 128, 169
Joux, 259
Judt, Tony, 43, 240
July Days, the, 65, 68
July Monarchy, the, xiv, 51, 57, 58, 60, 61, 73, 88

Kérillis, Henri de, 182, 188, 191
Kipling, Rudyard, 134
Kupferman, Fred, 184, 216

laïcité (secularism), 260
Lan, David, 244
Langlois, C.-V., 127
Languedoc, 19, 20, 244
Larousse, Pierre, 130, 131, 132
Larzac, Causse du Larzac, 22, 244
Latins, the Latin 'race', 134
Laval, Pierre, 180, 183, 187, 199
Lavisse, Ernest, 6, 156, 162
Le Bras, Hervé, 40
Le Havre, 34, 36, 45
Le Pen, Jean-Marie, 126
Le Play, Frédéric, 151
Le Roy Ladurie, Emmanuel, 287
Le Sant, Adélaïde, 66
Lebovics, Herman, 241
Leconte, Floreska, 76
legislature, 52
Leo X (Pope), 3, 120
Leroux, 65, 67, 71, 99
Liberation, the, xiv, 110, 128, 208, 209, 214, 216, 217, 219, 222, 245–7, 266, 269
Libre Parole, La, 115, 163
lieu de mémoire (site of memory), 132, 254, 283
Ligue de la Patrie Française, 163, 174
Limousin, 243, 278
Loire-Inférieure, 199–202
Loir-et-Cher, 60
Londinières, 34, 38
Lorient, 72
Loti, Pierre, 92, 132, 170
Lottman, Herbert, 216

Loubet, Emile, 162
Louis Napoleon, 56, 57, 73
Louis XVI, 21
Louverture, Toussaint, 256, 259
Lozère, 246, 247
Luchaire, Jean, 182

MacDonald, Robert, 130
Macé, Jean, 76
Mafeking, 141, 142, 150
Mahy, François de, 136
Maine-et-Loire, 199, 202, 203, 209
Maistre, Joseph de, 85, 99
Manche, 59
Manon des sources, 233
Manosque, 233, 239
Mantoux, Paul, 141
Marguerite de Navarre, 7
Marne, 56
Marseilles, 239
Marvejols, 246, 247
Massif Central, 19, 56, 246
Maurevert, Seigneur de, 11
Maurras, Charles, 113, 116, 235, 236, 244
Maussion, Clotide, 72
McPhee, Peter, 50, 52, 57
Meaux, 9, 12
Medici, Catherine de', 1–16 passim
medicine, 65, 88, 95, 169, 280
Mehenni, Ferhat, 257, 258
Méline, Jules, 117, 241
memory, 285–95 passim
Meredith, George, 129
Meyerowitz, Heinz, 239
Michelet, Jules, xvi, 6, 7, 74, 82, 90–3, 97, 98, 99, 130, 142, 294
Milice, the, 214, 216
Mill, John Stuart, 129, 169, 174
Millau, 18, 19, 20, 21, 22, 23, 24, 25, 26, 27, 28
Mission du bicentenaire de la Révolution française (MCRF), 256–8
Mistral, Frédéric, 244
Modern and Contemporary France, xvi, 125
modernization, 50, 61, 104, 241

Mongin, Olivier, 293
Monnet, Georges, 240
Montauban, 19, 20, 27
Morbihan, 54, 64, 71, 72
Morénas, François, 239
Mystères de Londres, Les, 132
Mystères de Paris, Les, 132
myth, 117, 122, 127, 128, 132, 223,
   229, 234, 242, 258, 290

Nant, 19, 207
Nantes, 17, 19, 20, 64–80 passim, 198,
   201, 206, 207, 289
Napoleon, 18, 28, 33, 34, 36, 56,
   57, 68, 70, 74, 75, 132, 134,
   259, 264, 272
National Assembly, 21, 22, 23, 24,
   60, 72
national character, 126, 129, 130, 142
National Guard, 53, 70, 72
National Review, the, 134, 141
National Revolution, 197, 199, 200,
   201, 202, 235
national superiority, 126
nationalism, 75, 104–23 passim, 126,
   141, 218, 235, 237, 241, 245
Nausée, La, 234
Neufchatel, 31–46 passim
newspapers, xv, 33, 58, 70, 94, 125,
   138, 188, 189, 261
Nîmes, 19, 28, 236
Nineteenth Century and After, The, 143
Noiriel, Gérard, 249, 250
Nora, Pierre, 249, 291
Nord, 56
Nordau, Max, 128
Normandy, 40, 43, 126
notables, 32, 34, 36, 38, 39, 46,
   52, 58, 72
Notre Temps, 182
Novick, Peter, 216

Observatoire interrégional du
   politique, 277
Occitanism, 244, 245
Occupation, the, 121, 122, 131,
   180–96 passim, 197–212 passim,
   213, 217, 238, 243, 244
Opportunistes, 117–19

Orleanism, 68
Orleanists, 66, 71, 114
Orléans, 7, 8, 66
Orwell, George, 288
Oswald, John, 84
Outram, Dorinda, 218

Panama scandal, the, 128
Paris, xiv, xv, xvi, 4, 9–14, 25, 26, 31,
   42, 52, 55, 56, 65, 66, 68, 69, 70,
   72, 74, 79, 89, 95, 106–8, 113, 114,
   128, 130, 132, 149–53, 157–60,
   162, 172, 180–4, 187–92, 234, 235,
   241, 257
Paris 1900 Exhibition, 152–64 passim
Paris Commune, the, 78, 106, 107,
   108, 113, 152
Paris International Assembly, 153
Paris-Match, 126
patriotism, 21, 75, 111, 129, 216,
   219, 245, 246
Paul VI (Pope), 269
Paulhan, Jean, 129
peasants, 4, 32, 43, 44, 78, 104, 203,
   205, 206, 237, 239, 241, 242, 244
Peer, Shanny, 241, 242
Péguy, Charles, 116, 177, 283, 291
Perdiguier, 67, 78
'Perfidious Albion', 127, 128, 132
Péri, Gabriel, 188
Pétain, Marshal Philippe, 122, 180,
   187, 199, 202, 236, 237, 266
Philip II, 7, 8, 13
Phlipponneau, Michel, 245
Pierre Pucheu, 200
Pius XII (Pope), 268, 269
Poincaré, Raymond, 91, 170, 173
Poissy, 2, 8
Poitou, 19
Polignac, 65
Pontivy, 64
Popular Front, the, 110, 238–42, 245
postcolonial minorities, 249–63 passim
Pouthas, Charles, xiv
poverty, 65, 66, 69, 70, 72, 79, 282
prejudice, 95, 96, 98, 125, 127, 138,
   162, 290
propaganda, 5, 14, 96, 126–8, 184, 188,
   191, 210, 236, 237, 243, 265

Protestantism, 17–30 passim, 112
Protestants, 2, 3, 17, 19, 21, 24, 25,
    29, 109, 117, 257
Proudhon, 67, 82
Provisional Government of 1848, 60
Pullar, Sir Robert, 156–159
Puritanism, 41
Pyrénées-Orientales, 50

Rabelais, 75
racial inter-marriage, 73
Ralliement, the, 106, 114, 176
Rancière, Jacques, 293
Reclus, Elisée, 91–3, 100, 151–3, 161
Reclus, Paul, 92, 152
Reformed Church, 19, 20
religion, 1, 5, 7, 76, 104–23 passim,
    204, 264–84 passim
remémorisation, 287
Rémond, René, xiv
Rennes, 72, 140, 150, 233, 234
Renoir, Jean, 238
Republican tradition, 123, 249–63
    passim
republicanism, 21, 64, 93, 104, 119,
    236, 241, 249, 251, 255, 261, 273
Rerum Novarum, 120
Resistance, the, 233
Réunion de l'Ouest, 68
Review of Reviews, the, 138, 142
Revocation of the Edict of
    Nantes, 20
Revue des deux mondes, La, 88, 134, 173
Revue illustrée des animaux, La, 92–4
Ribbentrop, Joachim von, 184–90, 192
Robespierre, Maximilien, 5, 64, 260
Rochette, 20, 21
Rodez, 23, 25
Rogers, Susan Carol, 242
Roland, Pauline, 74
Romains, Jules, 185, 186
'romantic history', 134
Rome, xiv, 127, 132, 267, 268,
    270, 282, 283
Ronsard, 7
Roquefort, 26
Rosanvallon, Pierre, 59
Roscoff, 151
Rouen, 33, 34, 38, 44, 45

Rousseau, Jean-Jacques, 75, 83, 84,
    97, 118
Royal Navy, 139
Rural Centre, the, 242, 243
ruralism, 233–48 passim

Saint Affrique, 18–29
Saint Helena, 132
Saint Jean-du-Bruel, 19, 20
Saint Rome-de-Tarn, 20, 26
Saint-Avertin, 200
Saint Bartholomew's Day Massacre,
    2, 11
Sainte-Beuve, 69
Saint-Nazaire, 209
Saint-Pierre, 59, 83, 84, 205
Saint-Simonianism, 66, 67
Saint-Simonians, 65–7, 69
Saint-Vallier, 53
Salisbury, Lord (Robert Cecil), 134, 161
Sand, George (Aurore Dupin), 74
Sarthe, 40, 53, 202
Saumur, 205, 206, 208, 209
scapegoating, 216, 217
Scapini, Georges, 184, 187
Scott, James C., 244
Scruton, Roger, 81
Second Empire, 104
Second Republic, the, 50, 51, 53–8,
    60, 273
Second Vatican Council, 267, 269, 275
Second World War, 246, 253, 266,
    273, 283
Seine-Inférieure, 31–49 passim
Sekula, Allan, 222
Service du Travail Obligatoire, 204
Seymour, David, 238
Siegfried, André, 32
Singer, Peter, 81
slavery, 256, 259, 261
Socialism, 64, 66, 67, 68, 71, 73, 77,
    110, 111, 117, 119, 120, 122,
    172, 182, 187, 240, 241
Society of the Rights of Man, the, 68
Sokal, Alain, 126
Solidarité, La, 73
SOS-Racisme, 259
Spain, 7, 10, 11, 12, 13, 264–84
    passim

Spanish Civil War, the, 237
Spencer, Colin, 82
Staël, Madame de, 74
Stead, W. T., 138
Stenger, Gilbert, 235
stereotypes, 125, 126, 130, 224
Sternhell, Zeev, 106
Stowe, Harriet Beecher, 74
Sun, the, 125
Sunday Times, The, 125
Sutherland, Nicola, 2, 6, 11
syphilis, 6, 70

Taine, Hippolyte, 33, 114, 134, 176
Tarn, 19
Tavannes, 7, 11, 12
Terre libre, La, 240
Thermidor, the coup of 9 Thermidor,
    26, 27, 37
Third Republic, the, 37, 91, 98,
    99, 105–9, 112, 113, 115,
    129, 169, 199
Times, The, xvi, 141
Tocqueville, Alexis de, xvi, 59
Todd, Emmanuel, 40
Todorov, Tzvetan, 127
Toulouse, 20, 21, 54, 171, 217
Tournay, 60
Tours, 72, 197, 200, 202, 203,
    205, 208
Trafalgar, battle of, 139, 142, 143
Transvaal, the, 140, 150

Underdown, David, 42, 46
Union des Femmes françaises,
    222

United States, the, 151, 174, 250, 251
universal suffrage, 61
Université Nouvelle, 152
University College, London, xviii

Valmont, 34
Vannes, 72, 77
Var, 53, 55, 56, 57, 240
Vatican, the, 12, 173, 267, 268,
    269, 282
vegetarianism, 81, 88, 93–8
Vendée, 56, 84, 169, 278
Vendredi, 239, 240
Verdun, 139, 184
Verne, Jules, 132, 133
Versailles Treaty, the, 142
Vichy regime, the, xiv, 112, 120,
    122, 197, 208, 266
Victoria, Queen, 140, 143, 150
Villebois de Mareuil, Colonel, 141
Villermé, 69, 70
Virgili, Fabrice, 214
Voltaire, 75, 83, 97

Waterloo, 18, 28, 126, 132, 142
Weber, Eugen, 51, 129, 242
Weil-Curiel, André, 182, 184

xenophobia, 2, 125–48 passim

Yacine, Kateb, 260
Yonne, 60

Zeitgeist, 176, 177, 286, 288
Zola, Emile, 119, 120, 163, 170,
    173, 174, 175